Primate ecology and conservation

Selected Proceedings of the Tenth Congress of the International Primatological Society, held in Nairobi, Kenya, in July 1984

Volume 2

Primate ecology and conservation

Edited by

JAMES G. ELSE

Institute of Primate Research, National Museums of Kenya

PHYLLIS C. LEE

Sub-Department of Animal Behaviour, Department of Zoology University of Cambridge

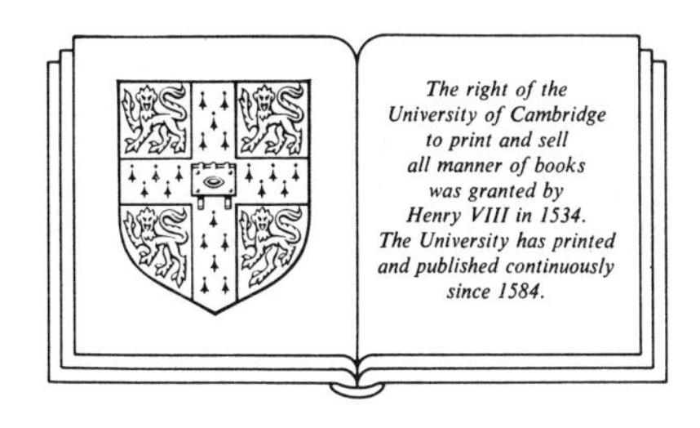

CAMBRIDGE UNIVERSITY PRESS

Cambridge

London New York New Rochelle

Melbourne Sydney

Published by the Press Syndicate of the University of Cambridge
The Pitt Building, Trumpington Street, Cambridge CB2 1RP
32 East 57th Street, New York, NY 10022, USA
10 Stamford Road, Oakleigh, Melbourne 3166, Australia

First published 1986

Printed in Great Britain at the University Press, Cambridge

British Library cataloguing in publication data

Primate ecology and conservation. –
(Selected proceedings of the tenth Congress of the
International Primatological Society; v. 2)
1. Primates – Ecology 2. Mammals –
Ecology
I. Else, James G. II. Lee, Phyllis C.
III. Series
599.8′045 QL737.P9

Library of Congress cataloging in publication data

Primate ecology and conservation.

(Selected proceedings of the Tenth Congress of the
International Primatological Society; v. 2)
1. Primates – Ecology – Congresses. 2. Wildlife conservation –
Congresses. 3. Mammals – Ecology – Congresses.
I. Else, James G. II. Lee, Phyllis C. III. Series: International
Primatological Society. Congress (10th : 1984 : Nairobi, Kenya).
Selected proceedings of the Tenth Congress of the International
Primatological Society; v. 2.
QL737.P9I53 1984 Vol. 2 599.8 s [599.8′045] 86-8287

ISBN 0 521 32451 3 hard covers
ISBN 0 521 31012 1 paperback

PN

CONTENTS

CONTRIBUTORS

F. Antinucci, *Institute of Psychology CNR, University of Rome, Rome 00157, Italy*

J. Altmann, *Allee Lab for Animal Behavior, University of Chicago, Chicago, Illinois 60637, USA*

J. Berkson, *Research Department, Zoological Society of San Diego, San Diego, California 92112, USA*

E. J. Brennan, *Institute of Primate Research, National Museums of Kenya, Box 24481, Karen, Kenya*

D. J. Chivers, *Sub-Department of Veterinary Anatomy, University of Cambridge, Cambridge CB2 1QS, UK*

J. G. Else, *Institute of Primate Research, National Museums of Kenya, Box 24481, Karen, Kenya*

A. A. Eudey, *Department of Anthropology, University of Reno, Reno, Nevada 89557, USA*

L. M. Fedigan, *Department of Anthropology, University of Alberta, Edmonton, Alberta T6G 2H4, Canada*

D. L. Forthman Quick, *Research Department, Los Angeles Zoo, Los Angeles, California 90027, USA*

J. U. Ganzhorn, *Abteilung Verhaltensphysiologie, Institut für Biologie III, University of Tübingen, D 7400 Tübingen 1, Federal Republic of Germany*

A. W. Goldizen, *Division of Biological Sciences, University of Michigan, Ann Arbor, Michigan 48109, USA*

K. M. Green, *Department of Zoological Research, Smithsonian Institution, Washington D.C. 20008, and Remote Sensing Systems Laboratory, Department of Civil Engineering, University of Maryland, Maryland 20732, USA*

M. J. S. Harrison, *Laboratoire d'Ecologie Générale, Museum National d'Histoire Naturelle, 91800 Brunoy, France*

J. P. Hearn, *Institute of Zoology, Regent's Park, London NW1 4RY, UK*

M. K. Izard, *Duke University Primate Center, Durham, North Carolina 27705, USA*

C. H. Janson, *Department of Zoology, University of Washington, Seattle, and State University of New York, Stony Brook, New York 11794, USA*

J. W. Kemnitz, *Wisconsin Regional Primate Center, Madison, Wisconsin 53715, USA*

A. Kortlandt, *Department of Animal Psychology and Ethology, University of Amsterdam, The Netherlands*

P. C. Lee, *Institute of Primate Research, National Museums of Kenya, and Sub-Department of Animal Behaviour, University of Cambridge, Madingley, Cambridge CB3 8AA, UK*

D. G. Lindburg, *Research Department, Zoological Society of San Diego, San Diego, California 92112, USA*

M. J. McFarland, *Department of Biology, Princeton University, Princeton, New Jersey 08544, USA*

A. M. MacLarnon, *Department of Anthropology, University College London, London WC1E 6BT, UK*

I. Malik, *Department of Zoology, University of Delhi, New Delhi 110011, India*

R. D. Martin, *Department of Anthropology, University College London, London WC1E 6BT, UK*

C. R. Menzel, *California Primate Center, University of California, Davis, California 95616, USA*

A. H. Mitchell, *Department of Anthropology, Yale University, New Haven, Connecticut 06529, USA*

R. A. Mittermeier, *Primate Program, World Wildlife Fund US, Washington D.C. 20009, USA*

R. P. Mukherjee, *Zoological Society of India, Calcutta 700 087, India*

L. T. Nash, *Department of Anthropology, Arizona State University, Tempe, Arizona 85285, USA*

M. M. Neu, *Wisconsin Regional Primate Center, Madison, Wisconsin 53715, USA*

L. Nightenhelser, *Research Department, Zoological Society of San Diego, San Diego, California 92112, USA*

G. W. Norton, *Department of Applied Biology, University of Cambridge, Cambridge CB2 3DX, UK*

J. J. Raemaekers, *Sub-Department of Veterinary Anatomy, University of Cambridge, Cambridge CB2 1QS, UK*

P. K. Seth, *Department of Anthropology, University of Delhi, Delhi 110007, India*

S. Seth, *Department of Anthropology, University of Delhi, Delhi 110007, India*

E. L. Simons, *Duke University Primate Center, Durham, North Carolina 27705, USA*

M. F. Stevenson, *Royal Zoological Society of Scotland, Murrayfield, Edinburgh EH12 6TS, UK*

S. C. Strum, *Department of Anthropology, University of California, San Diego, La Jolla, California 92093, USA*

M. F. Teaford, *Department of Cell Biology and Anatomy, School of Medicine, Johns Hopkins University, Baltimore, Maryland 21205, USA*

J. Terborgh, *Department of Biology, Princeton University, Princeton, New Jersey 08544, USA*

R. L. Tilson, *Biological Program, Minnesota Zoological Garden, Apple Valley, Minnesota 55124, USA*

E. Visalberghi, *Institute of Psychology CNR, University of Rome, Rome 00157, Italy*

D. Western, *Wildlife Conservation International (New York Zoological Society), P.O. Box 62844, Nairobi, Kenya*
P. C. Wright, *Duke University Primate Center, Durham, North Carolina 27705, USA*

PREFACE

The papers included in this volume were presented at the Tenth Congress of the International Primatological Society, held in Nairobi, Kenya, in July 1984. The contributions have been revised for publication, but the length has been kept short in order to allow as many as possible of the high-quality presentations to be included. A major emphasis of the Congress was to integrate the current research on diverse aspects of the biology and behaviour of primates, using specific questions of current interest rather than treating each field separately. In organising sections around questions rather than species or single topics we hope to promote the exchange of ideas between typically distinct fields within a broad general theme. Thus we have included a wide range of subjects and species in the different integrated sections.

In this volume, we have tried to draw together both the detailed and broader issues of primate ecology under investigation in the many different areas of the world with natural populations of primates, and new experimental and laboratory work. The setting of the Congress in Nairobi reflected the contribution of East Africa to studies of primates, in the past and the present, and in particular to studies of primates in the wild. We are also pleased to include a section (Part IV) on the New World primates, whose biology and behaviour are less well known than that of the Old World species, and those of Africa specifically.

The first papers deal with detailed laboratory studies of primate nutrition, food selection and food exploitation (Part II). This work is then related to new field studies of primate ecology that have used the same detailed and physiological approach. As our understanding of primate behaviour in relation to ecology increases, more studies have begun to focus on complex issues of phytochemistry, digestion and physiology and these approaches are presented in Part II. At the same time studies of primates in their natural environment continue to provide basic information on the diversity and adaptability of primates (Part III). The integration of our augmented knowledge of primate habitats with a greater understanding of physiology is increasingly allowing us to define and predict both the behavioural relations of primates with their habitat, and general patterns of social organisation. This, in turn, leads us towards an understanding of one of the greatest needs faced by primates in their natural habitats: that of conservation.

The problems of primate conservation are becoming increasingly severe as human populations expand and need additional land for development. We have approached this question in the three remaining sections of papers: Part V deals with direct confrontations between the needs of primates and those of the human populations; Part VI considers the status of primates in the wild, and the various new (and past) techniques for assessing populations and problems in conservation; Part VII integrates these approaches in a broader view.

The papers in this volume present both the advances that have been made recently in our understanding of the biology, behaviour and ecology of primates, and suggestions for further research, especially in the area of primate conservation.

We would like to thank all the session chairs and the external reviewers of the papers, whose valuable advice and suggestions for revisions have helped maintain high standards for the papers included in this volume. We thank the Government of the Republic of Kenya and the National Museums of Kenya for encouraging and hosting the Congress, and the participants from many parts of the world who made the Congress a success. We are especially grateful to Dr David Chivers for his support, comments on manuscripts and encouragement of the publication of the *Proceedings*. Dr Robin Pellew of Cambridge University Press has been untiring in his assistance. PCL would like to thank Professors P. Bateson and R. A. Hinde for providing facilities during the editing of these volumes.

James G. Else
P. C. Lee
Nairobi and Cambridge
September 1, 1985

Part I

Introduction

I.1

Current issues and new approaches in primate ecology and conservation

D. J. CHIVERS

Introduction

Primate ecology originated more than 50 years ago from the studies of Carpenter (1934, 1940) on howler monkeys and gibbons, but it has only really developed in the last 20 years following the 1960s boom in primate field studies (DeVore, 1965; Altmann, 1967; Jay, 1968). Attention was focussed on the ecological aspects of primate behaviour by the work of Crook & Gartlan (1966) and Eisenberg, Muckenhirn & Rudran (1972), when it became apparent that social behaviour was influenced by habitat, producing differences even between populations of the same species. This has led to a dichotomy in primate field studies, and in related captive and laboratory studies.

While many continue to quantify the details and dynamics of social behaviour (see *Proceedings* Volume 3; *Primate Ontogeny, Cognition and Social Behaviour* (CUP, 1986)), others have concentrated on the equally detailed study of vegetation, feeding and ranging behaviour and inter-specific relations in the community. Such studies are essential for contributing to the explanation of patterns of social behaviour observed, and for the conservation of primate populations, whether it be the protection of natural ecosystems or the management of disturbed habitats.

While concern about the over-exploitation of wildlife was first transformed into action over 80 years ago (Fitter & Scott, 1978), conservation has only developed in the last 10 years, certainly as a science (e.g Soulé & Wilcox, 1980; Sutton, Whitmore & Chadwick, 1983), as human destruction of natural ecosystems has escalated to the extent that it threatens the long-term survival of humans as well as of other animals. Because of their close relationship to humans, and their relatively large size (especially in forest ecosystems), primates have a special role to play in the development of strategies of environmental protection and management.

Let us consider briefly recent developments and future potentials of primate ecology and conservation, and their inter-relationships.

Primate ecology

Ecological study involves quantitative descriptions of:

1. The habit in terms of (a) plant form and species composition, distribution and abundance, with the collection of samples for identification and chemical analysis, and (b) faunal composition, especially birds and mammals but also key invertebrates (i.e. potential competitors for resources)
2. The size and composition of social groups in each primate species studied, and how this varies with habitat, activity and time
3. Activity patterns and budgets, with special reference to (a) movements vertically and horizontally through the habitat, and (b) food selection at different times of the day, month and year and in successive years
4. All aspects of foraging and ingestive behaviour, in relation to more detailed captive studies of food preparation and ingestion
5. The nature and frequency of intra- and inter-specific interactions, to develop an understanding of relationships within the community.

Such information leads to the formulation of ecological strategies for each species studied, identifying the relation between foraging strategy and social system, and estimating the intake of energy relative to its output (a topic that has been little investigated in the field, mainly because of the difficulties involved in its measurement). It is also increasingly important to document spatial and temporal variation in all these aspects of ecology, so as to explain more fully patterns of habitat use within and between species.

Captive studies are essential for pursuing the investigation of energetics and useful for explaining foraging strategies, mainly food selection, in terms of the content of nutrients, digestion-inhibitors and toxins in each food. Laboratory analysis reveals the constituents of each food and the digestive capabilities of each primate for different foods.

What progress have we made with such ecological studies, and in what directions are advances most needed? Crook & Gartlan (1966) equated small social groups with forest habitats and large social groups with more open savannah habitats; there were small groups again in the most open (arid) habitats. Eisenberg *et al.* (1972) linked dietary preferences with different habitats and social structure. Such generalities provoked more precise investigation on a large scale.

Kay (1973) was among the first of many to illustrate a closer correlation between body size and diet, with the smallest primates being insectivorous, the largest folivorous, and primates of intermediate size frugivorous; smaller frugivores had insects as the main supplement, large frugivores ate leaves as well as fruit. An even closer correlation was found between the biomass (kg/km^2) of a specific primate population and diet (proportions of animal matter, fruit and foliage) by Hladik & Chivers (1978). Home-range size and metabolic needs have been included in correlative studies (Harvey & Clutton-Brock, 1981; Mace *et al.*, 1983). One of the most thorough analyses has been that of Clutton-Brock & Harvey (1977); they established seven ecological categories of primate (three nocturnal, four diurnal) from a detailed analysis of measures of body size, arboreality, frugivory/folivory, home-range size and group size.

Even these refined techniques, however, do not produce conclusive results. The ecological categories tend to be too broad, some are more cohesive than others and some variables vary more than others between ecological categories; the data come from many sources and are not always truly comparable. Altmann pointed out as early as 1974 that a variety of confounding variables impeded this approach of correlation, and Richard (1985) stressed the value of examining the distribution of single variables to understand the functional significance of different forms of social organisation.

The sequence of events started by Crook & Gartlan (1966) was initiated by Hall's (1963) conclusion that group size represents adaptations to different environments; this followed on from the observation that smaller groups occurred in more impoverished environments. It is perhaps a pity that the former was pursued at the expense of the latter, since the inconclusive results would seem to stem from a failure to appreciate that it is habitat quality which is the key to primate socio-ecology, and that it can be as variable within habitat types as between them.

Abundant data are now available on the habitats and diets of different primates (e.g. Clutton-Brock, 1977; Davies, Caldecott & Chivers, 1984; and from community studies, see below). Dietary information is mainly in terms of the time spent eating different foods, which is the most accessible information to the field worker, and the most relevant to the ecological aspects of activity budgets. For understanding nutrition and energetics, however, the most relevant information is the weight of each food consumed, and progress is slowly being made to produce such data (e.g. Hladik & Hladik, 1972; Goodall, 1977; Hladik, 1977*a*, 1977*b*, 1978; Iwamoto, 1978; Chivers & Raemaekers: Chapter II.3). Such studies mark the start of the analysis of foods for nutrients – sugars, protein, fats, vitamins, minerals and trace elements – which led on to the analysis of foods for secondary compounds – digestion inhibitors and toxins – (McKey *et al.*, 1978, 1981; Oates, Waterman & Choo, 1980; Milton, 1980; Vellayan, 1981; Waterman & Choo, 1981; Wrangham & Waterman, 1982; Ganzhorn: Chapter II.1; Harrison: Chapter II.2).

Feeding behaviour can now be explained as a compromise between obtaining essential nutrients and avoiding harmful substances (although Waterman (1983) cautions us from over-emphasising the role of secondary compounds in food selection). Primates generally select a diet rich in protein and carbohydrate and low in fibre (and tannins or alkaloids). Thus, we are beginning to understand why many primates have such a varied diet, why they select the plant species and parts that they do, and why they eat what they do when preferred fruits are scarce. This requires knowledge of what is available in each habitat, when and in what amounts (see below). Consideration is now being given to passage rates through the gastrointestinal tract for different diets in different species (e.g. Milton, 1984), because of its significance for digestion, but little attention has been given to the role of taste in food selection (Hladik & Chivers, 1978), perhaps because of the experimental difficulties involved. The study of energy intake in relation to expenditure is also being developed (e.g. Waser & Homewood, 1979; Mace *et al.*, 1983).

This has shifted the research emphasis into the laboratory, but field workers are still pre-eminent in its development since it has proved difficult to involve laboratory workers, with the notable exception of Peter Waterman (but see Nordin, 1981; Kemnitz & Neu: Chapter II.6); nutritionists seem generally to be totally involved with human problems. The impetus for field work is still sustained by the questions posed by plant and food chemistry. For example, seed-eating by colobine monkeys in West Africa was thought to be a consequence of poor soils producing plants so well defended chemically that even the colobine monkeys with their sacculated stomachs could not find sufficient digestible leafy foods (McKey *et al.*, 1981); parallel situations have been found in South-east Asian colobines in North-east Borneo (Davies *et al.*, 1984) and in gibbons on the Mentawai Islands (Whitten, 1980). Now, however, Harrison (Chapter II.2) has found the same West African colobine monkey eating even more seeds in forest growing on rich soils. The answers will come from more extensive, problem-orientated field work.

While these dietary studies have been used to explain the evolution of primate social systems (e.g. Wrangham, 1980), they become more meaningful in a broader sense when considered in relation to the evolution of flowering plants (Ripley, 1984) and the co-evolution of plants and animals (Freeland & Janzen, 1974; Gilbert & Raven, 1975). The relationship between primates and their habitats needs further elucidation by primate ecologists, along with a clearer understanding of the inter-relationships between primate species in the same habitat, and between primates and other animals, particularly those occupying similar ecological niches.

Long-term community studies (e.g. Madagascar: Charles-Dominique *et al.*, 1980; Uganda: Struhsaker, 1978; Malaysia: Chivers, 1980; Peru: Terborgh, 1983; Panama: Leigh, Rand & Windsor, 1982) have shown how primates of different sizes, social structure and diet are integrated into a community within the same habitat, by minimising competition through the development of different strategies. In Malaysia, for example, lorises forage individually for insects and fruit by night, monogamous family groups of gibbons defend territories in which they exploit small fruit sources, multi-male groups of macaques forage widely for large fruit sources, and one-male groups of langurs exploit foliage intensively, consuming the seeds of fruit eaten, rather than dispersing them directly as do the more frugivorous primates. Such studies allow comparison of different species of the same genus in the same habitat (see also Hladik, 1977*a*; Fleagle & Mittermeier, 1980; Nash: Chapter III.4) and in different habitats (e.g. Lindburg, 1980; Caldecott & Fa, unpublished; Davies & Oates, unpublished), as well as of the same species in different habitats (e.g. Hrdy, 1977).

These studies of foraging strategies and niche separation are best studied in long-term research programmes in well-established field stations. There is a good geographical spread of such stations of differing longevity, intensity of study and continuity, except in the highly threatened forests of South America (Table 1). While the research returns from such long-established sites, with their ever-extending data-bases, are exponential, the inevitable focus on particular species, while productive in some respects, will be limiting in others. Hence the continuing need for studies in other sites, mainly to accumulate information on, and improve our understanding of, ecological variation in all the features mentioned above within and between species. As with the colobine monkeys, it is only by increasing the number of studies on feeding behaviour in relation to habitat, that the data from long-term stations can be fully explained and the ecology of the taxon properly understood.

Thus, there is a need to consolidate on, and improve the financial viability of, the network of long-term research stations for ecological research, and to develop problem-orientated research programmes at such stations and wherever else may be necessary, with the appropriate laboratory follow-up. In ecological terms the emphasis should be on plants (their diversity, production cycles of leaves, flowers and fruit, pollination and seed dispersal), and the diets and foraging strategies of primates (and other competing animals) – that is, the use of such resources by each species within the community framework.

Table 1. *New and old sites of long-term studies of primate socio-ecology*

Africa	Asia	America
Tiwai Island (Sierra Leone)	Polonnaruwa (Sri Lanka)	Hacienda la Pacifica and Santa Rosa National Park, Guanacaste (Costa Rica)
Tai Forest (Ivory Coast)	Aligarh	Barro Colorado Island (Panama)
Korup Forest Reserve (Cameroun)	Jodhpur	Raleighvallen-Voltzberg Reserve (Surinam)
Makokou	Anamalai Wildlife Sanctuary (India)	Fuframa, near Manaus
Lope-Okanda Reserve (Gabon)	Khao Yai National Park (Thailand)	Fazenda Montes Claros, Minas Gerais (Brazil)
Parc des Volcans (Rwanda)	Krau Game Reserve (West Malaysia)	
Kibale Forest (Uganda)	Sepilok Reserve, Sabah	
Amboseli National Park (Kenya)	Samunsam Reserve, Sarawak (East Malaysia)	
Gombe Stream Reserve	Gunung Leuser Reserve (Sumatra)	
Mahale Mountains	Tanjong Puting Reserve and Kutai Forest Reserve, Kalimantan (Indonesia)	
Mikumi National Park (Tanzania)		
Berenty Reserve (Madagascar)		

Primate conservation

Conservation, in the broadest sense, involves:

1. The protection of the full diversity of natural ecosystems;
2. The management of disturbed habitats for sustained yields of plant and animal products, which requires that the resident species flourish; and
3. The rescue of primates (and other animals) made homeless by habitat clearance for agriculture, for translocation and/or (as a last resort) for captive breeding for research on nutrition and reproduction (in particular), essential health research and/or reintroduction to semi-natural habitats for education (including tourism) and research.

This requires the identification of areas for total protection and for sustained management through ecological survey and study, and the appropriate political and financial lobbying for their establishment, followed by rigorous monitoring and management. From the primate viewpoint, zoogeographical (Marsh, in press) and socio-ecological (Raemaekers & Chivers, 1980; Chivers, in press) considerations are essential, to provide a regional overview to what will inevitably be a national activity. From the human viewpoint, long-term environmental stability and economic prosperity (Myers, 1979, 1983) are essential. At first glance the needs of primates and humans might appear incompatible, but this is clearly not so; recent events in Africa (droughts and famine), Asia (devastating floods, soil erosion and fires) and South America (droughts and famine, disruption of river-based economies) show that environmental mismanagement works to the detriment of both wildlife and people.

Tropical rain-forest is the most threatened primate habitat (see also Mittermeier: Chapter VI.7), because of its economic important (in the short term according to current practices) and because its inhabitants are more specialised in their arboreal adaptations than those primates exploiting more terrestrial niches. Since such primates are wholly or mostly arboreal, and because they are among the largest of forest animals, they are excellent indicators of the health of the forests concerned, and, because of their close relationship to humans, ideal focal ('flagship') species for publicising the threats to their habitats. This is just a means to an end, however, since it is the whole ecosystem (and its diversity) which one is seeking to conserve – for humans as well as primates, and all the other animals and plants on which they depend.

The long-term value of tropical forests has been widely publicised (e.g. IUCN/WWF campaigns; Myers, 1979, 1983); they are both environmental and economic, global as well as local (Table 2). However widely the problems are appreciated, the pressures are such that the appropriate solutions are not being implemented sufficiently rapidly. It is a matter of immediate economic necessity *versus* long-term stability, the transition from the former to the latter being difficult. The crisis is growing as developing countries seek to escape from the spiral of large incomes from the sale of timber and forest products in the short term with much uncertainty about the future, into a long-term strategy of survival and prosperity from sustained yields. However much income for development can be derived from sacrificing the forest, and whatever one's beliefs about the viability of monocultures in monsoonal or ever-wet regions, the costs in terms of environmental catastrophes and poor or lost yields are likely to be prohibitive (e.g. Myers, 1979, 1983; Furtado, 1980; Sutton *et al.*, 1983).

The solutions have three aspects (Table 2):

1. The establishment of national parks and reserves to give total protection to watersheds (thereby protecting ecosystems at higher altitudes) and representative areas of lowland ecosys-

Table 2. *Conservation of tropical rain-forests (from Chivers, in press)*

Values (long-term)	Pressures	Solutions
Water and soil balance Climate rainfall pattern atmospheric gas balance	Hunting Harvesting Farming	Total protection of watersheds and significant representatives of each ecosystem, especially those with high plant/animal diversity
40–50% world's plant and animal species genetic diversity pivotal plant/animal links	Pet trade Power water oil	Wide-ranging management of buffer zones to reserves for sustained yields
Sustained yields timber, canes, fibres gums, waxes, resins, foods – plant and animal medicines Education and research Recreation	Selective logging Clear-felling for timber for fuel for agriculture	Agro-forestry and agriculture in areas cleared of forest, with improved efficiency

tems (which tend to be the most abundant and diverse in terms of flora and fauna)
2. The efficient management of substantial buffer zones to these sanctuaries, and of other available tracts of forests and other ecosystems, for sustained yields of the great variety of plant and animal products available
3. The more efficient use of land already deforested for agriculture, with the development of agro-forestry where possible.

These approaches are already being developed in various tropical countries, but the international community must give more support in terms of funds and expertise, to make full use of the available manpower and, at the same time, reducing the excessive demand for products such as cheap timber and beef. If forest clearance can be reduced (preferably to zero), and protected and managed areas can be increased, there is still time to restore a long-term balance. For example, while forest clearance has continued in South-east Asia, the area of protected forests has increased in the last few years (Table 3), even though figures for protected forests may include significant areas of production forest.

Primate ecology and conservation

The ecologist is a significant member of the band of environmental scientists who have an important role to play in the long-term strategy of maintaining environmental stability and promoting the economic productivity of various ecosystems.

The primate ecologist is already identifying the ecological and behavioural needs of a wide variety of primates (although not necessarily the most endangered ones), which are central to the establishment of protected areas with the correct (and adequate) resources, and to the management of disturbed areas. Surveys are only really meaningful in the light of systematic data collected during long-term studies. Primate ecologists have a major contribution to make in identifying critical situations and in pressing for the appropriate remedies; for example, they are contributing to the action plans being developed for Africa (Oates, unpublished), Madagascar (Pollock, unpublished), Asia (Eudey, unpublished) and the Americas (Mittermeier, unpublished) as the Primate Specialist Group's (of the Species Survival Commission of IUCN) contribution to the World Conservation Strategy.

A start has already been made, somewhat belatedly, in assessing the tolerance of various primates to disturbance (Wilson & Wilson, 1975;

Table 3. *Protected forests in some South-east Asian countries (from Chivers, in press)*

Country	Total land area km^2	1975[a] Forested area km^2	%	Protected forests km^2	%	1985[b] Protected forests actual + proposed km^2	%	(proposed only) %
'Assam'	121900	47900	39	235	0.2	4937	4	(0)
Bangladesh	142776					4498	3	(1)
Burma	678033					11886	2	(1)
Thailand	514910	94452	18			41484	8	(0)
West Malaysia	128013	66950	52	8150	6	39138	31	(5)
East Malaysia	201727	128700	64	1740	1	71297	35	(2)
Indonesia								
Sumatra	473970	260000	55	18280	4	76597	16	(3)
Java	126501	28000	22	2422	2	12294	10	(4)
Kalimantan	539500	419000	78	11410	2	105724	20	(11)

[a] from Chivers (1977)
[b] from Protected Areas Data Unit, Conservation Monitoring Centre, Cambridge

Marsh & Wilson, 1981; Johns, 1983 and in progress; Johns & Skorupa, in press). In South-east Asia, for example, while orang-utans and proboscis monkeys are intolerant of disturbance, gibbons and langurs (and macaques) seem to survive well, at least in the long term, after light selective logging (10 trees/ha or 4% of trees), even though up to 45% of trees may be damaged in the process.

Although Wolfheim (1983) suggests that larger primates are more vulnerable to disturbance, because they need more food and space, occur at lower densities, breed more slowly, are more vulnerable to hunters and tend to be more recently derived and specialised, Johns & Skorupa (in press) show, by an elegant analysis, that diet is a stronger indicator of survival ability than is body size. The extent and rate at which disturbed forests regenerate exploitable foods is probably crucial, and the more rapid regeneration of foliage than fruit will favour the more folivorous primates. Complementing the study of primate survival in disturbed habitats, is the study of primates (and other animals) in forest patches of different sizes (Lovejoy *et al.*, 1983), which has important implications for reserve design and topical theories about island biogeography (e.g. Soulé & Wilcox, 1980).

Thus, there is plenty of scope for ecologists in disturbed habitats, and in helping to resolve the conflict between humans and primates in more open habitats fringing cultivated areas (see Part V). Ecologists take a back seat, however, in those aspects of conservation involving captive breeding, when the integration of reproductive physiology and behaviour becomes essential. The reintroduction of surplus captive animals, and the translocation of homeless animals, bring the ecologist to the fore again. These are conservation tools that have not been adequately considered or tested for primates as they have for other mammals, although there are important exceptions (Beck *et al.*, in press; Strum, in press). Costs and practical problems tend to be prohibitive, but we need to pursue actively all possibilities. In addition to educational and conservation publicity values, reintroduction does seem to be feasible for the smaller primates that are easy to breed and maintain in captivity. For larger primates, translocation offers the best prospects if suitable, unpopulated habitat is available. It is the social complexity of primates which tends to prevent their successful movement, especially from captivity, but there are exceptions which we need to identify without delay.

Finally, it should be emphasised that all primatologists, especially ecologists, must be prepared to devote considerable time and energy to applying the academic fruits of their labours to the conservation problems outlined above, or the subjects of their more erudite research will be lost for ever. Success ultimately lies in being able to influence commerce and politics in order to secure environmental stability and long-term productivity (see Part VII; Sutton *et al.*, 1983). We must step beyond the bounds of academic research and coordinate our efforts with sympathetic industrialists and economists in the political arena, where the crucial decisions are made.

References

Altmann, S. A. (1967) (ed.) *Social Communication among Primates*. Chicago: Chicago University Press

Altmann, S. A. (1974) Baboons, space, time and energy. *Am. Zoologist*, **14**, 221–48

Beck, B., Kleiman, D., Rettberg, B. & Dietz, J. (in press) Preparation of golden lion tamarins for reintroduction to the wild. In *Primates Conservation – The Road to Self-sustaining Populations*, ed. K. Benirschke & D. Lindburg. New York: Springer Verlag

Carpenter, C. R. (1934) A field study of the behavior and social relations of howling monkeys. *Comp. Psychol. Monogr.*, **10**, 1–168

Carpenter, C. R. (1940) A field study in Siam of the behavior and social relations of the gibbon *Hylobates lar*. *Comp. Psychol. Monogr.*, **16**, 1–212

Charles-Dominique, P., Cooper, H. M., Hladik, A., Hladik, C. M., Pages, E., Pariente, G. F., Petter-Rousseaux, A., Schilling, A. & Petter, J. (1980) (eds) *Nocturnal Malagasy Primates: Ecology, Physiology and Behaviour*. London: Academic Press

Chivers, D. J. (1977) The lesser apes. In *Primate Conservation*, ed. Prince Rainier & G. H. Bourne, pp. 539–98. New York: Academic Press

Chivers, D. J. (1980) (ed.) *Malayan Forest Primates: Ten Years' Study in Tropical Rain-forest*. New York: Plenum Press

Chivers, D. J. (in press). South-east Asian Primates. In *Primate Conservation – The Road to Self-sustaining Populations*, ed. K. Benirschke & D. Lindburg. New York: Springer Verlag

Clutton-Brock, T. H. (1977) (ed.) *Primate Ecology: Studies of Feeding and Ranging Behaviour in Lemurs, Monkeys and Apes*. London: Academic Press

Clutton-Brock, T. H. & Harvey, P. H. (1977) Primate ecology and social organisation. *J. Zool.*, **183**, 1–39

Crook, J. H. & Gartlan, J. S. (1966) Evolution of primate societies. *Nature*, **210**, 1200–3

Davies, A. G., Caldecott, J. O. & Chivers, D. J. (1984) Natural foods as a guide to nutrition of Old World primates. In *Standards in Laboratory Animal Management*, ed. UFAW, pp. 225–44. Potters Bar: Universities Federation for Animal Welfare.

DeVore, I. (1965) (ed.) *Primate Behavior*. New York: Holt, Rinehart & Winston

Eisenberg, J., Muckenhirn, N. A. & Rudran, R. (1972) The relation between ecology and social structure in primates. *Science*, **176**, 863–74

Fitter, R. & Scott, Sir Peter (1978) *The Penitent Butchers*. London: Collins

Fleagle, J. G. & Mittermeier, R. A. (1980) Locomotor behavior, body size and comparative ecology of seven Surinam monkeys. *Am. J. Phys. Anthropol.*, **52**, 301–22

Freeland, W. J. & Janzen, D. H. (1974) Strategies in herbivory by mammals: the role of plant secondary compounds. *Am. Naturalist*, **108**, 269–89

Furtado, J. I. (1980) (ed.) *Tropical Ecology and Development*. Kuala Lumpur: The International Society of Tropical Ecology

Gilbert, L. E. & Raven, P. H. (1975) (eds) *Coevolution of Animals and Plants*. Austin: University of Texas Press

Goodall, A. G. (1977) Feeding and ranging behaviour of a mountain gorilla group (*Gorilla gorilla berengei*) in the Tshibinda-Kahuzi Region (Zaïre). In *Primate Ecology*, ed. T. H. Clutton-Brock, pp. 450–79. London: Academic Press

Hall, K. R. L. (1963) Variations in the ecology of the chacma baboon. *Symp. Zool. Soc. Lond.*, **10**, 1–28

Harvey, P. H. & Clutton-Brock, T. H. (1981) Primate home range size and metabolic needs. *Behav. Ecol. Sociobiol.*, **8**, 151–6

Hladik, C. M. (1977*a*) A comparative study of the feeding strategies of two sympatric species of leaf monkeys: *Presbytis senex* and *Presbytis entellus*. In *Primate Ecology*, ed. T. H. Clutton-Brock, pp. 324–53. London: Academic Press

Hladik, C. M. (1977*b*) Chimpanzees of Gabon and chimpanzees of Gombe: some comparative data on the diet. In *Primate Ecology*, ed. T. H. Clutton-Brock, pp. 481–501. London: Academic Press

Hladik, C. M. (1978) Adaptive strategies of primates in relation to leaf-eating. In *Ecology of Arboreal Folivores*, ed. G. G. Montgomery, pp. 373–5. Washington, D.C.: Smithsonian Institution

Hladik, C. M. & Chivers, D. J. (1978) Ecological factors and specific behavioural patterns determining primate diet. In *Recent Advances in Primatology*. vol. 1, *Behaviour*, ed. D. J. Chivers & J. Herbert, pp. 433–44. London: Academic Press

Hladik, C. M. & Hladik, A. (1972) Disponibilités alimentaires et domaines vitaux des Primates à Ceylan. *Terre et Vie*, **26**, 149–215

Hrdy, S. B. (1977) *The Langurs of Abu: Female and Male Strategies of Reproduction*. Cambridge, Mass.: Harvard University Press

Iwamoto, T. (1978) Food availability as a limiting factor on population density of the Japanese monkey and gelada baboon. In *Recent Advances in Primatology*, vol. 1, *Behaviour*, ed. D. J. Chivers & J. Herbert, pp. 287–303. London: Academic Press

Jay, P. (1968) (ed.) *Primates: Studies in Adaptation and Variability*. New York: Holt, Rinehart & Winston

Johns, A. D. (1983) Ecological effects of selective logging in a West Malaysian rain-forest community. Ph.D. thesis, University of Cambridge

Johns, A. D. & Skorupa, J. P. (in press) Trends in the survival of rain-forest primates. *Int. J. Primatol.*

Kay, R. F. (1973) Mastication, molar tooth structure and diet in primates. Ph.D. thesis, Yale University

Leigh, E. G., Rand, A. S. & Windsor, D. M. (1982) (eds) *The Ecology of a Tropical Forest: Seasonal Rhythms and Long-term Changes*. Washington, D.C.: Smithsonian Institution

Lindburg, D. G. (1980) (ed.) *The Macaques: Studies in Ecology, Behavior and Evolution*. New York: Van Nostrand Reinhold

Lovejoy, T., Bierregaard, R. O., Rankin, J. & Schubart, H. O. R. (1983) Ecological dynamics of tropical forest fragments. In *Tropical Rain Forest: Ecology and Management*, ed. S. L. Sutton, T. C. Whitmore & A. C. Chadwick, pp. 377–84. Oxford: Blackwell Scientific Publications

Mace, G. M., Harvey, P. H. & Clutton-Brock, P. H. (1983) Vertebrate home range size and energetic requirements. In *The Ecology of Animal Movement*, ed. I. Swingland & P. J. Greenwood, pp. 32–53. Oxford: University Press
McKey, D. B., Waterman, P. G., Mbi, C. N., Gartlan, J. S. & Struhsaker, T. T. (1978) Phenolic content of vegetation in two African rain-forests: ecological implications. *Science*, **202**, 61–4
McKey, D. B., Gartlan, J. S., Waterman, P. G. & Choo, G. M. (1981) Food selection by black colobus monkeys (*Colobus satanas*) in relation to plant chemistry. *Biol. J. Linn. Soc.*, **16**, 115–46
Marsh, C. W. (in press) A framework for primate conservation priorities in Asian moist tropical forests. In *Conservation of Primates and Tropical Forests*, ed. C. W. Marsh, R. A. Mittermeier & J. S. Gartlan. New York: Allan R. Liss
Marsh, C. W. & Wilson, W. L. (1981) *A Survey of Primates in Peninsular Malaysian Forests*. Universiti Kebangsaan Malaysia and Cambridge: Cambridge University Press
Milton, K. (1980) *The Foraging Strategy of Howler Monkeys: A Study in Primate Economics*. New York: Columbia University Press
Milton, K. (1984) The role of food-processing factors in primate food choice. In *Adaptations for Foraging in Nonhuman Primates*, ed. P. S. Rodman & J. G. H. Cant, pp. 249–79. New York: Columbia University Press
Myers, N. (1979) *The Sinking Ark*. Oxford: Pergamon
Myers, N. (1983) *A Wealth of Wild Species: Storehouse for Human Welfare*. Boulder, Colorado: Westview
Nordin, N. (1981) Voluntary food intake and digestion in Malaysian primates with special reference to the intake of energy and protein. *Malaysian Appl. Biol.*, **10**, 163–75
Oates, J. F., Waterman, P. G. & Choo, G. M. (1980) Food selection by the South Indian Leaf-Monkey, *Presbytis johnii*, in relation to leaf chemistry. *Oecologia (Berlin)*, **45**, 45–56
Raemaekers, J. J. & Chivers, D. J. (1980) Socio-ecology of Malayan forest primates. In *Malayan Forest Primates*, ed. D. J. Chivers, pp. 279–316. New York: Plenum Press
Richard, A. F. (1985) *Primates in Nature*. New York: W. H. Freeman
Ripley, S. (1984) Environmental grain, niche diversification and feeding behaviour in primates. In *Food Acquisition and Processing in Primates*, ed. D. J. Chivers, B. A. Wood & A. Bilsborough, pp. 33–72. New York: Plenum Press
Soulé, M. E. & Wilcox, B. A. (1980) *Conservation Biology: An Evolutionary-Ecological Perspective*. Sunderland, Massachusetts: Sinauer Associates
Struhsaker, T. T. (1978) Food habits of five monkey species in the Kibale Forest, Uganda. In *Recent Advances in Primatology*, vol. 1, *Behaviour*, ed. D. J. Chivers & J. Herbert, pp. 225–48. London: Academic Press
Strum, S. (in press) Translocation of primates. In *Primate Conservation – The Road to Self-sustaining Populations*, ed. K. Benirschke & D. G. Lindburg. New York: Springer Verlag
Sutton, S. L., Whitmore, T. C. & Chadwick, A. C. (1983) (eds) *Tropical Rain Forest: Ecology and Management*. Oxford: Blackwell Scientific Publications
Terborgh, J. W. (1983) *Five New World Primates*. Princeton: Princeton University Press
Vellayan, S. (1981) Chemical composition and digestibility of natural and domestic food of the lar gibbon (*Hylobates lar*) in Malaysia. M.Sc. thesis, University Pertanian Malaysia

Waser, P. M. & Homewood, K. (1979) Cost-benefit approaches to terrestriality: a test with forest primates. *Behav. Ecol. Sociobiol.*, **6**, 115–19
Waterman, P. G. (1983) Distribution of secondary metabolites in rain forest plants: toward an understanding of cause and effect. In *Tropical Rain Forest: Ecology and Management*, ed. S. L. Sutton, T. C. Whitmore & A. C. Chadwick, pp. 167–79. Oxford: Blackwell Scientific Publications
Waterman, P. G. & Choo, G. M. (1981) The effects of digestibility-reducing compounds in leaves on food selection by some Colobinae. *Malaysian Appl. Biol.*, **10**, 147–62
Whitten, A. J. (1980) The Kloss gibbon in Siberut rain forest. Ph.D. thesis, University of Cambridge
Wilson, C. C. & Wilson, W. L. (1975) The influence of selective logging on primates and some other animals in East Kalimantan. *Folia Primatol.*, **23**, 245–74
Wolfheim, J. H. (1983) *Primates of the World: Distribution, Abundance and Conservation*. Seattle: University of Washington Press
Wrangham, R. W. (1980) An ecological model of female-bonded primate groups. *Behaviour*, **75**, 262–300
Wrangham, R. W. & Waterman, P. G. (1982) Feeding behaviour of vervet monkeys on *Acacia tortilis* and *Acacia xanthophloea*: with special reference to reproductive strategies and tannin production. *J. Anim. Ecol.*, **50**, 715–31

Part II

Nutrition, Diet and Feeding Adaptations

Introduction *D. J. Chivers*

Ecological studies of primates, in conjunction with the continued documentation of the intake of different types of food and patterns of ranging, are focussing more on forest and savannah plant phenology (the study of production cycles of various edible plant parts) and on plant chemistry (the analysis of primary and secondary compounds in each food). Spurred on by the development of nutrition studies in captivity, by the quantification (by weight) of each food ingested, and by studies of metabolic rates and energetics, primate ecologists have turned increasingly towards collecting comparable data from natural populations. The traditional quantification of diet in terms of the time spent eating each food continues, however, to have relevance in the analysis of activity budgets and provides baseline data for the more detailed studies of nutrition.

The study of nutrition, diet and feeding adaptations can be divided into examining food acquisition, best studied in the wild, and food processing, best studied in captivity. Complementary and integrated field and laboratory studies are essential, however, for a full understanding of each topic. While field studies of food acquisition are numerous, studies of food processing, particularly in terms of physiology, lag behind. Jaws and teeth have been well studied and masticatory processes clarified, and baseline data on the gastrointestinal tract are available. Little is known yet about passage rates of food types or of the digestive processes in each part of the gut in different primate species, with the exception of the sacculated stomachs of fore-gut fermenters. These areas deserve more attention from primate ecologists. The nutritional requirements of primates and their energetics remain to be fully investigated.

The papers in this section illustrate both the variety of approaches to these important subjects and recent developments. A wide range of primate species is sampled, from prosimians through New and Old World monkeys, to the apes. All these primates, including humans, are covered in Chapter II.7.

Field, captive and laboratory studies are integrated in the different papers in this section. Ganzhorn (Chapter II.1) and Harrison (Chapter II.2) deal with food acquisition and the importance of plant chemistry in the food choice of captive and wild primates. These papers highlight the potential for increasing our understanding of primate ecology by investigating the interaction between plant chemistry and primate digestive physiology. Chivers and Raemaekers (Chapter II.3) relate the acquisition of food among wild gibbons to their nutritional requirements, and these principles are then related to diets for captive animals. The importance of tool use in food processing is explored by Visalberghi and Antinucci (Chapter II.4), while the relations between tooth wear and food processing are discussed by Teaford (Chapter II.5). Both principles of food acquisition (in terms of hunger and preferences) and of food processing (in terms of caloric intake) are dealt with by Kemnitz and Neu (Chapter II.6). MacLarnon *et al.* (Chapter II.7) use a detailed investigation of the morphology of the gut among many different primate species to define and predict properties of digestion and food acquisition.

II.1

The influence of plant chemistry on food selection by *Lemur catta* and *Lemur fulvus*

J. U. GANZHORN

Introduction

Many different plant chemicals are known to influence food selection by herbivorous primates. Most of this information stems from studies of the Colobinae of Africa and Asia (Hladik, 1977; Oates, Swain & Zantovska, 1977; McKey, 1978; McKey *et al.*, 1978; Oates, 1978; Oates, Waterman & Choo, 1980; Moreno-Black & Bent, 1982) and the howling and spider monkeys of the Neotropics (Hladik *et al.*, 1971; Glander, 1975, 1978, 1981; Milton, 1979; Nagy & Milton, 1979; Gaulin & Gaulin, 1982; Estrada, 1984). There have also been a few studies on the effect of plant chemicals on food selection by omnivorous Old World primates such as vervet monkeys (Wrangham & Waterman, 1981), chimpanzees (Hladik, 1978; Wrangham & Nishida, 1983), and baboons (Hausfater & Bearce, 1976). Very little is known about the influence of plant compounds on the feeding behavior of Malagasy prosimians (C. M. Hladik, 1978, 1979; A. Hladik, 1980; C. M. Hladik, Charles-Dominique & Petter, 1980; Petter-Rousseaux & Hladik, 1980; Glander & Rabin, 1983).

Here I present the results of a captive study to assess the influence of plant compounds on food selection by two Malagasy prosimians, *Lemur catta* and *Lemur fulvus*, which are sympatric in the wild.

Materials and methods

A group of *Lemur catta* (11 individuals) and a group of *Lemur fulvus* (7 individuals) were studied in a 0.5 ha enclosure in a mixed pine forest at the Duke University Primate Center (USA). The animals were provisioned with monkey chow and fruit daily but supplemented their diet with plants growing in the enclosure. Only the feeding behavior on plant material growing in the enclosure is considered here.

Feeding behavior of these two groups was recorded using both scan and focal animal sampling. Scan samples were taken at intervals of 10 min during the winter, and 15 and 30 mins during the summer. Samples were distributed evenly across days. At each interval it was recorded whether an animal was resting, grooming, moving, feeding, or involved in other activities. If an animal was feeding, the plant species and plant part were registered. The average feeding bout of *L. catta* and of *L. fulvus* on natural food lasted for 155 ± 180 s and for 212 ± 231 s, respectively. (One feeding bout was defined as an uninterrupted period of feeding separated by more than 5 min from any other feeding event.) Since feeding bouts were short, consecutive 10-min interval samples can be considered independent from each other. Feeding on natural food accounted for 1356 of 21924 individual scan samples for *L. catta* and for 1056 of 18079 individual scan samples for *L. fulvus*. Focal animals were observed continuously from dawn to dusk on 14 days. This method was used to estimate the amount of food eaten per animal per day. The number and size of each food item eaten by the focal animal was recorded. The same items were collected the next day, dried and weighed. Plant species composition and availability of different plant parts were determined for the different seasons of the year. Plant species composition was measured by the number of parts of each species touching a pole of 5 m height. For the canopy above 5 m, a sighting was made through a 1 m pipe held vertically. The fraction of the sky not obscured by foliage is equivalent to the negative logarithm of the number of leaves that would touch the pole (MacArthur & MacArthur, 1961). These readings were taken at 520 spots chosen at random in the enclosure. Phenological data were taken weekly. The availability of different plant parts was calculated as the proportion of a plant species in the total plant biomass multiplied by the number of weeks that the part was present.

Chemical analyses were made on 74 parts of 24 different plant species for the following compounds: fiber and ash (Goering & van Soest, 1970, modified according to van Soest & Robertson, 1980, 1981, cited in Robbins, 1983); protein (Bradford, 1976); alkaloids (qualitatively in triple assays with Mayer's, Wagner's and Dragendorff's reagent (Cromwell, 1955), in which the sample was considered to contain alkaloids only if all three assays showed some precipitation, and quantitatively according to Oates *et al.*, 1977); tannins (qualitatively according to Wilson & Merrill, 1931, cited in Farnsworth, 1966; Fong *et al.*, unpublished laboratory manual; quantitatively according to Oates *et al.*, 1977); cyanogenic substances (Feigl & Anger, 1966); and minerals (Loewther, 1980) (for more details see Ganzhorn, 1985).

Feeding records were correlated with the availability of the different plant parts. The feeding records were then normalized for the availability of the different items and simple and semipartial correlations between the normalized feeding records and the concentrations of the plant compounds were calculated (Sokal & Rohlf, 1969). The analysis was done separately for the winter, when none of the deciduous trees had any leaves, and for the growing season.

Results

In the winter, *L. catta* chose plant items according to their availability ($r = 0.354$, $p < 0.05$, d.f. = 35). None of the analyzed plant compounds seemed to influence their food choice (Table 1).

During the summer, there was no significant correlation between the feeding records of *L. catta* and the availability of different plant parts, but the normalized feeding data were significantly correlated with the water and sodium content of their food items (water: $r = 0.275$, $p < 0.05$, d.f. = 51; sodium: $r = 0.348$, $p < 0.05$, d.f. = 42). However, water content is negatively correlated with sodium concentration ($r = 0.357$, $p < 0.05$, d.f. = 42). Thus, if semipartial correlations are calculated, neither water nor sodium is correlated significantly with *L. catta*'s choice (semipartial correlations: water: $r = 0.172$, $p > 0.05$, d.f. = 41; sodium: $r = -0.278$, $p > 0.05$, d.f. = 41). Although the lack of correlation with availability suggests that *L. catta* did select particular food items during the summer, the factors determining their choice were not revealed.

Feeding records of *L. fulvus* were not significantly correlated with the availability of different food items in either season (winter: $r = 0.318$, $p > 0.05$, d.f. = 35; summer: $r = 0.008$, $p > 0.05$, d.f. = 51). During the winter, the normalized feeding records were positively correlated with protein content (winter: $r = 0.347$, $p < 0.05$, d.f. = 35). No other plant compound significantly influenced food selection.

During the summer, the feeding records of *L. fulvus* were significantly correlated with water content ($r = 0.507$, $p < 0.001$, d.f. = 51) and with the concentrations of protein, tannins, potassium and magnesium (protein: $r = 0.436$, $p = 0.001$, d.f. = 51; tannin: $r = -0.283$, $p < 0.05$, d.f. = 51; potassium: $r = 0.365$, $p < 0.05$, d.f. = 43; magnesium: $r = 0.355$, $p < 0.05$, d.f. = 43). However, protein, potassium and magnesium concentrations were significantly correlated with water content. If these correlations between variables are taken into account by calculating semipartial correlations, feeding records of *L. fulvus* were only correlated significantly with water content and the concentration of protein (semipartial correlation for water: $r = 0.420$, $p < 0.01$,

Table 1. *Influences on food selection of* Lemur catta *and* Lemur fulvus, *analyzed as simple correlations*

	Lemur catta				*Lemur fulvus*			
	winter		summer		winter		summer	
	r	*p*	*r*	*p*	*r*	*p*	*r*	*p*
Availability	0.354	0.032	0.231	0.097	0.318	0.055	0.008	0.955
Water	−0.007	0.971	0.275	0.046	0.276	0.133	0.507	0.001
Fiber	0.067	0.697	−0.147	0.293	−0.066	0.702	−0.264	0.051
Ash	0.011	0.947	0.175	0.211	0.167	0.331	0.236	0.082
Protein	0.020	0.904	0.187	0.179	0.347	0.035	0.436	0.001
Tannin	0.297	0.074	0.158	0.258	−0.028	0.871	−0.283	0.036
Alkaloids	0.058	0.610	−0.001	0.992	−0.022	0.902	0.132	0.345
Sodium	−0.135	0.461	−0.348	0.024	−0.327	0.067	−0.096	0.535
Potassium	−0.235	0.195	0.149	0.341	0.139	0.447	0.365	0.014
Calcium	0.058	0.751	0.104	0.512	0.003	0.986	0.145	0.347
Magnesium	0.088	0.631	0.208	0.182	0.049	0.791	0.355	0.017

d.f. = 50; semipartial correlation for protein: $r = 0.397$, $p < 0.01$, d.f. = 50).

Though the semipartial correlation between the number of feeding records and tannin concentration was not significant after subtraction of the effects of water and protein (semipartial correlation for tannin: $r = -0.122$, $p > 0.05$, d.f. = 49), *L. fulvus* avoided plant parts with high tannin concentration. In the summer, natural food items eaten by *L. fulvus* contained on the average 6.2% tannin, but plant parts not eaten contained on the average 11.6% (Mann Whitney U-test: $z = 3.05$, $p < 0.01$).

Whether or not plant items with high tannin concentrations were avoided or not was dependent on the amount of natural food eaten per day. During the winter, when *L. fulvus* ate 15–20 times less natural food than during the summer, there was no difference between the tannin concentrations of rejected and consumed plant items (rejected items: $\bar{x} = 14.5\%$; consumed items: $\bar{x} = 18.2\%$; Mann Whitney U-test: $z = 1.24$, $p > 0.05$). Thus, tannins were avoided by this lemur species only when they ate substantial amounts of plant material.

Alkaloids also seemed to influence the feeding behavior of *L. fulvus*. When these animals had to choose between two parts of the same plant which contained different concentrations of alkaloids, in five out of six cases they chose the part with the lower alkaloid concentration (Table 2). However, they also ate leaves of clover (*Trifolium repens*) containing 4.8 μg of alkaloids. These leaves had very high concentrations of proteins, which apparently overrode the effect of the alkaloids.

Discussion

In contrast to *L. fulvus*, it is difficult to understand the basis of food selection in *L. catta*, whose choices were partially influenced by the availability of different food items. This species obtained less than 5% of its daily protein intake from plant material found in the enclosure. It might well be that the amount of natural food eaten by the captive *L. catta* was of such small nutritional importance that choice between items was of little consequence. But though there was still a tendency to choose food items according to their availability in the summer, feeding records of *L. catta* were not significantly correlated with the availability of different food items during this time of the year. Thus, *L. catta* selected specific food items, but, in contrast to *L. fulvus*, their choice was not based on any of the components which were analyzed.

The different responses of these two lemur species toward plant components may be part of altogether different feeding strategies (Ganzhorn, unpublished). In that study *L. fulvus* had longer feeding bouts on preferred food items and ate more fruit than *L. catta*. Furthermore the food passage time of *L. fulvus* was shorter than that of *L. catta*, thus reflecting a more frugivorous diet of *L. fulvus* (Chivers & Hladik, 1980; Milton, 1981). Finally, the habitat use of *L. fulvus* was correlated significantly with the distribution of their major natural food plants. Also their utilization of the area changed markedly after feeding sites had been translocated. On the other hand the use of the enclosure by *L. catta* was neither correlated with the abundance of their major natural food plants nor changed their ranging behavior as much as it did in *L. fulvus* after the translocation of the feeding site.

Thus, the feeding pattern of *L. catta* can be interpreted as an adaptation to an environment with unpredictable distribution of food

Table 2. *Influence of alkaloids on food selection by* L. catta *and* L. fulvus. *B = leaf blade, FB = flowerbud, FR = fruit, LB = leaf bud, ML = mature leaf, PE = petiole, SC = scale, ST = stem, YFR = young fruit, YL = young leaf, YS = young shoot*

Species	Plant part	Alkaloids (μg/g)	Selection ratio[a]	
			L. catta	*L. fulvus*
Cornus florida	FB	8.3	3.5	0
	YFR	0	0	4.3
Trifolium repens	ML	4.8	0	136.7
	ST	0	0	0
Liriodendron tulipifera	YL	38.0	13.3	6.7
	SC	0	0	27.3
	LB	22.8	100.0	0
Morus rubra	YL	19.2	1.7	0
	YFR	0	0	eaten[b]
	YS	0.8	0	8.3
Fraxinus americana	YL–B	14.9	1.3	4.1
	YL–PE	1.2	8.7	45.5
Ligustrum vulgare	ML	3.8	0	1.1
	FR	0	0	11.5
	BARK	8.7	0	0

[a] Selection ratio = number of feeding records/availability of plant parts (availability = % of plant species on total vegetation biomass × time of phenological presence of that plant part).

[b] *L. fulvus* fed on the young fruits of *M. rubra* extensively. However, by the time systematic activity records were taken this food item was totally consumed.

(e.g. arid regions). In those environments food items have to be eaten as they are encountered. Detrimental effects of plant chemicals may be avoided by eating only small amounts of many different plant species. *L. fulvus*, however, seems to be more adapted to areas with clumped food resources. Since they have long feeding bouts on any particular food item they need to discriminate plant parts more carefully to avoid poisoning. These different feeding strategies might possibly allow the two lemur species to coexist in the same area of Madagascar without obvious competition (Sussman, 1972, 1974).

Acknowledgements

I thank E. L. Simons and J. I. Pollock for their permission to work with the animals. Plants were identified by T. W. Johnson and R. L. Wilbur. Chemical analyses were done in the laboratories of G. Batzli, D. J. Fluke, K. E. Glander and D. Seigler, C. Chitterling carried out the mineral analysis. This paper benefited greatly from comments of K. E. Glander, P. H. Klopfer, J. F. Oates and P. C. Wright. The study and its presentation was financed by grants from the Deutscher Akademischer Austauschdienst and from the Reinhold-und-Maria-Teufel-Stiftung to the author and by grants to K. E. Glander (Department of Anthropology, Duke University, Durham, USA). This is Duke University Primate Center publication No. 260.

References

Bradford, M. (1976) A rapid and sensitive method for the quantification of microgram quantities of protein utilizing the principle of protein-dye-binding. *Anal. Biochem.*, **72**, 248–54

Chivers, D. J. & Hladik, C. M. (1980) Morphology of the gastrointestinal tract in Primates: comparison with other mammals in relation to diet. *J. Morphol.*, **166**, 337–86

Cromwell, B. T. (1955) The alkaloids. *Mod. Methods Plant Anal.*, **4**, 367–74

Estrada, A. (1984) Resource use by Howler Monkeys (*Alouatta palliata*) in the rain forest of Los Tuxtlas, Veracruz, Mexico. *Int. J. Primatol.*, **5**, 105–31

Farnsworth, N. R. (1966) Biological and phytochemical screening of plants. *J. Pharm. Sci.*, **55**, 225–76

Feigl, F. & Anger, V. A. (1966) Replacement of benzidine by copper ethylacetate and tetra base as spot-test reagent for hydrogen cyanide and cyanogen. *Analyst*, **91**, 282–4

Ganzhorn, J. U. (1985) Habitat use and feeding behavior in *Lemur catta* and *Lemur fulvus*. Dissertation, Universität Tübingen

Gaulin, J. C. S. & Gaulin, C. K. (1982) Behavioral ecology of *Alouatta seniculus* in Andean cloud forest. *Int. J. Primatol.*, **3**, 1–32

Glander, K. E. (1975) Habitat description and resource utilization: a preliminary report on mantled howling monkey ecology. In *Socioecology and Psychology of Primates*, ed. R. H. Tuttle, pp. 37–57. The Hague: Mouton

Glander, K. E. (1978) Howling monkey feeding behavior and plant secondary compounds. A study of strategies. In *The Ecology of Arboreal Folivores*, ed. G. G. Montgomery, pp. 561–74. Washington: Smithsonian

Glander, K. E. (1981) Feeding patterns in mantled howling monkeys. In *Foraging Behavior*, ed. A. C. Kamil & T. D. Sargent, pp. 231–57. New York: Garland Press

Glander, K. E. & Rabin, D. P. (1983) Food choice from endemic North Carolina tree species by captive prosimians. *Am. J. Primatol.*, **5**, 221–9

Goering, H. K. & Soest, P. J. van (1970) Forage fiber analysis. *Agric. Handb.*, No. 379. USDA: Agricultural Research Service

Hausfater, G. & Bearce, W. H. (1976) Acacia tree exudates: their composition and use as food source by baboons. *East Afr. Wildl. J.*, **14**, 241–3

Hladik, A. (1980) The dry forest of the west coast of Madagascar: climate, phenology, and food availability for Prosimians. In *Nocturnal Malagasy Primates*, ed. P. Charles-Dominique, H. M. Cooper, A. Hladik, C. M. Hladik, E. Pages, G. F. Pariente, A. Petter-Rousseaux, J. J. Petter & A. Schilling, pp. 3–40. New York: Academic Press

Hladik, C. M. (1977) A comparative study of the feeding strategies of two sympatric species of leaf monkeys: *Presbytis senex* and *Presbytis entellus*. In *Primate Ecology*, ed. T. H. Clutton-Brock, pp. 481–501. New York: Academic Press

Hladik, C. M. (1978) Adaptive strategies of primates in relation to leaf eating. In *The Ecology of Arboreal Folivores*, ed. G. G. Montgomery, pp. 373–95. Washington: Smithsonian

Hladik, C. M. (1979) Diet and ecology of prosimians. In *The Study of Prosimian Behavior*, ed. G. A. Doyle & R. D. Martin, pp. 307–57. New York: Academic Press

Hladik, C. M., Charles-Dominique, P. & Petter, J. J. (1980) Feeding strategies of five nocturnal prosimians in the dry forest of the west coast of Madagascar. In *Nocturnal Malagasy Primates*, ed. P. Charles-Dominique *et al.*, pp. 41–73. New York: Academic Press

Hladik, C. M., Hladik, A., Bousset, J., Valdebouze, P., Viroben, G. & Delort-Laval, J. (1971) Le régime alimentaire des Primates de l'île de Barro-Colorado (Panama). *Folia Primatol.*, **16**, 85–122

Loewther, J. R. (1980) Use of a single sulfuric acid-hydrogen peroxide digest for the analysis of *Pinus radiata* needles. *Commun. Soil Sci. Plant Anal.*, **11**, 175–88

MacArthur, R. H. & MacArthur, J. (1961) On bird species diversity. *Ecology*, **42**, 594–8

McKey, D. (1978) Soils, vegetation and seed eating by black colobus monkeys. In *The Ecology of Arboreal Folivores*, ed. G. G. Montgomery, pp. 423–37. Washington: Smithsonian

McKey, D., Waterman, P. G., Mbi, C. N., Gartlan, J. S. & Struhsaker, T. T. (1978) Phenolic content of vegetation in two African rain forests: ecological implications. *Science*, **202**, 61–4

Milton, K. (1979) Factors influencing leaf choice by howler monkeys: a test of some hypotheses of food selection by generalist herbivores. *Am. Naturalist*, **114**, 362–78

Milton, K. (1981) Food choice and digestive strategies of two sympatric primate species. *Am. Naturalist*, **117**, 496–505

Moreno-Black, G. S. & Bent, E. F. (1982) Secondary compounds in the diet of *Colobus angolensis*. *Afr. J. Ecol.*, **20**, 29–36

Nagy, K. A. & Milton, K. (1979) Energy metabolism and food consumption by wild howler monkeys (*Alouatta palliata*). *Ecology*, **60**, 475–80

Oates, J. F. (1978) Water-plant and soil consumption by guereza monkeys (*Colobus guereza*): a relationship with minerals and toxins in the diet. *Biotropica*, **10**, 241–53

Oates, J. F., Swain, T. & Zantovska, J. (1977) Secondary compounds and food selection by Colobus monkeys. *Biochem. Syst. Ecol.*, **5**, 317–21

Oates, J. F., Waterman, P. G. & Choo, G. M. (1980) Food selection by the South Indian leaf monkey, *Presbytis johnii*, in relation to leaf chemistry. *Oecologia*, **45**, 45–56

Petter-Rousseaux, A. & Hladik, C. M. (1980) A comparative study of food intake in five nocturnal prosimians in simulated climatic conditions. In *Nocturnal Malagasy Primates*, ed. P. Charles-Dominique *et al.*, pp. 169–79. New York: Academic Press

Robbins, C. T. (1983) *Wildlife Feeding and Nutrition*. New York: Academic Press

Sokal, R. R. & Rohlf, F. J. (1969) *Biometry*. San Francisco: Freeman

Sussman, R. W. (1972) An ecological study of two Madagascan primates: *Lemur fulvus rufus* Audebert and *Lemur catta* L. Ph.D. thesis, Duke University

Sussman, R. W. (1974) Ecological distinctions in sympatric species of Lemur. In *Prosimian Biology*, ed. R. D. Martin, G. A. Doyle & A. C. Walker, pp. 75–108. London: Duckworth

Wrangham, R. W. & Nishida, T. (1983) *Aspilia spp.* leaves: a puzzle in the feeding behavior of wild chimpanzees. *Primates*, **24**, 276–82

Wrangham, R. W. & Waterman, P. G. (1981) Feeding behaviour of vervet monkeys on *Acacia tortilis* and *Acacia xanthophloea*: with special reference to reproductive strategies and tannin productions. *J. Anim. Ecol.*, **50**, 715–31

II.2

Feeding ecology of black colobus, *Colobus satanas*, in central Gabon

M. J. S. HARRISON

Introduction

A major characteristic of Colobine monkeys is the ruminant-like digestive system, where food is digested by bacterial fermentation in the enlarged forestomach. The ability to digest cellulose means that plant structural carbohydrates, a major component of leaves, can be used as the main source of energy for these primates. It is not surprising, then, that most field-studies of the African genus *Colobus* and the Asian genus *Presbytis* record foliage as comprising between 50% and 80% of the diet. More recent studies, however, have shown that labelling all Colobines as 'leaf-eaters' is premature: Davies & Bennett (unpublished) (see Davies, Caldecott & Chivers, 1984) recorded 37% and 34% for foliage in the diets of *Presbytis rubicunda* and *P. melalophos* in Malaysia, and McKey *et al.* (1981) recorded 43% for foliage in the diet of black colobus (*Colobus satanas*) in Cameroon.

McKey's work at Douala-Edea in Cameroon showed striking differences in the feeding behaviour of *C. satanas* compared with red colobus (*C. badius*) and black-and-white colobus (*C. guereza*) in East Africa. *C. satanas* fed predominantly on seeds; they ate a low proportion of leaves and were highly selective of the leaves they did eat, choosing these from rare plants and avoiding most of the common species. Chemical analysis of plants showed that overall levels of tannins and phenolics in the vegetation were high while those of nutrients were low (Gartlan *et al.*, 1980). *C. satanas* selected leaves relatively richer in nutrients and lower in digestion-inhibitors such as fibre and tannin. McKey proposed that *C. satanas* adopted a feeding strategy based on seeds, which are nutritious and low in digestion-inhibitors, in response to the high levels of secondary compounds in leaves. These in turn may be related to the poor quality of the soil: high levels of chemical defence may have been selected because replacement of leaves eaten by herbivores would involve greater cost to plants growing on poor soils than to plants growing on soils rich in nutrients (Janzen, 1974).

The population of *C. satanas* studied by McKey inhabits the coastal forest of the marine sedimentary basin in Cameroon, where soils are extremely sandy and low in plant nutrients. Analysis of soils and plant secondary compounds at Kibale Forest in Uganda, where *Colobus* populations eat high proportions of leaves, shows that the fine textured soils are richer and the vegetation is indeed lower in chemical defences (Gartlan *et al.*, 1980). There are, however, alternative hypotheses to account for differences between these two sites, relating to their differences in altitude and in plant-species diversity.

The present study was aimed at further evaluation of hypotheses relating soils, phytochemical profiles of different forests, and feeding strategies in *Colobus* monkeys. A site was chosen where *C. satanas* occurs in continental lowland rainforest, outside the influence of the coastal sedimentary basin. Here soils would be different, and differences in plant secondary chemistry could be expected. The Lopé Reserve in central Gabon was chosen as there is considerable circumstantial evidence to support the prediction that the continental forest there grows on richer soil and is lower in leaf chemical defences than its coastal forest counterpart.

1. Over 100 km inland of the sandy coastal basin, the Lopé has clay soils that are *a priori* more suitable for retention and transport of plant nutrients.
2. There are visible levels of insect damage to leaves, suggesting a relative lack of chemical defenses.
3. Deciduous species seem relatively common; trees are thus not as selected for long-lived leaves as they might be in poor-soil environments where leaves are costly to replace.
4. Streams are not tea-brown in colour, as they are in areas where humic acids (derived from tannins and other phenolics) are leached from the leaf-litter (Janzen, 1974).
5. There are much lower levels of rainfall in the Lopé: 1600 mm annually, compared with 4000 mm at Douala-Edea in Cameroon. This has implications for the deciduous behaviour of trees, as well as the degree of leaching of soil nutrients.
6. Much vegetation in the Lopé grows at a forest–grassland interface, where rapidly growing pioneer or light-loving species may be favoured. Rapid plant growth may place

demands on a tree's resources that conflict with expenditure on chemical defence.

If foliage in the Lopé forest is not as inhospitable to a folivore as it is in Douala-Edea's coastal forest, the prediction is that *C. satanas* in the Lopé would revert to a more typical pattern of *Colobus* folivory, including a higher proportion of leaves in the diet and being less selective in their choice of these leaves. Smaller home-ranges and higher population density might also be expected if leaf quality was no longer a limiting factor. I studied the Lopé forest and its black colobus between August 1983 and April 1984, and shall present here some of the main features of the feeding and ranging behaviour of the colobus, and some preliminary results from the analysis of soils and vegetation.

Study site and methods

The study was carried out in the de Brazza hill-range in the Réserve de la Lopé-Okanda, south of the River Ogooué in central Gabon. A single group of between 9 and 13 colobus was studied for 9 months. Observations were made during monthly, 5-day sample periods to record feeding and ranging patterns. Soils, leaves, and seeds were systematically collected for chemical analysis, and the structure, composition, and phenology of the vegetation was sampled. Methods were directly comparable to those used in the studies of Struhsaker (1975), Oates (1977), and McKey *et al.* (1981) on *Colobus* in Uganda and Cameroon.

Results

Main features of the soil and vegetation

The soils sampled in the de Brazza hills are sandy clay loams, with a mean textural breakdown of 51% sand, 21% silt, and 28% clay (Table 1). Compared to the very sandy soils of the coastal basin (86% sand and 8% clay at Douala-Edea), clay soils have lower porosity and higher ion retention properties, providing a more suitable medium for plant nutrients. Similar soil texture was found in the fertile soils of

Table 1. *Soil texture and acidity in three tropical rain forests*

	% sand	% silt	% clay	pH
Douala-Edea, Cameroon	86	6	8	3.9
Lopé, Gabon	51	21	28	4.5
Kibale, Uganda	59	28	13	5.6

Kibale Forest in Uganda. Soils at the Lopé are also less acid than at Douala-Edea, with a mean pH of 4.5, which is close to the pan-tropical average (Proctor *et al.*, 1983). This falls between the very acid soils of Douala-Edea, with a mean pH of 3.9, and those of Kibale with a pH of 5.6.

The predominant plant families in the Lopé, based on the 20 most common species in terms of biomass, are Mimosaceae (represented by 5 species), and Caesalpiniaceae, Myristicaceae, and Burseraceae (each with 3 species) (Table 2). Although the vegetation of Lopé and Douala-Edea has a considerable number of species in common, only a single species (*Strombosiopsis tetrandra*) comes within the top 20 species in both sites. At Douala-Edea, Burseraceae are rare, and Mimosaceae are apparently not represented at all. Thus the major components of vegetation are quite different in these two forests.

Species diversity is greater in the Lopé forest. Counting trees over 50 cm GBH in a 1.45 ha sample area, there are about 90 species in the Lopé and 53 at Douala-Edea. Only 58% of individual trees are from the top 20 species at Lopé, compared to 78% at Douala-Edea.

Main features of colobus feeding and ranging behaviour

There are several striking features of the diet (Table 3). First, seed-eating by *C. satanas* in the Lopé is more predominant than it is in Douala-Edea, making up on average 60% of the diet. The seed component varied between 32% and 82% of the monthly diet. Secondly, intake of mature leaves was remarkably low, averaging only 3% of the

Table 2. *Dominant plant families and species diversity in two tropical rain forests*

	Lopé, Gabon	Douala-Edea, Cameroon
Dominant tree families (no. species in top 20 species)	Mimosaceae (5) Caesalpiniaceae (3) Myristicaceae (3) Burseraceae (3)	Euphorbiaceae (4) Caesalpiniaceae (2) Olacaceae (2) Ochnaceae (1)
No. species in 1.45 ha sample area	90[a]	53
% of all individual trees in top 20 species	58%	78%

[a] not all determined at present.

diet, with a maximum in any one month of 11%. Young leaves averaged 23% of the diet. While this represents data from only 9 months of an annual cycle, bias is unlikely as the missing period covered 6 weeks in both wet and dry seasons (May, June, July), when colobus were observed eating both young leaves and seeds and there were no dramatic changes in plant phenology or food availability (Tutin, personal communication). There was a strong negative correlation between the intake of seeds and young leaves ($r = -0.89$, $p < 0.001$), suggesting these to be the major alternative elements in the diet. However, except for one month with low seed-availability, seeds dominated the diet in every month, even when edible young leaves were available in abundance.

The total home-range area visited during the study was 184 ha. Of this total, 84 ha was visited during 5-day sample periods.

Discussion

At Douala-Edea, the low quality of leaves placed constraints on food-choice, led to a switch in dietary focus from leaves to seeds, and demanded increased ranging in search of suitable leaf items (McKey & Waterman, 1982). The working hypothesis in the present study was that in the Lopé this constraint would be released, and with a wider choice of suitable leaves the colobus would be able to eat more foliage and reduce the extent of their ranging.

Results were contrary to expectation. *C. satanas* in the Lopé were even less folivorous and more wide-ranging than at Douala-Edea. The total leaf portion of the diet at Lopé was 26%, compared with 43% at Douala-Edea. This was due to the virtual absence of mature leaves from the former diet; while young leaf intake was equivalent, the mature leaf component was only 3% at Lopé, compared with 20% at Douala-Edea. The total known home-range of McKey's study group at

Table 3. *Diet and range-size of* C. satanas *in two different forests*

	Proportion of diet (mean % range)			Range-size (ha)	
	seeds	young leaves	mature leaves	5-day samples	total
Lopé, Gabon	60 (32–82)	23 (5–56)	3 (0–11)	84	184
Douala-Edea, Cameroon	53 (20–85)	20 (5–40)	20 (5–40)	60	69

Douala-Edea was 69 ha, well under half the total of 184 ha for the Lopé study group. Both groups were of similar size.

There are three possibilities to account for this behaviour. First, the leaf chemistry of vegetation in the Lopé may be even more unsuitable for folivores than at Douala-Edea. Food-choice may thus be even more limited, and wider ranging more necessary. In this case, the relation between soil and phytochemistry is more complicated than present evidence suggests. Alternatively, chemical defence of the vegetation at Lopé may indeed be less restrictive for a folivore, and the colobus could survive on a much more leafy diet, but seeds may be positively selected to provide the richest source of nutrition. Seeds from Douala-Edea are rich in fats but lower in nitrogen than leaves, and there *C. satanas* select seeds with the highest nitrogen content. Seeds from the Lopé may have higher contents of fats and nitrogen. A third possibility is that differences between Lopé and Douala-Edea in nutrient and secondary chemistry may be small, and seeds may simply be far more widely available at Lopé, both seasonally and in absolute abundance.

Analysis of Lopé plant chemistry is currently in progress, and should help discriminate between these possibilities. Whatever the case, the predominance of seed-eating in *C. satanas* has important implications for our understanding of the evolution of folivory in *Colobus* monkeys. The current feeding adaptations of *C. satanas* in continental forest may relate to selection pressures acting during the Pleistocene, when the species was restricted to moist coastal forest refugia. Seed-eating evolved in such conditions may now persist through phylogenetic inertia, in the absence of the original constraints. Alternatively, seed-eating may be a general characteristic of the black-and-white colobus group, at least in West Africa, with *C. guereza* as the exception. More data are needed from other black-and-white colobus populations before this can be assessed, and before general statements can be made about the extent of folivory in this species-group.

Acknowledgements

I gratefully acknowledge The Royal Society (UK) and the Centre National de Recherche Scientifique (France) for financial support, and the Ministère des Eaux et Forêts (Gabon) for permission to work in the Reserve de la Lopé-Okanda.

References

Davies, A. G., Caldecott, J. O. & Chivers, D. J. (1984) Natural foods as a guide to nutrition of old world primates. In *Proceedings of Universities Federation of Animal Welfare Symposium*. London: Zoological Society

Gartlan, J. S., McKey, D. B., Waterman, P. G., Mbi, C. N. & Struhsaker, T. T. (1980) A comparative study of the phytochemistry of two African rainforests. *Biochem. Syst. Ecol.*, **8**, 401–22

Janzen, D. H. (1974) Tropical blackwater rivers, animals, and mass fruiting by the Dipterocarpaceae. *Biotropica*, **6**(2), 69–103

McKey, D. B., Gartlan, J. S., Waterman, P. G. & Choo, G. M. (1981) Food selection by black colobus monkeys (*Colobus satanas*) in relation to plant chemistry. *Biol. J. Linn. Soc.*, **16**, 115–46

McKey, D. B. & Waterman, P. G. (1982) Ranging behaviour of a group of black colobus (*Colobus satanas*) in the Douala-Edea Reserve, Cameroon. *Folia Primatol.*, **39**, 264–304

Oates, J. F. (1977) The guereza and its food. In *Primate Ecology*, ed. T. H. Clutton-Brock, pp. 276–321. London: Academic Press

Proctor, J., Anderson, J. M., Chai, P. & Vallack, H. W. (1983) Ecological studies in four contrasting lowland rain forests in Gunung Mulu National Park, Sarawak. *J. Ecol.*, **71**, 237–60

Struhsaker, T. T. (1975) *The Red Colobus Monkey*. Chicago: Chicago University Press

II.3

Natural and synthetic diets of Malayan gibbons

D. J. CHIVERS AND J. J. RAEMAEKERS

Introduction

The nine species of gibbons, or lesser apes, and their distribution in S.E. Asia (mainly allopatric) are summarised by Chivers & Gittins (1978); their socio-ecology and evolutionary biology are reviewed in relation to current problems of conservation by Chivers (1977); all are monogamous and territorial. We are concerned here with the diets and feeding strategies of the three species occurring in the centre of their range, in Peninsular Malaysia, and mostly with contrasts between the sympatric lar gibbon, *Hylobates lar*, and the much larger siamang, *H. syndactylus*; comparative data will be presented for the ecological equivalent of the lar gibbon in the north of the Malay Peninsula, the agile gibbon, *H. agilis*. The diet of these wild gibbons, in terms of food type, quantities and timing will be compared with examples of diet and feeding behaviour in captive gibbons.

The data on free-ranging gibbons were obtained mostly during four field studies in three localities in the Malay Peninsula. DJC studied the siamang in lowland forest around Kuala Lompat in the Krau Game Reserve from January 1969 until May 1970, and in the hill forest of Ulu Sempam, north of Bukit Fraser, with 2-month follow-up visits in 1972, 1974 and 1977 (Chivers, 1974; Chivers, Raemaekers & Aldrich-Blake, 1975). JJR compared the diets and feeding and ranging strategies of lar gibbon and siamang at Kuala Lompat from March 1974 until March 1976 (Raemaekers, 1977, 1978*a*, 1978*b*, 1979). MacKinnon & MacKinnon (1978) carried out intensive sampling of the feeding and ranging behaviour of all primates at Kuala Lompat for 6 months in 1973. Gittins (1979) studied the ecology of the agile gibbon on Gunung Bubu, Perak, and at its border with lar gibbons in Ulu Mudah, Kedah, from March 1974 until March 1976.

Data on captive gibbons have been obtained from Brambell (1976) and Doherty & MacNamara (1976), or kindly supplied by the zoos of London, Twycross, San Francisco, Los Angeles, Milwaukee and Washington, D.C.

Natural diets

Types of food

The natural food of gibbons may be divided conveniently into three categories: the reproductive and vegetative parts of plants, and animal matter. Plant reproductive parts comprise fruit and flowers; drab catkins or conspicuous flowers, large or small, closed or open, are all eaten. The bulkier, less colourful flowers probably provide a similar range of nutrients to the pulps of drier fruit. The nectar content of conspicuous animal-pollinated flowers yields useful amounts of protein, free amino acids, lipids and ascorbic acid, in addition to dissolved sugars (Hladik *et al.*, 1971).

Gibbons, particularly the smaller ones, seldom eat ripe fruit, and rarely digest seeds. From their viewpoint a fruit is a parcel of succulent pulp sandwiched between unwanted coat and seed. The seed is usually swallowed and excreted whole, or otherwise rejected from the mouth while feeding. Succulent fruit are rich in free sugars, the most quickly assimilated and usable source of energy. The distinction between fruit pulp and seed is important, although the categories are repeatedly lumped in studies of primate feeding behaviour. Edible seeds often belong to commoner species than do attractive pulps, and this directly affects the carrying capacity of the habitat for pulp and seed-eaters, and hence their population densities. Moreover, seeds contain different proportions of nutrients (e.g. sometimes a high fat content).

Most vegetative plant parts of trees and climbers are eaten: young leaves, including both unopened shoots and opened leaves lacking mature size, colour or texture; mature leaves; leaf stems; and stems of soft-stemmed climbers. No bark is eaten. Leaves contain more protein and less fibre when young (Hladik, 1978). In some leguminous leaf shoots protein may account for half the dry weight. Gibbons, especially the siamang, concentrate on young rather than mature leaves (see below), despite the perennial vastly greater abundance of the latter. Young leaves must be selected either for their higher protein content, for their lower fibre content, for their lower content of toxic secondary compounds, or for all three reasons. The colon is relatively bulky (Chivers & Hladik, 1980) and some bacterial fermentation is likely to occur, just as it does in humans. The surprising efficiency of fibre breakdown found in experiments on humans (Southgate *et al.*, 1976), however, is belied by their inability to live on a very high-fibre diet (McCance & Lawrence, 1929). There is therefore some doubt as to how much use the gibbons can make of fibre.

Animal food is almost exclusively invertebrate, and mostly insects. It is obtained in three ways: (1) the ape chances upon a mass or column of termites or social ants and stops to mop them up for a few minutes; (2) the ape visits trees of certain species infested with caterpillars or insect galls; (3) the ape catches prey during slow-moving bouts of foraging among mature, and sometimes dead, leaves. The success rate of the last method is about one catch each minute; many of the catches are spiders. The main attraction of invertebrates as food is presumably the greater range of amino acids found in the proteins of individual animals than in the proteins of individual plant species (Hladik *et al.*, 1971).

Dietary proportions

The proportions of total feeding time spent on the major food types by the three Malayan gibbons are summarized in Table 1. Feeding time is a less informative measure of diet than weight of intake, but is far easier to employ in the prevailing observation conditions.

Efforts were made to sample the rate at which gibbons ate different foods (Chivers, 1974; Raemaekers, 1979). Although there was much variation between particular foods defined by plant part and species, there was less differece between overall scores for leaves and fruit. Animal matter is eaten much more slowly than plant matter, because the items are so much smaller and usually more dispersed. Consequently the proportion of animal matter in the diet is over-represented by feeding time (Table 1); but then it is also disproportionately valuable in the quantity of protein yielded.

The 5 kg gibbons, *H. lar* and *H. agilis*, eat relatively more fruit and less green plant matter than the 10 kg *H. syndactylus*. Moreover, reports from studies of *H. lar* and other small gibbons elsewhere may suggest a lower intake of leaves than reported here (Carpenter, 1940; Ellefson, 1974; Whitten, 1980).

The small species also occupy larger home ranges, travel further about them each day visiting more food sources, and have less group cohesion than the siamang. These differences in dietary proportions and ranging behaviour can be related to the siamang's stockier body

Table 1. *Features of natural diet and ranging behaviour of Malayan gibbons*

Species	Alert period start stop	Alert period duration (min)	Feeding time duration (min)	Feeding time % of alert period	Food (%) fruit	Food (%) leaves	Food (%) insects	Food (%) figs	Feed visits per day	Feed visits duration (min)	Day range (m)	Home range (ha)
H. syndactylus												
Ulu Sempam[a]	06.37–16.48	610	276	45	49	50	1	41	12.0	23	800	15
Kuala Lompat[a]	06.42–17.00	643	354	56	41	57	2	24	13.3	27	969	34
Kuala Lompat[b]	06.28–17.32	664	345	52	47	45	8	23	8.5	41	640	28
Kuala Lompat[c]	06.42–16.32	591	300	51	42	43	15	22	12.6	24	723	47
H. lar												
Kuala Lompat[b]	–		204		60	36	4	23			1850	54
Kuala Lompat[c]	06.43–15.18	553	228	43	57	29	13	22	14.4	16	1521	57
H. agilis												
Gunung Bubu[d]	06.41–15.59	550	196	36	61	39	1	17	12.0		1335	29

Observers: [a]D. J. Chivers; [b]J. R. & K. S. MacKinnon; [c]J. J. Raemaekers; [d]S. P. Gittins shape, which should be associated with a lower metabolic rate (MacKinnon, 1977; Rijksen, 1978; Raemaekers, 1979). With their relatively simple gastrointestinal tract, gibbons almost certainly obtain fewer calories per unit time and weight from leaf than from fruit pulp, which have similar water, but different nutrient, contents. The siamang's predicted lower metabolic rate matches its leafier diet. Since even tender young leaves are commoner than fruit, e.g. at Kuala Lompat over 12 months 50% of 900 trees bore young leaves of species eaten by gibbons in that year, but only 8% of eaten species bore fruit (Raemaekers, 1977), travel in search of food can be reduced. This is desirable for the siamang, because it weighs twice as much as the small species, but travels more slowly without a proportionately longer 'arm-stride' to offset the extra cost of shifting its greater weight.

Temporal variation in diet

Many tropical mammals show peaks of feeding and travelling after dawn and before dusk, and rest during the heat of the day; *H. lar* and *H. syndactylus*, however, tend to feed for about half of each hour from 0700 to 1600 inclusive (Fig. 1). *H. lar* does not feed for so long overall and stops about an hour earlier. Considering the extent of the

Fig. 1. Number of minutes of feeding by adult males in each hour of the day at Kuala Lompat (120 days for each species: 13 142 feeding observations for siamang and 6852 for lar gibbons).

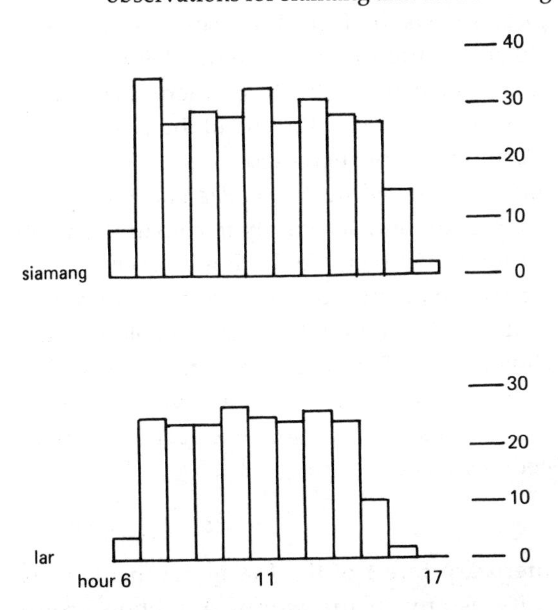

dietary overlap with sympatric monkey species (MacKinnon & MacKinnon, 1978) which exhibit the more usual feeding pattern, such temporal differences could be a means of reducing behavioural competition. *H. agilis* is very similar to the other two gibbons in its daily pattern (Gittins, 1979).

Chivers (1975) reported that, while the consumption of other food categories varied little through the day, the time which the siamang spent eating fruit decreased and that spent eating leaves increased. Fruit filled about 60% of feeding time at dawn, but only about 20% by mid-afternoon. There was a secondary peak of fruit-eating at the end of the day. The pattern was confirmed in *H. syndactylus* and demonstrated in *H. lar* as well (Raemaekers, 1978*b*). The special attraction of figs for the first feeding bouts of the day seems to be that they are the most efficient food source in terms of time and energy to restore blood sugar levels after a long overnight fast of 12–14 hours. The emphasis on leaves later in the day may simply be a reciprocal consequence of taking up the morning with fig-eating – although there are several other explanations, such as that leaves may contain more sugar after several daylight hours of photosynthesis and less effort is required to harvest them.

Day-to-day fluctuations in dietary proportions can be quite marked. The proportions of food types in the diet also vary considerably between months. The nutrient intake from these foods may however vary less (Hladik, 1977). Variation between months is expected because almost all tropical forests in fact show some seasonality (Whitmore, 1975; Raemaekers, Aldrich-Blake & Payne, 1980).

Despite the almost year-round availability of figs, there seems to be less fruit consistently in the rainy season at the end of the year. At this time it seemed that *H. lar* apparently increased its search effort to obtain adequate high-energy fruit, while *H. syndactylus* conserved energy by decreasing search effort and selectivity to consume a wide range of low-energy leaf foods (Chivers, 1974; Raemaekers, 1977). Chivers (1974) argued that the siamang obtained consistent proportions of fruit, leaves and animals throughout the year by exploiting the plant community as a whole, rather than its component species. This is possible because of the different periodicity of floral cycles – annual, biennial, sporadic, and non-annual, presumably endogenous, longer or shorter than a year (Medway, 1972).

Diet diversity

The lowland dipterocarp forest of the Malay Peninsula is as diverse a plant community as any in the world. A gibbon group occupying 25–60 ha has access to perhaps 500 species of trees and climbers. Towards the extremes of their geographical distribution gibbons occur in less diverse habitats, but the range of foods available is still great, since they inhabit only tall, dense forest. In the face of such diversity, it is not surprising that the gibbons at least nibble a huge range of foods; nevertheless, most of their feeding time is spent on a small proportion of these.

A compact method of expressing the variation in dietary diversity through time is to compare monthly values of an information index, such as the Shannon–Weaver index (Wilson & Bossert, 1971), which accounts for how many foods were eaten and the spread of feeding time across them. When this is done it can be seen that the apes eat a more restricted diet in the wettest weather, when ranging was reduced, and the most varied diets were eaten in the driest weather, when ranging was increased (Fig. 2) (Raemaekers, 1977).

Chivers (1974) identified three kinds of food species – primary, secondary and tertiary – which Raemaekers (1977) refined into four classes of food use.

1. *Staples*, eaten in most months, because of asynchronous activity in individuals of a common plant species (e.g. *Gnetum*), or a rarer one producing food over a long period (e.g. *Derris* and *Sloetia*). The former staples are mostly young leaves of climbers, since few trees bear young leaves for long; they occur near much-frequented sleeping trees, and one may

Fig. 2. Variation between months in dietary diversity for siamang and lar gibbons at Kuala Lompat (identified plant foods only). Primary monsoon: November to January; secondary monsoon: April to May.

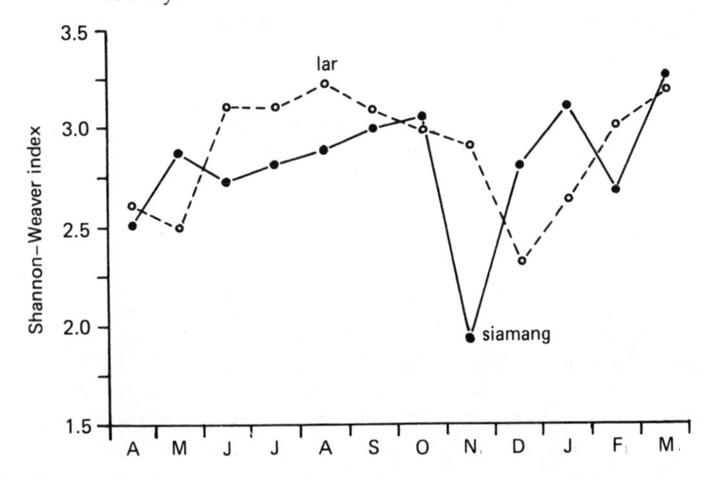

surmise that the certainty of a large evening meal in a known place gives the apes the freedom to be more exploratory during the day at the risk of not finding much food (Raemaekers, 1978*a*).

2. *Periodicals*, eaten in about 6 months of the year, borne by rare, dispersed species with shorter periods of production than the rare staples, and some asynchrony (e.g. *Diospyros* and *Polyalthia*).
3. *Monthly specials*, eaten intensively in only 1–3 consecutive months, produced by one individual of a rare species, or a few in synchrony (e.g. *Artocarpus*).
4. *Nibbles*, eaten briefly and occasionally as a result of the gibbon's continuous sampling of the vegetation to find out what is edible.

Diet selection

Gibbons are extremely selective feeders compared with other primates, which no doubt accounts for their low biomasses of around 50 kg/km^2. While *H. syndactylus* has been classed as a folivore, and *H. lar* as a frugivore (Chivers, 1974; Ellefson, 1974), on the basis of the amount of each food category ingested, Raemaekers (1977) argues that it is more useful for some purposes to think of *H. lar* as a fruit-pulp seeker, and *H. syndactylus* as a fig seeker.

The plant part most eaten is not necessarily that most sought after. When opportunities taken are expressed as percentages of those provided, it is clear that both gibbon species selected figs more strongly than other fruit and flowers overall, and these more strongly than young leaves overall (Table 2).

Table 2. *Selection for plant parts by siamang and lar gibbons*

Food category	Group	Feeding opportunities provided	Feeding opportunities taken	Taken as % provided
Fig fruit	Siamang	26	10	39
	Lar	25	10	40
Other fruit	Siamang	71	7	10
	Lar	97	9	9
Flowers	Siamang	19	2	10
	Lar	15	2	13
Young foliage	Siamang	308	6	2
	Lar	273	7	3

Apart from species and food type, the gibbons also select for large source size within their food species. Available data confirm that it pays them to travel until they reach a big source rather than to stop more frequently at smaller sources.

The intake of food is limited by chewing rate, especially for leaves, and collection rate, especially for fruit. Since the siamang eats more leaves and has a greater chewing capacity, because of its relatively larger palate and more efficient cheek teeth (Kay, 1975), and the lar gibbon eats more fruit and moves faster, the diets of both species are broadly consistent with maximising the rate of food intake.

The siamang and lar gibbons differ in the rate of intake of some foods, but not of others, and the direction of the difference varies, e.g. *H. lar* eats figs faster, and *H. syndactylus* eats leaves faster. In some food trees the adult female may eat twice as fast as the male, and on average she feeds for 50 min longer each day (Chivers, 1974). There may also be differences in the intake of different food types: Raemaekers (1977) notes that extra feeding by the female may be directed mainly at insects, and suggests that this increase in protein intake relates to pregnancy and lactation.

Synthetic diets

Timing

Although captive gibbons are only fed once or twice a day (Table 3), they do not usually gorge themselves immediately, so that they have continual access to food throughout the day in a manner comparable to wild gibbons. Zoos do not always, however, provide the bulk of food in the morning.

Variety

While zoos do not study the composition of wild diets or try to match them biochemically or energetically, since such data are only just becoming available, they do make strenuous efforts in many cases to cater for the requirements of each species in terms of variety and balance (e.g. Brambell, 1976; Doherty & MacNamara, 1976; Kerr, 1976). Kerr stresses the need to cater especially for newly imported animals and discusses the various diets under the headings of 'natural', commercial, semi-purified and elemental (or chemically defined). While natural and commercial diets may be cheap and palatable, they are often lacking in nutritional balance, and may be of unknown digestibility and toxicity. A wealth of captive study and experimentation has led to satisfactory solutions in most cases. Primates adapt

Table 3. *Synthetic diets of some captive gibbons*

	Twycross	London	San Francisco	Los Angeles	Milwaukee	Washington
Species	7 *H. syndactylus* 2 *H. concolor* *H. pileatus* 3 *H. agilis* *H. lar* 1 *H. muelleri*	5 *H. lar*	7 *H. syndactylus* 2 *H. concolor*	3 *H. syndactylus* 2 *H. pileatus* 2 *H. hoolock*	6 *H. syndactylus* 3 *H. lar*	3 *H. syndactylus* 8 *H. concolor*
Feeding time	evening	? evening	morning (pick all day)	14.00	once/day	09.00 and 15.00
Commercial foods						
Monkey chow			yes	yes	yes	mornings
Primate diet		25% 0.45 kg/day = 2875 kJ/day				mornings
Primate pencils	mornings					
Esotone tables		1				

Other foods		75% 1.35 kg/day = 2250 kJ/day			
Hard-boiled egg	alt. mornings		yes	yes		
Brown bread	1 slice		yes		yes	
Kale						yes
Cabbage						yes
Beans						yes
Lettuce	½			yes	yes	
Carrot	1		yes	yes	yes	
Apples	1–2		yes	yes	yes	yes
Bananas	4		yes	yes	yes	yes
Oranges	when available			yes	yes	yes
Yams (cooked)			yes	yes	yes	
Other	occ. some tomato, grapes, cucumber, pear, celery		mirror plant, occ. bird	cooked horse meat	sunflower seeds, peanuts, celery, onions, pineapple	

especially well to a change of environment, and the frugivorous apes adapt easily to a captive diet of fruit and vegetables balanced with animal protein.

Captive gibbons receive a wide variety of fruit and vegetables, often according to availability, supplemented with one or more commercial foods, of which Purina monkey chow and BP Nutrition primate diet are the most popular (Table 3). Monkey chow contains not less than 25% crude protein and 5% crude fat, and not more than 3.5% crude fibre, 3% added minerals and 6% ash, with a wide range of vitamins and trace elements, providing 4.5 kcal/g of food (Doherty & MacNamara, 1976). Primate diet contains 12% moisture, 4% dry matter of fibre, 23% protein, 42% carbohydrate and 6% fat, yielding 2.8 kcal/g of food, supplemented with vitamins A, D_3 and E, calcium and phosphorus (Brambell, 1976).

Quantity

Some zoos do not weigh out diets very precisely, presumably on the reasonable assumption that higher primates will eat what they need, and that deficiencies can be spotted easily and remedied. Other zoos pay detailed attention to the quantities of nutrients, energy and weight of food supplied to each animal, although it can never be certain that these are the precise needs of the animal, or that it will consume these quantities, especially where it is caged with others.

At London Zoo it is assumed that each animal requires $140W^{0.75}$ kcal/day, (where W is body weight), with extra for pregnancy, lactation and competition. Gibbons are classed as general feeders in contrast to the other dietary groups of predators, coprophages, fruit bats and plant feeders (Brambell, 1976). From an assumed total body weight of 23.5 kg, the five lar gibbons are considered to have a combined metabolic weight of 15.3 kg, which requires 9180 kJ/day. This is supplied as 1.8 kg of food, composed of 25% primate diet, yielding 2875 kJ/kg, and 75% fruit and vegetables, yielding 2250 kJ/kg (M. R. Brambell, personal communication). This works out at 0.36 kg/gibbon, but three are immature.

The National Zoological Park in Washington D.C. has three pairs of concolor gibbon, one with an infant, plus a single female, and a pair of siamang with a juvenile. The concolors each receive 0.53 kg/day and the siamang 0.71 kg/day (M. Roberts, personal communication). At the Bronx Zoo in New York, by comparison (Table 4), where there were three immature lar gibbons and two immature siamang (Doherty & MacNamara, 1976), the former each received 0.84 kg/day and the latter

2.23 kg/day (if the clearly erroneous figures for the siamang have been read correctly). At both London and Washington, however, the smaller supply of food contained a much higher proportion of commercial foods.

Health

The National Zoological Park feeds apples, bananas and oranges to the apes, regarding them as empty calories which are justified in view of the year-round outdoor housing of the gibbons (M. Roberts, personal communication); fruit supplies are increased during winter and other times of stress. They have noted the gibbons' delight at fresh browse, and the large leaf consumption of siamang under these conditions. Others have noted that too much fresh fruit promotes loose stools and loss of condition (E. H. Haimoff, personal communication).

Seasonal looseness of stools in siamang has led the National Zoological Park to improve their winter diet (M. Roberts, personal communication), by feeding 10% more kale and cabbage twice daily, and some bamboo, with the complete omission of oranges. Since the availability of fresh browse in the summer seemed to be linked with

Table 4. *Diets of gibbons at New York and Washington zoos*

	Bronx Zoo, New York		National Zoological Park, Washington, D.C.	
% by weight	*H. lar*	*H. syndactylus*	*H. syndactylus*	*H. concolor*
Primate diet	18	11	31	42
Monkey chow	4	6	12	16
Bananas	0	18	2	3
Apples			2	3
Carrots	18	22	8	11
Cabbage			16	8
Beans			8	5
Celery	0	3		
Yams	18	15		
Grapes	8	3		
Oranges	0	15	2	3
Kale			20	10
Eggs	4	3		
H grass	12	22		
Total weight (g/animal per day)	843	2231	714	532

their better health in that season, the Zoo has recently followed the example of Frankfurt Zoo by feeding throughout the rest of the year cut and trimmed leaves of maple, beech, willow and mulberry, which were sealed and frozen when available without loss of palatability.

It is clear from the descriptions of the behaviour of captive gibbons, supplied by the zoos who responded to our request for information, that care taken in providing an adequate and balanced diet leads to physiologically and socially healthy animals, as best proven by the occurrence of breeding in the monogamous pairings.

Concluding discussion

We have presented data on the time spent eating various foods by gibbons in the wild, and on amounts of various foods fed in captivity. To appreciate the relevance of field observations we have to be able to (1) transform time data into weights; and (2) identify differences in activity patterns, and hence in energy requirements, between the wild and captive states.

Data on feeding time are difficult to collect in the field. DJC determined the consumption by weight of the fig, *Ficus virens*, and of the fruit of *Eugenia* sp., as 2.1 and 2.4 g/min, respectively (Chivers, 1974). When eating these figs, mean daily consumption of food by siamang was consistently between 0.26 and 0.29 kg; at other times it was of the order of 0.15–0.18 kg. Gittins (personal communication) concluded that agile gibbons consumed about 0.21 kg/day of figs, from an intake of 7 g/min and a daily feeding time of 30 min; another species of fruit was eaten at the rate of 7 g/min of edible material. From Hladik's finding that primate species tend to consume food daily to the order of 15% of their body weight, Gittins predicted that figs accounted for about 25% of the daily intake of 0.825 kg in agile gibbons. Similar calculations for lar gibbons and siamang led to a predicted daily intake of 0.825 and 1.575 kg, respectively.

Siamang feed for 300 min or more each day, and lar and agile gibbons for about 200 min, with varying proportions of fruit, leaves and animal matter (Table 1). Estimates of the weight of each type of food consumed can be derived both from the proportions that each contributes to the diet in terms of feeding time, and from the actual time spent eating each food. From Hladik's (1977) study of primates in Sri Lanka, we can derive the relation between feeding time and food weight from scores for each of a variety of foods, both as proportions and rate of consumption in g/min. Thus, food intake by weight is 0.17 g/min for insects, 0.58 g/min for flowers, 0.85 g/min for leaves and 1.20 g/min for fruit.

From Hladik (1977) we can deduce consumption rates of 5 g/min for fruit, 4.1 g/min for leaves, 3.7 g/min for flowers and 0.8 g/min for insects (wet weights). Adjustments of JJR's (1977) dry weights for 67% water in flowers, 72% in fruit and 77% in leaves (Hladik, 1977) leads to scores of 5 g/min for fruit intake, 3.5–4.5 g/min for leaves and 1 g/min for flowers. Since the latter confirm the figures derived from Hladik's study, the wet weight values provide the basis for our quantification by weight of the diets of free-ranging gibbons (Table 5).

Table 5. *Estimates of food intake by weight of Malayan gibbons*

	H. syndactylus			*H. lar*	*H. agilis*
	K. Lompat			K. Lompat	G. Bubu
	1969–70	1975–76	1970	1975–76	1975–76
Feeding time (min/day)	354	310	276	217	196
% time spent feeding on:					
fig fruit	24	22	41	22	17
other fruit	8	14	6	28	41
all fruit	32	36	47	50	58
leaves	58	43	50	29	39
flowers	9	6	2	7	3
insects	2	15	1	13	1
mins/day spent feeding on:					
fig fruit	85	68	113	48	33
other fruit	28	43	14	61	80
all fruit	113	112	127	109	114
leaves	205	133	138	63	76
flowers	32	19	6	15	6
insects	7	47	3	28	2
kg/day (dry wt) eaten:					
fig fruit	0.116	0.093	0.154	0.080	0.055
other fruit	0.043	0.065	0.021	0.073	0.096
all fruit	0.158	0.158	0.175	0.153	0.151
leaves	0.217	0.141	0.146	0.047	0.057
flowers	0.048	0.029	0.009	0.008	0.003
kg/day (wet wt) eaten:					
fig fruit	0.413	0.330	0.549	0.286	0.197
other fruit	0.152	0.234	0.076	0.261	0.343
all fruit	0.565	0.564	0.625	0.547	0.540
leaves	0.945	0.613	0.636	0.205	0.248
flowers	0.145	0.086	0.027	0.023	0.009
insects	0.006	0.038	0.002	0.022	0.002
	1.661	1.301	1.290	0.797	0.799

The results indicate that siamang usually consume 1.25–1.7 kg/day of food, and lar gibbons *ca* 0.8 kg/day wet weight. The mean consumption rates are slightly below and above the estimate of food intake from body weight of 1.575 kg/day for siamang and just below the estimate of 0.825 kg/day for the smaller gibbons. The discrepancies could indicate either errors in the calculations or that siamang have more trouble finding adequate food than the smaller gibbons, despite their alternative strategy of switching to young leaves rather than increasing their search effort for fruit. The smaller quantity of food fed to captive gibbons (Table 4) could reflect either the higher nutrient quality of the foods and/or the lower level of activity, and hence of energy requirements, in the captive situation.

It is difficult to carry this discussion to a more satisfactory conclusion until more data are available on energy intake and expenditure, nutritional content of foods and quantities ingested both in the field and in captivity. Nevertheless, we hope that the comparative data collated here will assist in such investigations, as well as contributing to the healthy maintenance of gibbons in captivity.

Acknowledgements

We thank those who assisted us in our field work in Malaysia, especially the Department of Wildlife and National Parks in Kuala Lumpur and Temerloh, and the Department of Zoology, University of Malaya. Funding was received from the Commonwealth Scholarship scheme, the Science Research Council, the Goldsmiths' Company (London), the New York Zoological Society, the Boise Fund (Oxford), the Royal Anthropological Institute (London), the Royal Society (London), and the Medical Research Council. We thank the Departments of Physical Anthropology and Anatomy, University of Cambridge, for the accommodation and help they have provided.

We are very grateful to those who responded to our request for information on captive gibbons: Meg Muschamp of Twycross Zoo, Michael Brambell of the London Zoo (now at Chester), Saul Kitchener of the San Francisco Zoo, Michael Crotty of the Los Angeles Zoo, Sam La Malfa of the Milwaukee Zoo, and Miles Roberts of the Washington Zoo.

References

Brambell, M. R. (1976) Diets of mammals at the London Zoo. In *Handbook Series in Nutrition and Food*, section G, vol. II, pp. 381–7. Florida: CRC Press

Carpenter, C. R. (1940) A field study in Siam of the behavior and social relations of the gibbon, *Hylobates lar*. *Comp. Psychol. Monogr.*, **16**, 1–212

Chivers, D. J. (1974) The siamang in Malaya: a field study of a primate in tropical rain forest. *Contrib. Primatol.*, **4**, 1–335

Chivers, D. J. (1975) Daily patterns of ranging and feeding in siamang. In *Contemporary Primatology*, ed. S. Kondo, M. Kawai and A. Ehara, pp. 362–72. Basel: Karger

Chivers, D. J. (1977) The lesser apes. In *Primate Conservation*, ed. Prince Rainier & G. H. Bourne, pp. 539–98. New York: Academic Press

Chivers, D. J. & Gittins, S. P. (1978) Diagnostic features of gibbon species. *Int. Zoo Yearbook*, **18**, 157–64

Chivers, D. J. & Hladik, C. M. (1980) Morphology of the gastro-intestinal tract in primates: comparison with other mammals. *J. Morphol.*, **166**, 337–86

Chivers, D. J., Raemaekers, J. J. & Alrich-Blake, F. P. G. (1975) Long-term observations of siamang behaviour. *Folia Primatol.*, **23**, 1–49

Doherty, J. G. & MacNamara, M. C. (1976) Mammal diets: New York Zoological Park, Bronx, New York. In *Handbook Series in Nutrition and Food*, section G, vol. II, pp. 389–416. Florida: CRC Press

Ellefson, J. O. (1974) A natural history of the white-handed gibbon in the Malayan Peninsula. In *Gibbon and Siamang*, ed. D. M. Rumbaugh, pp. 1–136. Basel: Karger

Gittins, S. P. (1979) The behaviour and ecology of the agile gibbon, *Hylobates agilis*. Unpublished Ph.D. dissertation, University of Cambridge

Hladik, A. (1978) Phenology of leaf production in a rain forest of Gabon: distribution and composition of food for folivores. In *Ecology of Arboreal Folivores*, ed. G. G. Montgomery, pp. 441–64. Washington: Smithsonian

Hladik, C. M. (1977) A comparative study of the feeding strategies of two sympatric species of leaf monkeys, *Presbytis senex* and *Presbytis entellus*. In *Primate Ecology*, ed. T. H. Clutton-Brock, pp. 324–53. London and New York: Academic Press

Hladik, C. M., Hladik, A., Bousset, J., Valdebouze, P., Viroben, G. & Delort-Laval, J. (1971) Le régime alimentaire des primates de l'île de Barro Colorado (Panama). *Folia Primatol.*, **16**, 85–122

Kay, R. F. (1975) The functional adaptations of primate molar teeth. *Am. J. Phys. Anthropol.*, **43**, 195–216

Kerr, G. R. (1976) Diets (natural and synthetic): subhuman primates. In *Handbook Series in Nutrition and Food*, section G, vol. II, pp. 577–87. Florida: CRC Press

McCance, R. A. & Lawrence, R. D. (1929) The carbohydrate content of foods. Special Report Series, Medical Research Council, No. 135

MacKinnon, J. R. (1977) A comparative ecology of the Asian apes. *Primates*, **18**, 747–72

MacKinnon, J. R. & MacKinnon, K. S. (1978) Comparative feeding ecology of six sympatric primate species in West Malaysia. In *Recent Advances in Primatology*, vol. 1, *Behaviour*, ed. D. J. Chivers & J. Herbert, pp. 305–21. London: Academic Press

Medway, Lord (1972) Phenology of a tropical rainforest in Malaya. *Biol. J. Linn. Soc.*, **4**, 117–46

Raemaekers, J. J. (1977) Gibbons and trees: comparative ecology of the siamang and lar gibbons. Unpublished Ph.D. dissertation, University of Cambridge

Raemaekers, J. J. (1978*a*) Changes through the day in the food choice of wild gibbons. *Folia Primatol.*, **30**, 194–205

Raemaekers, J. J. (1978*b*) The sharing of food sources between two gibbon species in the wild. *Malay. Nat. J.*, **31** (3), 181–8

Raemaekers, J. J. (1979) Ecology of sympatric gibbons. *Folia Primatol.*, **31**, 227–45

Raemaekers, J. J., Aldrich-Blake, F. P. G. & Payne, J. B. (1980) The forest. In *Malayan Forest Primates*, ed. D. J. Chivers, pp. 29–61. New York: Plenum

Rijksen, H. J. (1978) *A Field Study of Sumatran Orang-utans: Ecology, Behaviour and Conservation*. Wageningen: H. Veenman & B. U. Zonen

Southgate, D. A. T., Branche, W. J., Hill, M. J., Drasar, B. S., Walters, R. L.,

Davies, P. S. & McLean Baird, I. (1976) Metabolic responses to dietary supplements of bran. *Metabolism*, **25**, 1129–35
Whitmore, T. C. (1975) *Tropical Rain Forests of the Far East*. Oxford: Clarendon
Whitten, A. J. (1980) Ecology and behaviour of the Mentawaian gibbon, *Hylobates klossii*. Unpublished Ph.D. dissertation, University of Cambridge
Wilson, E. O. & Bossert, W. H. (1971) *A Primer of Population Biology*. Stamford, Conn.: Sinauer

II.4

Tool use in the exploitation of food resources in *Cebus apella*

E. VISALBERGHI AND F. ANTINUCCI

Introduction

Both in captivity and in the wild, black-capped capuchins (*Cebus apella*) crack various kinds of nuts or hard-shelled mollusks in order to eat their contents. This task is accomplished in a variety of ways, two of which involved tool-using behaviors. They can hold the fruit with their hand(s) and strike it against a hard surface, or they can pound on it with a hard object. The first technique, which has been labelled *proto tool use* (Parker & Gibson, 1977), is very common and widespread, and has recently been described and studied systematically in the field (Izawa & Mizuno, 1977; Izawa, 1979; Terborgh, 1983). However, there are only anecdotal reports of the second technique – that involving *true tool use* – either in the wild (quoted in Jennison, 1927, and Hill, 1960) or in captivity (Romanes, 1883; Kluver, 1933; Nolte, 1958; Vevers & Weiner, 1963; Tobias, 1965).

The present study originated from casual observations of a group of *Cebus apella* housed in the Rome Zoo. It was noticed that one monkey of this group attempted to crack a nut open by pounding on it with whatever object was accidentally available in its cage: an empty can, a cardboard box, a piece of rubber and even a potato. No other cagemate was ever observed to do so.

An experiment was set up in which the tool-using behavior could be explored and evaluated systematically. For this purpose the tool-using capuchin (called subject C) was given the task of cracking nuts of two different sizes, hazelnuts and walnuts. He was offered a choice among three tools, similar in size, a stone, a block of wood and an empty plastic container which, because of their weights (500, 120, and 40 g, respectively) were of different efficacy as pounding objects. In order to evaluate the efficacy of the tool-using strategy, subject C's performance was compared on the same task to his own performance when no pounding object was at his disposal and to that of a cagemate (called subject R) who had not been seen engaging in tool-using behavior.

Methods

Subjects C and R were adult *Cebus apella* males. They were born and housed in an indoor–outdoor cage (approximately 12.3 m^2) with other conspecifics. Each subject was tested alone.

In each trial the three tool objects – stone, wood and plastic – were aligned on the ground near, and parallel to, the front wire mesh at distances of approximately 40 cm from one another. A nut was then given to the subject from either the far left side or the far right side of the front wire mesh. Each trial lasted 5 min or until a solution occurred, whichever came first. Twelve trials – six hazelnuts and six walnuts – were presented per session, alternating the kind of nut on each trial. Subject C completed a total of eight sessions, and subject R a total of five sessions. Subsequently, C was given five additional sessions without tools, following the same procedure.

The one–zero sampling method (Altmann, 1974) was used to score, in 10-s intervals, the occurrence of pounding behavior, tooth-action and the proper use of tools. Time to solution was recorded, as well as the final action responsible for the cracking of the nut.

Results and discussion

When tools were given (tool condition), C used them appropriately in all but two trials, while R never did so. Obviously, when tools were not given (no-tool condition), C was in the same situation as R.

Subject C's mean times to solution were 13 s for the hazelnuts and 16 s for the walnuts in the tool condition, and 62 s for the hazelnuts and 40 s for the walnuts in the no-tool condition. R's times to solution were 42 s and 79 s for the hazelnuts and the walnuts, respectively. A comparison of these times clearly shows that using a tool confers a definite advantage. The times employed in this task when tools are not used are from 2.5 to 5 times longer than when tools are used.

Furthermore, the different characteristics of the two kinds of nut do not affect the performance when tools are used: there is no statistical difference between the opening times of the two kinds of nut in the tool condition (Mann–Whitney *U* Test).

The data show differences in the opening times, both between the two kinds of nut and the two subjects, when tools are not used. Subject

R performed significantly better in the hazelnut than in the walnut trials (Mann–Whitney *U* Test, $p < 0.05$). Subject C's times show an inverse tendency, though the difference between nut type is not significant. Comparison between the two subjects shows that their opening times differed significantly in the walnut trials (Mann–Whitney *U* Test, $p < 0.05$), but not in the hazelnut trials. In order to explore further these differences, the strategies employed by the two subjects in cracking open the nuts must be considered.

Non-tool-using strategies

Subject R used only the action of his teeth to open the nuts. Subject C used two kinds of action: tooth-action, as did R, and pounding. When pounding, the monkey would hold the nut in his hand(s) and strike it with a rapid succession of blows (generally from 5 to 10) against a hard surface. In most trials the monkey switched back and forth between tooth-action and pounding. Table 1 shows a number of measures relative to the differential deployment of these two techniques in the hazelnut and walnut trials (i.e. the number of trials in which at least one interval of tooth-action or pounding occurred, the mean number of tooth-action and pounding intervals, and the number of trials where the final action responsible for the cracking was tooth-action or pounding).

Obviously, the effectiveness of the two techniques depends on the size of the nut. The larger the nut, the more difficult it is to insert between the premolars and the less the pressure that the jaws can exert on it. Larger nuts, however, can be more easily held and more effectively pounded on a hard surface. Hazelnuts, being smaller, are easier to crack through tooth-action than the larger walnuts.

Table 1. *Utilization of tooth-action (T) and pounding (P) by subject C in the no-tool condition*

	Number of trials with at least one interval		Mean number of intervals		Cracking action	
	T	P	T	P	T	P
Hazelnut ($N = 27$)	27	19	4.5	5.0	27	0
Walnut ($N = 26$)	16	26	1.9	3.7	14	12

As shown in Table 1, C appeared to employ the two techniques according to their relative efficiency with respect to the two kinds of nut: more tooth-action with the hazelnuts and more pounding with the walnuts.

Subject R did not have pounding in his behavioral repertoire. Therefore, his performance was successful with the hazelnuts but extremely poor with the walnuts. His solution time in the walnut trials was approximately twice as long as his solution time in the hazelnut trials.

Tool-using strategies

When tools were available, C used them appropriately. Once C had taken hold of the nut, he would typically bring it close to one of the tools, position it on the ground, and then proceed to lift the tool with both hands and strike a blow onto the nut. The tool was then lifted again to check the result. If the nut had cracked, C would discard the tool and eat the content. If either the nut had resisted the blow or the blow had missed the target, a second, a third and sometimes even a fourth blow would be made. If still unsuccessful, he would change tools.

Did C discriminate among the three tools according to their relative efficiency? Table 2 presents a number of measures that capture the differential use of the three tools.

By and large, these measures show that C used the stone twice as much as the wood, which in turn was utilized twice as much as the plastic. This pattern is obviously consistent with the relative efficiency of the three tools as pounding objects.

C's first choice of tool can offer another indication of his ability to discriminate the most efficient percussor. The procedure employed in

Table 2. *Utilization of the stone, wood, and plastic tool by subject C in the hazelnut (H; N = 46) and walnut (W; N = 30) trials*

	Stone		Wood		Plastic	
	H	W	H	W	H	W
Number of trials with at least one interval	36	21	16	13	7	5
Mean number of intervals	0.98	1	0.50	0.57	0.22	0.17
Final action	31	17	12	10	0	0

the experiment was such that C had in turn each of the three tools as the closest, medium and farthest from the position where he received the nut. If C had not been actively selecting among them, the tool used after receiving the nut would have been the one closest to him. An analysis of the first tool chosen, combining hazelnut and walnut trials together, was, therefore, performed.

Table 3 shows, for each tool, the number of times in which, being in the closest position, the tool was in fact chosen, or discarded, and, if discarded, in favor of which other tool. As can be seen, the data clearly indicate an active choice of the stone tool. The stone was practically never discarded, as opposed to the wood and, even more, the plastic. When the wood or the plastic was discarded, it was nearly always in favor of the stone tool.

Conclusion

The comparison between the tool *versus* the no-tool condition shows the importance of tool use in feeding strategies. Tool use not only reduces the amount of time (and hence, presumably, effort) needed to gain access to the highly nutritional content of a nut but also brings about a levelling of the different 'resistances' offered by the nuts.

A monkey's success in cracking open a nut without the use of an instrument, i.e. only through the action of his body parts, is determined both by the muscular strength of the animal and by the physical properties of the fruit, i.e. its hardness and size. By contrast, the rupture force that might be exerted by striking a blow with a hard percussor is so widely superior to the one that might be exerted

Table 3. *First choice of tool by subject C*

	In closest position	Chosen	Discarded	In favor of	
Plastic	23	5	18	4	Wood
				14	Stone
Wood	23	12	11	11	Stone
				0	Plastic
Stone	20	17	3	3	Plastic
				0	Wood

through direct actions, that differences both among fruits and in the physical strength of different animals are reduced considerably. As a result, when tools are used, success in cracking nuts is much more constant and stable, depending only on the accuracy of the striking action. Access to food resources is, therefore, less and less controlled by environmental variables (physical properties of the nuts) or by the physical characteristics of the animal.

Obviously, a successful tool-using strategy requires not only that the animal be capable of performing the appropriate action with the tool (i.e. striking the nut), but also that the appropriate tool objects be available. It is in connecting these two factors that the value of 'intelligent' tool use emerges. When this capacity is present, any suitable object can be put freely into the role of percussor, allowing the exploitation of a wide range of possibilities offered by the environment and at the same time a selection of the most efficient one can also be made.

References

Altmann, J. (1974) Observational study of behavior: sampling methods. *Behaviour*, **49**, 337–67

Hill, W. C. O. (1960) *Primates*, vol. IV, *Cebidae*, part A. Edinburgh: Edinburgh University Press

Izawa, K. (1979) Foods and feeding behavior of wild black-capped capuchin (*Cebus apella*). *Primates*, **20**, 57–76

Izawa, K. & Mizuno, A. (1977) Palm-fruit cracking behavior of wild black-capped capuchin (*Cebus apella*). *Primates*, **18**, 773–92

Jennison, G. (1927) *Natural History: Animals*. London: Black

Kluver, H. (1933) *Behavior Mechanisms in Monkeys*. Chicago: Chicago University Press

Nolte, A. (1958) Beobachtungen über das Instinktverhalten von Kapuzineraffen (*Cebus apella* L.) in der Gefangenschaft. *Behaviour*, **12**, 183–207

Parker, S. T. & Gibson, K. (1977) Object manipulation, tool use and sensorimotor intelligence as feeding adaptations in cebus monkeys and great apes. *J. Hum. Evol.*, **6**, 623–41

Romanes, G. J. (1883) *Animal Intelligence*. New York: D. Appleton & Co.

Terborgh, J. (1983) *Five New World Primates*. Princeton: Princeton University Press

Tobias, P. V. (1965) Australopithecus, *Homo abilis*, tool-using and tool-making. *South Afr. Archeol. Bull.*, **20**, 167–92

Vevers, G. & Weiner, J. (1963) Use of a tool by a captive capuchin monkey (*Cebus apella*). *Symp. Zool. Soc. Lond.*, **10**, 115–18

II.5

Dental microwear and diet in two species of *Colobus*

M. F. TEAFORD

Introduction

Variations in dental microwear have recently been correlated with variations in feeding behavior for a number of mammals (see for example Rensberger, 1978; Walker, Hoeck & Perez, 1978; Kay, 1981; Ryan, 1981; Teaford & Walker, 1983). More recent analyses have documented interspecific differences in primate dental microwear apparently related to the presence or absence of hard objects in the diet (Teaford & Walker, 1984; Teaford, 1985). Can finer microwear differences be related to more subtle dietary differences in other primate species? To begin to answer that question, a sample of 10 *Colobus guereza* and 10 *Colobus badius* was examined from the Smithsonian Institution. Recent behavioral work (e.g. Clutton-Brock, 1975; Struhsaker, 1975; Oates, 1977; Marsh, 1981) indicates that *C. badius* has a more varied diet than *C. guereza*, and that *C. badius* probably incorporates higher proportions of fruit into its diet. Based on previous work, *a priori* expectations include a more variable microwear in *C. badius*, with perhaps a higher incidence of microscopic pits than in *C. guereza*.

Methods

Scanning electron microscope techniques and statistical analyses were essentially the same as those described by Teaford (1985) and by Teaford & Walker (1984), the only exception being one additional statistical analysis (No. 5 below). Microwear features were counted and measured directly from SEM micrographs (using a digitizer and minicomputer system), and two micrographs were taken of each of two facets (facets 3 and 9, Kay & Hiiemae, 1974) for each individual. All micrographs were taken on an ETEC OMNISCAN scanning electron microscope. In total, five sets of comparisons were run for this sample: (1) interfacet comparisons of the relative proportions of microscopic pits and scratches (using Chi-square analyses); (2) interspecific comparisons of the relative proportions of microscopic pits and scratches on homologous facets (using Chi-square analyses); (3) interspecific comparisons of the number of microwear features per field of examination (using the nonparametric Mann–Whitney test); (4) interspecific comparisons of the average width of microscopic pits (using the Mann–Whitney test); and (5) interspecific comparisons of the variation in the proportion of pits between individuals (using a variance ratio test).

Results

The number and percentages of pits and scratches are presented in Table 1, from which one point should be evident: for each species, there are relatively more pits on facet 9 than on facet 3 (both

Table 1. *Pits and scratches on facets 3 and 9 of the maxillary second molars of* Colobus guereza *and* Colobus badius

	Facet 3	Facet 9
Colobus guereza		
Number of pits (%)	303 (15.5)	380 (18.3)
Number of scratches (%)	1656 (84.5)	1694 (81.7)
Total number of features	1959	2074
Colobus badius		
Number of pits (%)	232 (14.1)	376 (22)
Number of scratches (%)	1408 (85.9)	1335 (78)
Total number of features	1640	1711

Table 2. *Average percentage of pits for individual* Colobus guereza *and* Colobus badius

	Facet 3	Facet 9
Colobus guereza		
Average % of pits (std dev.)	15.8 (6.8)	18.4 (6)
Colobus badius		
Average % of pits (std dev.)	14.7 (10.3)	21.2 (18.5)

results significant, $p < 0.025$, chi-squared tests). In addition, although there are no significant interspecific differences in the proportions of pits and scratches for facet 3, the interspecific differences for facet 9 are significant ($p < 0.01$); i.e. *Colobus badius* exhibits relatively more pits on facet 9 than does *Colobus guereza*. Table 1 also shows that *C. guereza* exhibits absolutely more microwear features, and thus more features per field of examination, than does *C. badius* ($p < 0.02$, Mann–Whitney test). Average widths are not presented because there were no significant interspecific differences in the average width of pits on either facet. As can be seen from Table 2, the relative proportion of pits certainly varied between individuals; more importantly, on facet 9 *C. badius* showed greater variation between individuals than did *C. guereza* (as evidenced by the variance ratio test, $p < 0.01$). The differences in variance on facet 3 were not significant.

Discussion

As in previous work (Gordon, 1982; Teaford & Walker, 1984; Teaford, 1985), a crushing facet (9) has again proven to be more heavily pitted than a shearing facet (3). Interestingly, although *Colobus guereza* exhibited more microwear features than did *Colobus badius*, the interspecific differences in microwear shape were concentrated on the crushing facet. In other words, on facet 9, *C. guereza* exhibited relatively more scratches and relatively fewer pits than did *C. badius*. By itself, this would indicate that *C. guereza* is probably ingesting more tiny, abrasive particles in its diet. But this would also ignore the important variability in the *C. badius* sample – i.e. some *C. badius* show a great deal of enamel pitting while others show relatively little enamel pitting. In light of previous work, this variability probably reflects the variable diet of *C. badius*. Given the inter-individual variability in the *C. badius* sample, the interspecific differences in the shape of microwear features on facet 9 must be viewed as suggestive at best, while any further conclusions await analyses based upon larger samples.

Acknowledgements

I wish to thank Dr Richard Thorington for allowing access to specimens in his care at the Smithsonian Institution. This reseach was supported by National Service Award 1-F32-DE05312.

References

Clutton-Brock, T. H. (1975) Feeding behaviour of red colobus and black and white colobus in East Africa. *Folia Primatol.*, **23**, 165–207

Gordon, K. D. (1982) A study of microwear on chimpanzee molars: implications for dental microwear analysis. *Am. J. Phys. Anthropol.*, **59**, 195–215

Kay, R. F. (1981) The ontogeny of premolar dental wear in *Cercocebus albigena* (Cercopithecidae). *Am. J. Phys. Anthropol.*, **54**, 153–5

Kay, R. F. & Hiiemae, K. M. (1974) Jaw movement and tooth use in recent and fossil primates. *Am. J. Phys. Anthropol.*, **40**, 227–56

Marsh, C. W. (1981) Diet choice among red colobus (*Colobus badius rufomitratus*) on the Tana River, Kenya. *Folia Primatol.*, **35**, 147–78

Oates, J. F. (1977) The guereza and its food. In *Primate Ecology: Studies of Feeding and Ranging Behaviour in Lemurs, Monkeys and Apes*, ed. T. H. Clutton-Brock, pp. 275–322. New York: Academic Press

Rensberger, J. M. (1978) Scanning electron microscopy of wear and occlusal events in some small herbivores. In *Development, Function and Evolution of Teeth*, ed. P. M. Butler & K. A. Joysey, pp. 415–38. New York: Academic Press

Ryan, A. S. (1981) Anterior dental microwear and its relationship to diet and feeding behavior in three African primates (*Pan troglodytes*, *Gorilla gorilla gorilla*, and *Papio hamadryas*). *Primates*, **22**, 533–50

Struhsaker, T. T. (1975) *The Red Colobus Monkey*. Chicago: Chicago University Press

Teaford, M. F. (1985) Molar microwear and diet in the genus *Cebus*. *Am. J. Phys. Anthropol.*, **66**, 363–70

Teaford, M. F. & Walker, A. C. (1983) Dental microwear in adult and still-born guinea pigs (*Cavia porcellus*). *Arch. Oral Biol.*, **28**, 1077–81

Teaford, M. F. & Walker, A. C. (1984) Quantitative differences in dental microwear between primate species with different diets and a comment on the presumed diet of *Sivapithecus*. *Am. J. Phys. Anthropol.*, **64**, 191–200

Walker, A. C., Hoeck, H. N. & Perez, L. (1978) Microwear of mammalian teeth as an indicator of diet. *Science*, **201**, 908–10

II.6

Differential influence of various sweet solutions on caloric regulation by rhesus monkeys

J. W. KEMNITZ AND M. M. NEU

Introduction

Rhesus monkeys will drink large quantities of sucrose solutions and reduce their subsequent food intake as if to compensate for calories in the sucrose solution (Kemnitz *et al.*, 1981 and in press). The monkeys also drink large volumes of aspartame-sweetened water, which is essentially non-caloric, without an effect on their subsequent eating behavior (Kemnitz *et al.*, 1981). These data, as well as other lines of evidence (Moran & McHugh, 1981; Hansen, Jen & Kribbs, 1981; Hansen, 1982), suggest that rhesus monkeys are quite sensitive to the energy content of their food and that they modulate their intake to regulate their energy balance.

The present study was designed to examine in more detail the influence of sugar consumption on total caloric regulation and to extend the observations to solutions of sugars other than sucrose. Four sugars, sucrose, maltose, glucose and fructose, were utilized. Glucose and fructose are monosaccharides, while maltose and sucrose are disaccharides consisting of glucose plus glucose and glucose plus fructose, respectively. Solutions of these sugars were presented in several concentrations, and aspartame solution and plain water were used as control fluids. Two general questions were asked. First, do rhesus monkeys ingest these various sugar solutions differentially, as indicated by the total amount they consume and the distribution of drinking during a time-limited test? Secondly, do these sugars affect subsequent food intake differentially? The latter question was of particular interest because of a recent report that human subjects exhibit an exaggerated feeling of satiety after ingestion of fructose solutions (Rodin & Spitzer, 1983). Thus, the use of fructose was suggested to be a valuable dietary aid for reducing total caloric intake.

Methods

Subjects and maintenance conditions

Eight male rhesus monkeys (*Macaca mulatta*), ranging in age from 6 to 7 years and in weight from 8.9 to 12.4 kg at the beginning of the experiment, were used. They were individually caged (76 cm × 71 cm × 79 cm high) in the same room, which was maintained at approximately 21°C and lighted from 06.00 to 18.00 hours. Purina Monkey Chow® (St Louis, Missouri, USA) was available to them between 09.30 and 17.30 hours. Tap water was always available.

Testing procedure

Test fluids were presented in bottles with sipper tubes attached to the front wall of the cage during one 2-hour session per day (12.30–14.30 hours), 6 days per week. One fluid was used for each test. Fructose (Sigma Chemical Co., St Louis, Missouri, USA), glucose (Columbus Chemical Industries, Inc., Columbus, Wisconsin, USA), sucrose (EM Science, Gibbstown, New Jersey, USA), and maltose (Difco Laboratories, Detroit, Michigan, USA) were mixed in concentrations of 0.2M, 0.4M, 0.6M, 0.8M, and 1.2M. Plain water and aspartame solution (4.7×10^{-3}M, G. D. Searle and Co., Skokie, Illinois, USA) were used as control fluids. The monkeys were tested once with each sugar solution and twice with each control fluid, making a total of 24 tests per monkey. Each week, two of the eight monkeys were tested with each sugar. The order of presentation of the sugars was randomized for each monkey, and the order of sugar concentrations and control solutions within each week was also randomized. Randomization was restricted, however, to the extent that at least one and no more than two monkeys were tested with a given control fluid or sugar concentration on any particular day.

Food and tap water were available during the tests. Food intake was measured (Kemnitz *et al.*, 1984) during the 3-h interval before the fluid test, during the 2-h test, and at hourly intervals during the remaining 3 hours of food availability. Fluid intake was measured at 10, 20, 30, 45, 60, 90, and 120 min of the test by making a mark on the side of the bottle at the level of the fluid at that moment.

Statistical analysis

Data are presented as the mean ± standard error of the mean. Differences among means were evaluated by analysis of variance with Neuman–Keuls tests and differences between means by Student's

t-test for dependent samples. Pearson's Product Moment Correlation Coefficients were also evaluated by Student's t-test.

Results

Fluid intake

The monkeys drank large volumes of the sugar solutions at every concentration, but particularly the 0.2M and 0.4M solutions (Fig. 1). Intake of sugar solutions was 2.5–6 times greater than that of plain water (116 ± 29 ml). The monkeys also consumed a large quantity of aspartame-sweetened water and the volume of aspartame intake (347 ± 79 ml) was not statistically different from average intake of sugar solutions.

Collapsing the data across concentrations of each sugar, it is apparent that the monkeys drank more fructose than the other sugars during the test (Fig. 2). Rate of drinking was highest for all fluids during the first 10 min of the test, and the average rate of fructose consumption was slightly, but consistently, greater than those of the other sugars throughout the test. Patterns of intake of fructose and aspartame were approximately parallel and were characterized by higher rates (1.6 and

Fig. 1. The total volume of fluids consumed during the 2-hour tests.

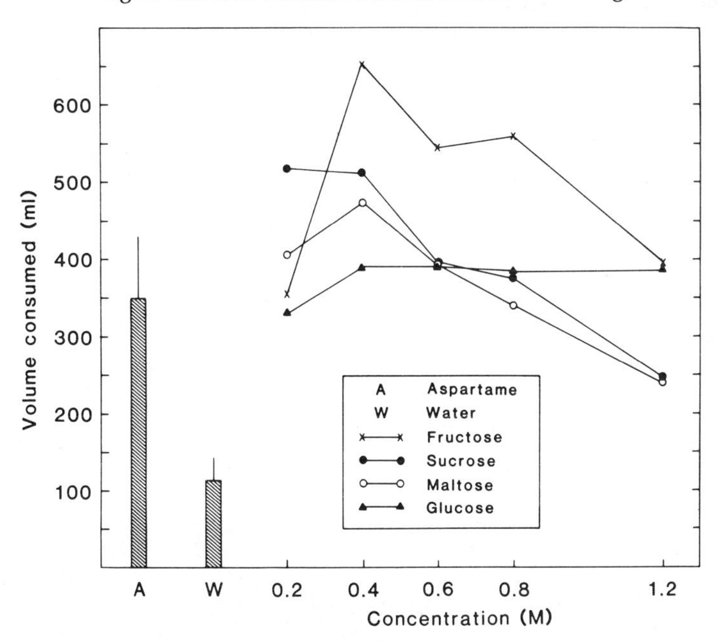

1.8 ml/min, respectively) during the second hour of the test than those of sucrose, glucose, and maltose (0.7–1.0 ml/min).

Individual differences in drinking behavior were prominent. In general, however, average intakes of all sweet solutions were positively correlated. That is, if an individual monkey drank a relatively large amount of one sweet solution, it was likely to drink relatively large quantities of the others as well. In contrast, intake of sweet solutions was not significantly correlated with water intake and it was not significantly correlated with baseline food intake or body weight.

Food intake

Consumption of maltose, sucrose, and glucose significantly inhibited intake of solid food during the 24-h interval beginning with the start of the drinking test. When compared to average food intake on days when plain water was the test solution (293 ± 12 g) intake was significantly lower when maltose (211 ± 22 g, $p < 0.01$), glucose (214 ± 15 g, $p < 0.01$) and sucrose (226 ± 20 g, $p < 0.05$), but not fructose (253 ± 21 g) were presented. Food intake following aspartame trials was similar to that following water trials (285 ± 28 g).

Fig. 2. Pattern of drinking during the tests. Note that patterns for sucrose (S), glucose (G), and maltose (M) are similar to each other, and that those for fructose (F) and aspartame (A) are approximately parallel. W = water.

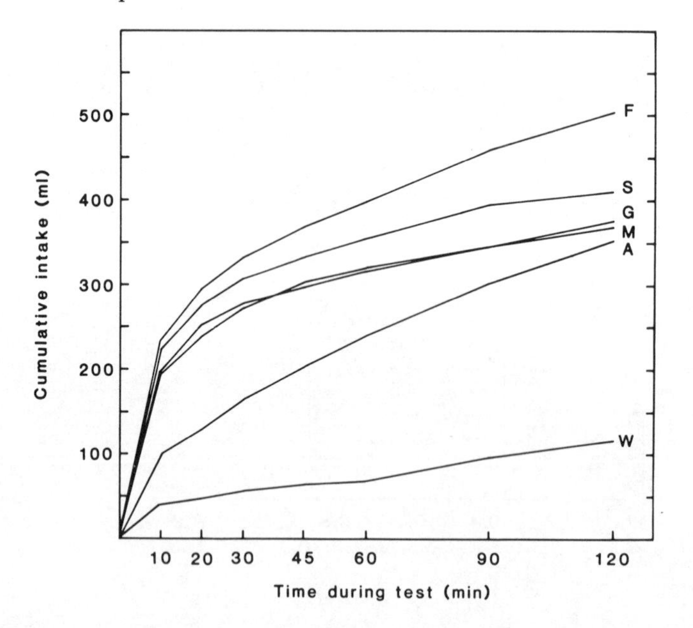

The temporal pattern of suppression of eating is illustrated in Fig. 3. The cumulative suppression of food intake (i.e. the difference in intake between water trials and trials in which sweet solutions were presented) during the 8 hours of access to food after the beginning of the test is plotted for each solution. Data for the sugars have been averaged across concentrations. During the 2 hours when the test fluids were available, intake was, on average, 18 g lower for trials with sugar solutions than for trials with water or aspartame solution. A greater inhibition of eating was apparent during the hour after the solutions were removed: 28 g for both maltose and sucrose, and 26 g for glucose, but only 17 g for fructose. Thereafter, the suppression related to fructose-drinking was clearly less than that for the other sugars (Fig. 3). Eating was slightly reduced shortly after aspartame-drinking, but compensation for this early suppression occurred during the following morning.

The magnitude of the reduction in food intake was related to the amount (by weight) of some of the sugars consumed. Average suppression of food intake was regressed on average solute intake for each sugar at every concentration according to the rectilinear, least-squares model. Correlations between these variables were +0.98 for sucrose, +0.97 for maltose, and +0.64 for glucose but only +0.33 for fructose.

Fig. 3. Cumulative difference in food intake between water trials and trials in which sweet solutions were presented during the 8 h of food availability after the beginning of the drinking tests.

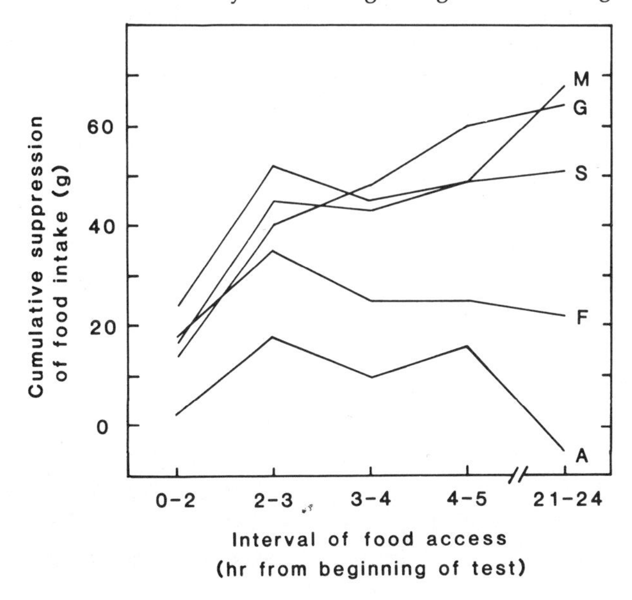

Since the caloric density of sugar (4 kcal/g) and Purina Chow (4.18 kcal/g by bomb calorimetry; Ralston Purina Co., 1981) are approximately equal, a slope of 1.0 in the regression equation would indicate accurate compensation for sugar intake by subsequent reduction of eating. The obtained slopes were 0.82 for sucrose, 1.15 for maltose, and 0.54 for glucose. The calculations for fructose generated a slope of 0.24. In other words, the monkeys regulated total caloric intake fairly accurately following ingestion of glucose and more accurately following ingestion of sucrose and maltose, but not well at all following ingestion of fructose.

Discussion

The monkeys drank large quantities of the sweet solutions. Sucrose, maltose, glucose, frucose, and aspartame were all consumed in greater quantities than plain water. In general, the monkeys drank more of the lower concentrations of sugar solutions and more of the fructose solutions than of the other sugars. These observations are consistent with our earlier observations of sucrose and aspartame consumption by rhesus monkeys (Kemnitz *et al.*, 1981 and in press) and correspond to observations of others concerning sweetness and pleasantness of sugar solutions as reported by human subjects (Moskowitz, 1971; Moskowitz *et al.*, 1974).

There were similarities and differences among fluids in the temporal pattern of consumption. The greatest rate of drinking of all fluids occurred during the first 10 min of the test and then steadily decreased. The rates of drinking sucrose, glucose, and maltose solutions were markedly lower during the second hour of the test, whereas the rates of drinking fructose and aspartame solutions were comparable and approximately double those for the other solutions during this interval. Whatever the factor that operated to inhibit drinking during the later phases of the test, it was less effective when fructose and aspartame were the substances ingested. The pattern of gastric emptying is more rapid for fructose than for some other sugars and this may be an important factor in modulating further drinking (Moran & McHugh, 1981). Direct comparisons of the rate of passage of aspartame, sucrose, and maltose through the stomach in the rhesus monkey have not been reported, so the importance of the pattern of gastric emptying in this context cannot be fully evaluated at the present time.

During the 3 hours after the beginning of the drinking test, food intake was lower when sweet solutions were used than when plain water was the test fluid. Food intake following fructose and aspartame trials was greater than following water trials, so that by the beginning of the next drinking test net suppression of eating was not statistically significant. Suppression of food intake following consumption of sucrose, maltose, and glucose, on the other hand, was statistically reliable. Furthermore, the magnitude of the reduction in food intake was highly correlated with the caloric consumption of these sugars. These data suggest that the monkeys were quite sensitive to the caloric content of the sucrose, maltose, and glucose solutions and they were adjusting their intake of solid food to compensate for the calories taken in the form of these sugars. The observation that food intake was not reduced following ingestion of large volumes of aspartame solution, which is essentially noncaloric, is consistent with this hypothesis and also demonstrates that gastric distension alone was not an important factor.

The absence of caloric compensation following fructose consumption remains to be explained. Fructose has been shown to differ from glucose and xylose in the rate at which it is emptied from the stomach, and the rate of gastric emptying might provide an early signal for the regulation of intake (Moran & McHugh, 1981). In the present study, however, the difference between fructose (as well as aspartame) and the other sugars as they affect food intake was most apparent beginning 3 hours after presentation of the fluids, indicating that the basis for the difference lies in absorptive or post-absorptive events. Fructose ingestion does not increase plasma glucose levels as do the other sugars used here (which all contained glucose), and it does not elicit an insulin response (Dunnigan & Ford, 1975). Plasma glucose and insulin levels have long been implicated in certain aspects of the control of eating behavior and it seems reasonable that these factors may have contributed to the effects reported here. In any event, the present data do support earlier reports of the capacity for accurate caloric homeostasis under certain circumstances in rhesus monkeys (Hansen *et al.*, 1981; Kemnitz *et al.*, 1981; Moran & McHugh, 1981; Hansen, 1982; Kemnitz *et al.*, in press), but they are inconsistent with the suggestion that fructose consumption provides a heightened sense of satiety compared to other sugars (Rodin & Spitzer, 1983).

Acknowledgements

This project was the basis for a Senior Honors Thesis submitted by Monika M. Neu to the Department of Psychology, University of Wisconsin-Madison. The authors are grateful to Professor R. W. Goy for his support during the course of the study and to Mary Schatz and Jackie Kinney for secretarial assistance. Annette Ripper of G. D. Searle and Co. provided the aspartame. The project was supported by an award from the Trewartha Honors Undergraduate Research Fund and by grant RR00167 from the National Institutes of Health. This is publication 24-016 of the Wisconsin Regional Primate Research Center.

References

Dunnigan, M. G. & Ford, J. A. (1975) The insulin response to intravenous fructose in relation to blood glucose levels. *J. Clin. Endocrinol. Metab.*, **40**, 629–35

Hansen, B. C. (1982) Induction and remission of obesity in monkeys: behavioral and physiological correlates. In *Changing Concepts of the Nervous System*, ed. A. R. Morrison & N. P. L. Strick, pp. 609–20. New York: Academic Press

Hansen, B. C., Jen, K.-L. C. & Kribbs, P. (1981) Regulation of food intake in monkeys: response to caloric dilution. *Physiol. Behav.*, **26**, 479–86

Kemnitz, J. W., Eisele, S. G., Lindsay, K. A., Engle, M. J., Perelman, R. H. & Farrell, P. M. (1984) Changes in food intake during menstrual cycles and pregnancy of normal and diabetic rhesus monkeys. *Diabetologia*, **26**, 60–4

Kemnitz, J. W., Gibber, J. R., Eisele, S. G. & Lindsay, K. A. (in press) Relationship of reproductive condition to food intake and sucrose consumption of female rhesus monkeys. In *Proceedings of the IXth Congress of the International Primatological Society*, vol. I. *Current Perspectives in Primate Social Dynamics*, ed. D. M. Taub and F. A. King. New York: Van Nostrand Reinhold, Inc.

Kemnitz, J. W., Gibber, J. R., Lindsay, K. A. & Brot, M. D. (1981) Preference for sweet and the regulation of caloric intake. *Am. J. Primatol.*, **1**, 313–14

Moran, T. H. & McHugh, P. R. (1981) Distinctions among three sugars in their effects on gastric emptying and satiety. *Am. J. Physiol.*, **241**, R25–R30

Moskowitz, H. R. (1971) The sweetness and pleasantness of sugars. *Am. J. Psychiatry*, **84**, 387–406

Moskowitz, H. R., Kluter, R. A., Westerling, J. & Jacobs, H. L. (1974) Sugar sweetness and pleasantness: evidence for different psychological laws. *Science*, **184**, 583–5

Ralston Purina Co. (1981) *Lab Chows®: The Control Factor*. St Louis, Missouri: Ralston Purina Co.

Rodin, J. & Spitzer, L. B. (1983) The effects of type of sugar ingested on subsequent eating behavior. Presented at the IVth International Congress on Obesity. New York. Abstract No. 84

II.7

Gastro-intestinal allometry in primates and other mammals including new species

A. M. MACLARNON, D. J. CHIVERS
AND R. D. MARTIN

Introduction

Measurements of the four major compartments of the gut (stomach, small intestine, caecum, and colon) in mammals have been subjected to allometric analysis in a number of previous studies (Chivers & Hladik, 1980; Martin *et al.*, 1985; MacLarnon *et al.*, in press). The effects of body size were successfully isolated, and remaining variation in gut size was shown to reflect known dietary habits quite clearly. In particular, fore-gut fermenting folivores, mid-gut fermenting folivores and faunivores formed three distinct groups. Several methods of multivariate analysis were used, yielding useful information on their performance and relative usefulness. The major steps in these analyses were:

1. The surface areas of the four gut compartments were shown generally to scale compatibly with predictions from Kleiber's Law of basal metabolic scaling (Kleiber, 1961).
2. Quotient values, or indices, were calculated for each species to represent the degree to which each gut compartment surface area is larger or smaller than expected, given the body weight of that species.
3. Using multivariate clustering techniques, overall relationships between species were determined on the basis of values for all four gut indices.

New data have been collected for seven additional mammal species: six New World monkeys (3 *Cacajao calvus*, 2 *C. melanocephalus*, 3 *Cebus apella*, 2 *Saimiri sciureus*, 3 *S. vanzolinii*,* 1 *Alouatta seniculus*) and the

* *Saimiri vanzolinii* is treated as a separate species from *S. sciureus* (Ayres, 1985).

Old World badger (*Meles meles*, 7 specimens). The results of further analyses are presented here.

1. Methods of analysis previously found to be most effective are used to analyse the relative gut compartment sizes of the new species.
2. A modification involving combination of the surface areas of the caecum and colon is tried, enabling the inclusion of caecumless species in the optimum type of multivariate analysis.
3. The necessity for careful calculation of allometric indices as a first step in analysis is demonstrated by presentation of results from a non-allometric attempt to analyse the same data.

Methods

All of the raw data were collected by DJC and C. M. Hladik by methods fully described in Chivers & Hladik (1980). Only data from fresh, adult specimens are used here, and only wild-caught specimens wherever possible. The new total sample comprises 80 mammal species (48 primate, 32 non-primate; 58 wild-caught, 22 captive). All specimens of the new species were wild-caught.

Best-fit lines, calculated by major axis techniques, were previously fitted to bivariate log–log plots of species averages of surface area *versus* body weight for each gut compartment. With the possible exception of the colon, the slope values were all compatible with 0.75, the value predicted for scaling according to Kleiber's Law of metabolic scaling (Kleiber, 1961). Lines with fixed slopes of 0.75 were fitted to the data and individual species deviations from these were used to calculate indices of relative gut compartment surface areas (Martin *et al.*, 1985). The equations for the original sample are used again here to calculate gastric (GQ), small intestinal (SQ), caecal (CQ), and colonic indices (LQ), for the new species. New equations, calculated for the entire new sample, would have had a negligible effect on the results obtained.

Two types of indices have previously been used. In Martin *et al.* (1985) absolute indices were used; in MacLarnon *et al.* (in press) both these and log indices were used and their effects on analyses compared. Both index types represent the residual variation in species gut compartment sizes after the removal of body size effects, i.e. the distances of individual species' points from the lines of slope 0.75 fitted to a logarithmic plot for each compartment surface area against body weight. Absolute indices are calculated by converting these distances into antilogarithmic form; log indices retain the logarithmic values. Log indices were found to be preferable, mainly because their calculation treats positive and negative deviations from the fitted slopes, i.e. relatively large and small gut compartments, as equally important. However, caecumless species have to be omitted from analyses using log indices as the log of zero is undefined. In an attempt to overcome this remaining problem a new method is tried here, involving the addition of the surface areas of the caecum and colon.

Comparison of species according to values of all four gut indices requires the use of multivariate methods. Two methods were used previously to represent the Euclidean distances between all pairs of species determined for their four indices: (1) dendrograms, producing hierarchical clusters; and (2) multidimensional scaling (MDScal) plots, 2-dimensional scatter plot representations. MDScal plots were found to be the more useful for both absolute and log indices as they do not impose a hierarchical structure, they are more informative and more stable to sample changes. None of the methods of analysis used requires any prior grouping of species, e.g. according to dietary classification. Species distributions are produced entirely on the basis of their relative gut compartment sizes and can thus be interpreted in the absence of prior assumptions, for example with respect to dietary preferences.

We have shown that gut compartment surface areas scale to body weight in a negatively allometric fashion (i.e. with scaling factors of less than 1) and in fact with exponent values of about 0.75 (Martin *et al.*, 1985). It is therefore to be expected that calculation of simple, non-allometric indices by dividing gut compartment surface areas by body weight would produce values too high for small-bodied species and too low for larger-bodied ones, the part played by body weight having been systematically underestimated and overestimated, respectively. The residual correlation of index values with body size should then be detectable in the results of multivariate analysis. The results of MDScal analysis on non-allometric indices of this type are presented here.

Results

Fig. 1 shows the results of MDScal analysis on the four gut surface area log indices for the total new sample (11 caecumless species are necessarily excluded, for reasons explained above). There are three major peripheral groups isolated from a central cluster of species. In Fig. 1 the fore-gut fermenting folivores lie to the bottom left, the mid-gut fermenting folivores to the top left. Specialist faunivores fall

Fig. 1. MDScal plot generated using log indices for four gut compartment surface areas (GQ, SQ, CQ, LQ) for 69 mammal species. Key: 1 = *Arctocebus calabarensis*; 2 = *Avahi laniger*; 3 = *Cheirogaleus major*; 4 = *Euoticus elegantulus*; 5 = *Galago alleni*; 6 = *G. demidovii*; 7c = *Lepilemur mustelinus*; 8 = *L. leucopus*; 9 = *Loris tardigradus*; 10 = *Microcebus murinus*; 11 = *Perodicticus potto*; 12 = *Saguinus geoffroyi*; 13 = *Aotus trivirgatus*; 14c = *Ateles belzebuth*; 15 = *Saimiri oerstedii*; 16 = *Cebus capucinus*; 17 = *Alouatta palliata*; 18c = *Lagothrix lagothricha*; 19 = *Miopithecus talapoin*; 20 = *Cercopithecus cephus*; 21 = *C. neglectus*; 22 = *C. nictitans*; 23 = *Cercocebus albigena*; 24 = *Macaca sylvanus*; 25 = *M. sinica*; 26 = *M. fascicularis*; 27 = *Papio sphinx*; 28c = *Erythrocebus patas*; 29 = *Colobus polykomos*; 30 = *Presbytis entellus*; 31 = *P. cristata*; 32 = *P. obscura*; 33 = *P. melalophos*; 34c = *P. rubicunda*; 35c = *Nasalis larvatus*; 36c = *Pygathrix nemaeus*; 37c = *Hylobates pileatus*; 38c = *H. (Symphalangus) syndactylus*; 39c = *Pongo pygmaeus*; 40 = *Pan troglodytes*; 41 = *Gorilla gorilla*; 42c = *Homo sapiens*; 43c = *Felis domestica*; 44c = *Canis familiaris*; 45* = *Mustela nivalis*; 46 = *Vulpes vulpes*; 47 = *Atilax paludinosus*; 48* = *Nandinia binotota*; 49 = *Poiana richardsoni*; 50 = *Genetta servalina*; 51* = *Mustela* sp.; 52*c = *Ailurus fulgens*; 53*c = *Nasua narica*; 54c = *Genetta* sp.; 55c = *Panthera tigris*; 56c = *Sus scrofa*; 57c = *Capra hircus*; 58c = *Ovis aries*; 59 = *Cervus elaphus*; 60c = *Equus caballus*; 61c = *Halichoerus grypus*; 62* = *Phocaena phocaena*; 63* = *Tursiops truncatus*; 64 = *Sciurus vulgaris*; 65 = *Epixerus ebii*; 66 = *Heliosciurus rufobrachium*; 67 = *Sciurus carolinensis*; 68 = *Oryctolagus cuniculus*; 69* = *Potamogale velox*; 70* = *Manis tricuspis*; 71 = *Dendrohyrax dorsalis*; 72* = *Bradypus tridactylus*; 73c = *Macropus rufus*; 74 = *Cacajao calvus*; 75 = *C. melanocephalus*; 76 = *Cebus apella*; 77 = *Saimiri sciureus*; 78 = *S. vanzolinii*; 79 = *Alouatta seniculus*; 80* = *Meles meles*. * = caecumless species not included in Fig. 1. c = captive specimens; only wild-caught specimens were used for all other species.

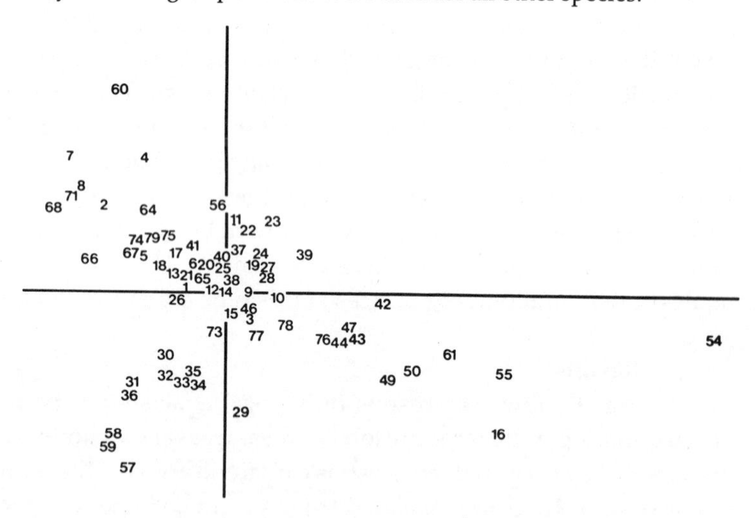

to the bottom right, the insectivores lying nearer to the central cluster than the carnivores. (The terms 'insectivore' and 'carnivore' are used here as *dietary* labels signifying invertebrate-eaters and vertebrate-eaters, respectively). Central cluster species are typically frugivorous to some extent, the more folivorous species falling at the bases of the two distinct folivorous offshoots.

The new *Cacajao* species (Fig. 1: 74, 75) lie on the edge of the central cluster towards the mid-gut fermenting offshoot. Fontaine (1981) studied these species in a semi-natural environment and described them as frugivore–folivores. Milton (1984) suggested, on the basis of similar food passage rates and dental morphology, that *Cacajao* may have a natural diet of fruit and seeds similar to *Chiropotes*. The present evidence suggests that *Cacajao* may be more folivorous than frugivorous, with some mid-gut adaptation for leaf digestion. The new *Alouatta* species, *A. seniculus* (Fig. 1: 79), falls between the *Cacajao* species and close to *A. palliata* (17), in accord with the fact that howler monkeys are reported to be the most folivorous of the New World monkeys (Milton, 1980).

The new *Cebus* species, *C. apella* (76), falls in the faunivorous offshoot, and makes interesting comparison with the most extreme primate member of this group, the *C. capucinus* (16) included in the original sample. *Cebus* species are probably all fairly insectivorous. Terborgh (1983) studied *C. apella* and *C. albifrons*, finding that 40% of their time was spent foraging for insects. He found similar results for *Saimiri sciureus* (77); it falls amongst other insectivorous primates, including the other new *Saimiri* species, *S. vanzolinii* (78), and *S. oerstedii* (15), at the inner margin of the faunivore group. The extreme position of the single *C. capucinus* specimen may be aberrant, especially as intraspecific gut size variation is relatively great (Martin *et al.*, 1985), but *C. apella* does lie in the same direction and the more outlying position of *C. capucinus* may reflect a particularly faunivorous diet for this species.

The other new species, the European badger (*Meles meles*), is caecumless and therefore excluded from Fig. 1. To enable inclusion of caecumless species in log index analysis, the surface areas of the caecum and colon were added together. Their combined surface area (*CC*) scales to body weight (*W*) compatibly with Kleiber's Law as 0.75 lies within the 95% confidence limits of the slope (0.71–1.06) of the best-fit line:

$$\underset{(\text{cm}^2)}{\log_{10} CC} = 0.87 \underset{(\text{g})}{\log_{10} W} - 0.62 \quad (r = 0.74, N = 80)$$

Caecum + colon surface area indices (CCQ) were therefore calculated as for the four separate gut compartments, from a line of fixed slope 0.75 fitted through the sample means:

$$\underset{(\mathrm{cm}^2)}{\log_{10} CC} = 0.75 \underset{(\mathrm{g})}{\log_{10} W} - 0.18$$

In all but three species (*Lepilemur mustelinus*, *Oryctolagus*, *Genetta* sp.), the caecum is absolutely smaller than the colon, and in most species its size is relatively insignificant and has little influence on the combined index. If, as has been suggested (Chivers & Hladik, 1980), the caecum and colon (or part of the colon) to some extent form a single functional unit, the loss of independent information on the caecum may not be important for functional interpretation of relative gut sizes.

The results of MDScal analysis on the reduced array of three gut compartment indices (GQ, SQ, CCQ) are shown in Fig. 2. The species distribution has much in common with Fig. 1, but has only two major offshoots from the central cluster instead of three. The mid-gut fermenting offshoot of Fig. 1 is compressed into the more frugivorous central cluster in Fig. 2. The fore-gut fermenting group remains constant and distinctive (29–36, 57–59). The faunivore offshoot, to the

Fig. 2. MDScal plot generated using log indices for three gut compartment surface areas (GQ, SQ, CCQ), combining caecum + colon surface areas, for 80 mammal species. (For key see Fig. 1.)

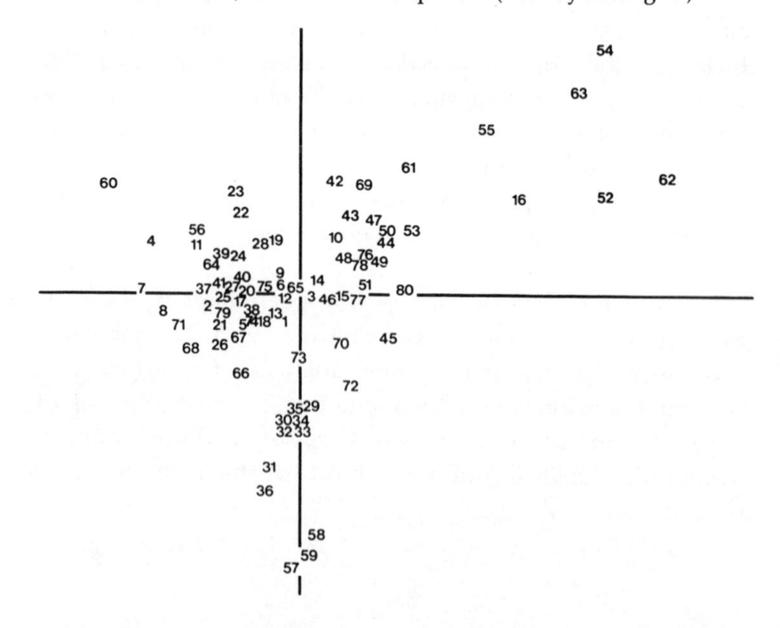

upper right of Fig. 2, gains some extra caecumless species compared with Fig. 1, but otherwise its constituent species remain constant, although some fall closer to the central cluster in Fig. 2 than they did in Fig. 1. All the sample members of the Carnivora (43–55, 80) lie in or towards the faunivorous offshoot, along with carnivores and insectivores from other orders. Man (42) lies a little to one side of this group in Fig. 2, compared with more definite membership in Fig. 1. Both suggest some adaptation to faunivory, but since gut size may be influenced by actual diet (Martin *et al.*, 1985) interpretation is particularly difficult in this case. Amongst the members of the group other than the Carnivora, the insectivorous species lie at its inner end, the carnivorous ones further out. Gut adaptation for insectivory is apparently similar to, but less extreme than, that for carnivory. *Manis* (70) has an insectivorous diet and a muscularly specialised stomach which compensates for its lack of teeth (Chivers & Hladik, 1980). It probably ingests about 50% debris along with ants and termites like other insectivores of its body size (McNab, 1984), which presumably requires an enlarged stomach. *Manis* lies between the other insectivorous species and the fore-gut fermenters on Fig. 2, clearly reflecting a combination of insectivory and stomach specialisation. The sample Carnivora range from exclusive carnivores e.g. *Panthera* (55), to *Nandinia* (48), which is primarily frugivorous (Charles-Dominique, 1978). To some extent the more carnivorous species lie further out along the faunivorous offshoot, but so does *Ailurus* (52) which is quite vegetarian. It is therefore difficult to interpret the position of the new member of the Carnivora, *Meles* (80), which primarily eats earthworms (Kruuk & Parish, 1981), but it does lie near other insectivorous species.

Of the species which fell in the mid-gut fermenting offshoot in Fig. 1, the horse (60) remains distinctive in Fig. 2; it has the sample's largest CCQ (and LQ), and smallest GQ. The orang-utan (39) (and to some extent the gibbons (37, 38)) moves from lying close to the faunivore offshoot in Fig. 1, to a position more understandably compatible with its frugivory in Fig. 2, the result of its small CQ, large LQ and resultant fairly large CCQ. The more folivorous central cluster species tend to fall on its left-hand side or lower edge. The two *Cacajao* species do not appear as folivorous as implied in Fig. 1.

Fig. 3 shows the results of MDScal analysis on simple, non-allometric indices of gut compartment surface areas divided by body weight. The distribution of species clearly shows the effects of the persisting influence of body size. There is marked contrast with Figs 1 and 2, in which the species distributions are apparently free from such an effect.

In general, small-bodied species fall towards the bottom of Fig. 3 and large-bodied species towards the top. Small body weight is the only common distinguishing feature of the most obvious offshoot from the central cluster, the line of species towards the bottom of the plot. The six bottom species (1, 67, 66, 6, 51, 45) are amongst the 10 lightest species of the whole sample (two of the others, 5 and 10, fall close by). These species are from three orders; two have relatively generalised diets (66, 67), two are carnivorous (45, 51) and two insectivorous (1, 6). The most extreme fore-gut fermenters lie to the far right (57–59), but the large stomach of the colobine monkeys barely suffices to distinguish some of them from the central cluster (e.g. 29, 30, 35) (cf. Figs 1 and 2). Similarly, only some of the most extreme mid-gut fermenters

Fig. 3. MDScal plot generated using non-allometric, simple ratio indices (surface area/body weight), for gut compartment surface areas for 80 mammal species. (For key see Fig. 1.)

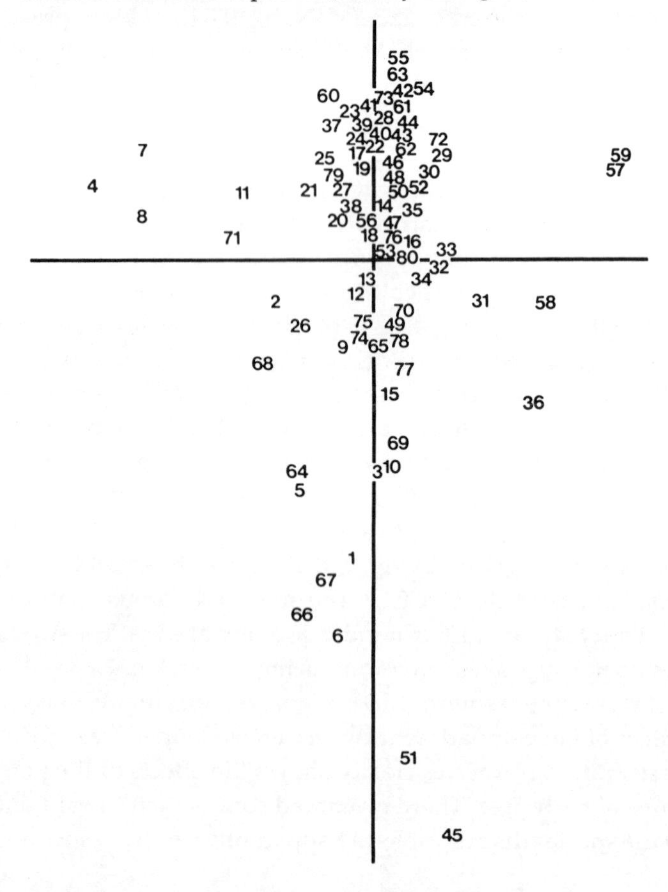

stand out clearly (4, 7, 8, 71), joined by *Perodicticus* (11). The extremely enlarged colon of the horse (60) results in no particular distinction here (cf. Figs 1 and 2); its large body weight overrides other influences to pull it to the top of the central cluster.

Conclusions

As a result of allometric analyses of gut compartment surface areas in mammals the following conclusions have been reached.

1. The surface areas of the stomach, the small intestine, the caecum and the combined caecum + colon surface area scale to body weight compatibly with an exponent of 0.75, in agreement with predictions from Kleiber's Law of basal metabolic scaling. The colon surface area apparently scales slightly differently, but for the sake of consistency, colonic indices (see 2) were also calculated assuming agreement with Kleiber's Law.
2. Allometric indices in log form were used to express residual variation in species gut compartment surface areas after the removal of the effects of body size in accordance with the relations established as in (1).
3. Multidimensional scaling (MDScal) plots of allometric log indices for each of the four separate gut compartments distinguish fore-gut fermenting folivores, mid-gut fermenting folivores and, a little less clearly, faunivores. The results suggest that the two new *Cacajao* species are possibly rather more folivorous than frugivorous; the new *Cebus* species and *Saimiri* species have indices consistent with their insectivory, the new *Alouatta* with its folivory.
4. Combining caecum + colon surface areas to enable inclusion of caecumless species in log index MDScal analysis produced some significant differences in the species distribution compared with that for four separate compartment indices. The former may be more correct for interpretation of functional adaptation if the caecum and colon commonly function as a single unit. Care must therefore be taken in the interpretation of MDScal plots with regard to their evident sensitivity to change in the constituent variables.
5. Members of the order Carnivora generally have relative gut size indices similar to insectivorous and carnivorous species of other orders, despite having a wide range of dietary preference in the sample species, including non-faunivory. The new member of the Carnivora, *Meles meles*, has gut size indices

comparable to those of other insectivorous species, but they do not clearly distinguish it as an insectivore amongst members of the Carnivora.

6. MDScal plots based on allometric indices exhibit species distributions that are manifestly free of body-size influence. In contrast, non-allometric indices (simple ratios of gut compartment surface areas to body weight) generate an unintelligible MDScal plot in which the species distribution confounds the effects of body size and of individual dietary adaptation. This type of approach clearly hinders meaningful interpretation and these results thus demonstrate the necessity for proper allometric extraction of the effects of body size as a prelude to multivariate analysis, at least in this case.

Acknowledgements

Specimens of the six new species of New World monkeys were provided by J. M. Ayres of INPA, Manaus and the Sub-Department of Veterinary Anatomy, Cambridge; the European badger specimens were provided by the Veterinary Investigation Centre, Cambridge. F. L. Brett gave much advice on the use of multivariate methods of analysis and considerable assistance in their execution by computer.

References

Ayres, J. M. (1985) On a new species of squirrel monkey, genus *Saimiri*, from Brazilian Amazonia (Primates, Cebidae). *Pap. Avul. Mus. Zool. Univ. São Paulo*, **36**, 147–64

Charles-Dominique, P. (1978) Ecologie et vie sociale de *Nandinia binotota* (Carnivores. Viverridés): comparaison avec les prosimiens sympatriques du Gabon. *Terre et Vie*, **32**, 477–528

Chivers, D. J. & Hladik, C. M. (1980) Morphology of the gastrointestinal tract in primates: comparisons with other mammals in relation to diet. *J. Morphol.*, **166**, 337–86

Fontaine, R. (1981) The Uakaris, genus *Cacajao*. In *Ecology and Behavior of Neotropical Primates*, ed. A. F. Coimbra-Filho & R. A. Mittermeier, pp. 443–93. Rio de Janeiro: Brazilian Academy of Sciences

Kleiber, M. (1961) *The Fire of Life: An Introduction to Animal Energetics*. New York: John Wiley

Kruuk, H. & Parish, T. (1981) Feeding specialisation of the European badger, *Meles meles*, in Scotland. *J. Anim. Ecol.*, **50**, 773–88

MacLarnon, A. M., Martin, R. D., Chivers, D. J. & Hladik, C. M. (in press) Some aspects of gastro-intestinal allometry in primates and other mammals. In *Morphogenèse du Crâne et Origenes de l'Homme*, ed. M. Sakka. Paris: CNRS

McNab, B. K. (1984) Physiological convergence amongst ant-eating and termite-eating mammals. *J. Zool.*, **203**, 485–510

Martin, R. D., Chivers, D. J., MacLarnon, A. M. & Hladik, C. M. (1985) Gastro-intestinal allometry in primates and other mammals. In *Size and Scaling in Primate Biology*, ed. W. L. Jungers, pp. 61–89. New York: Plenum

Milton, K. (1980) *The Foraging Strategy of Howler Monkeys*. New York: Columbia University Press
Milton, K. (1984) The role of food-processing factors in primate food choice. In *Adaptation for Foraging in Non-human Primates*, ed. P. S. Rodman & J. G. H. Cant, pp. 249–79. New York: Columbia University Press
Terborgh, J. (1983) *Five New World Primates*. Princeton, N.J.: Princeton University Press

Part III

The Use of Time and Space

Editors' introduction

Studies of the biology, behaviour and ecology of primates in the wild have been major subjects of research over the past 20 years. Despite much excellent previous field work, new studies continue to add to our overall knowledge of the diversity, adaptability and specialisations of primates. Such research provides essential information on the natural conditions and the range of ecological and social skills used by primates in exploiting their habitats.

Our understanding of how primates live in the wild is still incomplete, particularly when we consider the detailed questions that have been raised by field and laboratory studies. The studies of Seth and Seth (Chapter III.1) and Malik (Chapter III.2) on perhaps the best-known laboratory species, the rhesus monkey, illustrate both the breadth of flexibility of this species, and its specialisations to local ecological conditions such as food limitation, seasonal variation, and differences in habitats. While these papers address broad questions of the ecology of the rhesus monkey in the wild, they also consider more specific questions of how the observed ecological diversity can be related to variation in social behaviour; a topic of recurrent interest in primate studies.

Basic research on primates in the wild is becoming increasingly important for the conservation of threatened and endangered species. Until we have more high-quality information on the density, distribution, ecological requirements and social dynamics of these threatened species, such as that given by Mukherjee (Chapter III.3) for the hoolock gibbon, proposing priorities and programmes for their conservation will be inadequate.

More specific questions of the relations between ecological and social variables are also addressed. The relation between food distribution and exploitation, and patterns of social spacing are dealt with by Nash (Chapter III.4) and Menzel (Chapter III.5). In these two papers, one on ranging and food distribution in sympatric species of prosimians and the other on experimental manipulations of ranging patterns in social groups of titi monkeys, the factors influencing the spatial organising of primates with respect to food distribution and exploitation are presented. In the final paper by Norton (Chapter III.6), the social and cognitive issues influencing the use of spatially distinct resources are considered. He raises the problem of studying group decision processes and of their importance in food exploitation.

The papers in this section demonstrate our increasing ability to integrate and interpret the results of laboratory and field studies into a more complete understanding of the proximate and ultimate variables influencing the way primates allocate time to different activities, and acquire food in their physically and socially diverse and often fluctuating environments.

III.1

Ecology and behaviour of rhesus monkeys in India

P. K. SETH AND S. SETH

Introduction

Free-ranging rhesus monkeys (*Macaca mulatta*) in India, live in a wide variety of habitats, including areas of human settlement, forests, semi-deserts, and mangrove swamps (Seth & Seth, 1983; Seth, Seth & Shukla, 1983). Although conditions differ from one habitat to another, and within habitats through time, the macaques seem equally 'at home' in all these environments. They usually prefer the vicinity of cultivated fields, especially near small water pools, and trees near streams. They are less common in dense forests. In the cis-himalayan regions, they are found at elevations varying from 500 to 1500 m.

Of late, understanding the links between the environment, species-related feeding strategies, sex-related feeding strategies and social strategies have been emphasised in primatology. Thus, if we are to gain a more complete understanding of primate sociobiology, we must study the environmental conditions tending to place a premium on such strategies (Seth & Seth, 1986).

Methods

A few selected natural rhesus populations were observed for 12000 h between 1981 and 1984. Three approaches were used in observations: (1) overall troop movements, usually with individuals foraging as the troop moved through trees or over the ground; (2) foraging while troop members remained in a small circumscribed area such as a group of trees; and (3) resting, including activities such as sleeping, grooming, and play by infants and juveniles.

This paper reports on about 350 free-ranging rhesus groups living in different ecozones with varying degrees of human interference in the north and northwest belts of the Indian subcontinent. These groups have been under constant surveillance since May 1981 and were chosen for their year-round accessibility, their abundant populations and good conditions for visibility. In these groups, several or all individuals could be viewed in the course of a single observation cycle. Dawn-to-dusk observations were made for 15 days at a stretch on each group. In order to observe uninhibited feeding and behavioural strategies, no food was offered to the monkeys during routine observation cycles nor were they provisioned after establishing rapport.

In the present surveillance programme, emphasis was placed on the similarities and differences in specific activities between groups in diverse ecozones, using the same census and observation techniques (Seth & Seth, 1983). Reconnaissance trips were made within a radius of 300 km from Delhi, to locate groups comparable in terms of biome (arid or tropical) and habitat type (following an eight-point habitat classification scale). Potential study groups were selected for accessibility, visibility and approachability, as well as on the terrain type – both on and off established tracks within the forest through temple habitats (Puri, 1960; Ripley, 1965; Kaul, 1977).

Identification

Census of all groups was done by visual inspection, facilitated by the identification of individual group members. Complete counts were made at least three times during each observation schedule. Individual characteristics, such as permanent injury marks and scars or missing digits of hands or feet, were recorded for each animal to minimise the possibility of recounts. Age and sex were recorded separately to account for all individuals. The final troop counts were the result of many cross checks.

Age–sex classes used for classification and identification are presented below:

Adult male: more than 6 years old with a red scrotum

Adult female: more than 3 years old with red skin colour on the rump

Subadult male: 4–6 years old with pink scrotum

Adolescent female: around 3 years old with pinkish swellings on the hindquarters

Juvenile: J1 – over 1 year old
J2 – 2 or more years old
J3 – 3 years or older (only males)

Infant: below 1 year old
Newborn

Study areas and habitat

Surveys of the free-ranging rhesus monkeys were carried out in 23 districts in the States of Uttar Pradesh, Haryana, Himachal Pradesh, Jammu, Punjab, Rajasthan, and in the Union Territory of Delhi (Table 1).

The extreme zonation of the biotic characteristics of these study areas reveals a striking and often awesomely beautiful landscape which changes as one travels westwards and southwards. Each biome is in fact a varied aggregation of smaller communities. Data in this report are based on a study of eight basic habitats in the hills, plains, and arid ecozones of seven Indian states in north and northwestern India. These habitats are temple grounds and parklands surrounded by forested tracts, cities and towns with a relatively high population density, villages and ponds surrounded by extensive agricultural cultivation in an intermontane valley, roadsides and canal banks, and finally deciduous forests, in both the subtropical and arid biomes.

The rhesus groups could be classified in their habitats following an eight-point scale reflecting the degree of human interaction: temple (1); urban (2); village (3); village + pond (4); pond (5); roadside (6); canalside (7); and forest (8). However, the expanding human population (by far the greatest enemy) is encroaching on the habitat of freely moving rhesus monkeys through proliferating roads and farms, resulting in a massive squeeze on rhesus habitats. Although the populations studied are free ranging, they rely on people for at least some of their food. The urban and temple monkeys live in intimate contact with humans and face innumerable problems compared to their forest counterparts.

Results

Habitat and home range

As the complexity of the habitat increases, based on the eight-point habitat classification, the home range of the rhesus monkeys, except those in temples, decreases with little or no effect on average group size or adult sex ratios. Group sizes and sex ratios are highest in the middle of the habitat classification scale. This increase in ecological diversity is the result of deliberate changes in the environment.

Table 1. *Ecology of the area surveyed*

	Delhi	Haryana State				
		Rohtak	Jind	Bhiwani	Hissar	Ambala
Category	Densely inhabited, mixed terrain	Flat cultivated	Flat cultivated	Flat cultivated	Flat cultivated	Plains and Siwalik hills (South–West)
Height above sea level	200–240 m	218 m	210 m	235 m	235 m	1499.3 m (Karoh Peak)
Temperature:						
annual mean	24.8 °C	24.9 °C	25.0 °C	30 °C	30 °C	–
range	5–40 °C	8–44 °C	10–45 °C	5–43 °C	5–43 °C	–
Average rainfall	75.00 cm	75.00 cm	65.00 cm	42.91 cm	42.91 cm	71.12 cm (plains) 144.94 cm (hills)

Vegetation	Peepal, neem, sheesham, banyan, acacia and variety of horticultural plants (Forests: acacia & jal)	Peepal, banyan, sheesham, neem, acacia, janti, kaindu, jal (Forests: eucalyptus & sheesham)	Peepal, sheesham, banyan, neem, acacia, kaindu, kadam (Forests: eucalyptus & sheesham)	Peepal, acacia, jand, banyan, rohira, dhak, hingo, karil, jal, dhaman grass	Banyan, acacia, jand, peepal, dhak, hingo, karil, rohira, jal, dhaman grass	Sal, sheesham, chil (*Pinus longifolia*), harrar (*Terminalia chibula*)
Crops	Wheat, paddy, bajra & vegetables	Wheat, gram, jawar, bajra, sugar cane, paddy	Wheat, gram, bajra, paddy	Paddy, wheat, potato, sugar cane, sarson	Wheat, paddy, potato, sugar cane, sarson	Wheat, gram, barley, maize, paddy, pulses
Animals	Domesticated animals	Domesticated animals and sambhar, nilgai, jackal	Domesticated animals	Domesticated animals	Domesticated animals	Cattle, tiger, sambhar, chital, antelope, hyaena, dog

Table 1. *Continued*

	Uttar Pradesh State					Jammu
	Ghaziabad	Bulandshahr	Saharanpur	Dehradun	Nainital	
Category	Flat cultivated, Western Gangetic basin	Flat cultivated, Western Gangetic basin	Foothills of Siwalik, dense tropical forests, Gangetic basin	Foothills of Siwalik, dense tropical forests, Gangetic basin	Kumaon hills, mixed deciduous forests	Mountainous
Height above sea level	210–225 m	221.6m	300–583 m	600–2000 m	300–2300 m	305 m
Temperature:						
annual mean	–	–	30 °C	20 °C	20 °C	–
range	7–38 °C	–	3–44 °C	0–42 °C	0–35 °C	6.5–39 °C
Average rainfall	80 cm	61.14 cm	120–154 cm	150–214.99 cm	100–250 cm	107 cm

Vegetation	Sheesham, jamun, neem, peepal, mango, papaya, kikar	Sheesham, neem, peepal, siras, dhak, kikar	Sal, jamun, mango, banyan, lantana, peepal, rohini, leechi, karonda (Forests: sal & mixed)	Sal, lantana, rohini, jamun, neem, mango, sheesham, leechi, karonda, manjh (Forests: sal, eucalyptus & pine)	Sheesham, kikar, catacheu, sal, oak, pine, lantana (Forests: mixed & pine)	*Cedrus*, *Pinus*, elm, sandal wood, cypress, thuja, bhojpattar, postil
Crops	Wheat, paddy, potato, sugar cane	Wheat, maize, jwar, bajra, paddy, gram, cotton, sugar cane, barley	Wheat, paddy, sugar cane	Wheat, paddy sugar cane, potato	Wheat, paddy, beans, maize, sugar cane	Wheat, paddy, barley, beans, maize, walnut, apple
Animals	Domesticated animals	Cattle, camel, wild pig, jackal, wolf, hyaena	Domesticated animals & tiger, elephant, deer, wild cat, snakes	Domesticated animals & tiger, panther, elephant, jackal, wild cat, snakes	Domesticated & wild animals, wolf	Cattle, khakkar, leopard, marmots, musk deer, porcupine

Table 1. *Continued*

	Rajasthan State		Himachal Pradesh State		Punjab State
	Alwar	Jaipur	Simla	Solan	Patiala
Category	Plateaux plains and deserts	Plains and deserts	Mountainous	Mountainous	Flat cultivated and submountainous tract
Height above sea level	142 m	431 m	609.6–2590.8 m	1462 m	–
Temperature:					
annual mean	–	–	–	–	–
range	8–48 °C	3–45 °C	−4–27 °C	–	1–46 °C
Average rainfall	57.77 cm	70.06 cm	110.6 cm	–	69.34–85.60 cm
Vegetation	Dry, deciduous forest, dhak, salar, khair, chheela, kikkar, ber, bamboo, har, singar, lodsiali	Dry tropical thorn forest	Deodar, ban, oak, chil, walnut, peach, spruce, sal	Deodar, kail, silver fir, spruce, chil, sal, oak	Acacia, beri, sheesham, peepal, neem, dhak, jand, kreer, jal

Crops	Bajra, jwar, maize, wheat, barley, rice, pulses, sugar cane	Rice, wheat, cereals, pulses, sugar cane, cotton	Paddy, rabi, cotton, maize, black gram, barley, apple	Paddy, wheat, ginger, haldi (turmeric), kachalu, tobacco, potato, black gram, barley	Wheat, maize, paddy, sugar cane
Animals	Tiger, panther, spotted deer, sambar, nilgai, wolf, hyaena, jackal, fox, porcupine, hedgehog, hare, peacock, night birds, camel, cattle	Cow, sheep, goat, dogs, buffaloes, camel	Peafowl, dogs, sambhar, barking deer, ghooral, musk deer, hare, pheasants, partridges	Cattle, buffaloes, horses, ponies, donkeys, mules, sheep,goats, camels	Deer, nilgai, fox, jackal, wild cat, hare, peacocks, quail, pigeon, sand grouse

Temple monkeys always obtain their food from devotees or tourists visiting the temples. Urban rhesus often plunder houses or shops for food requirements. Food-sharing has not been observed. Temple groups have home ranges between 0.06 and 2.30 km^2, while urban groups have smaller home ranges, followed by village groups. Roadside groups also have very small home ranges, between 0.01 and 1 km^2. The home range of the canal side rhesus groups varies from 0.07 to 3.0 km^2. On average, forest groups of rhesus have larger home ranges than those in other habitats. Urban groups in the foothills of the Himalayas, village groups in Hissar district and most of the roadside groups of monkeys have relatively small home ranges varying between 0.03 and 0.17 km^2.

Interestingly, the home-range size is neither an index of group size nor of sex composition, but probably reflects the richness of the environment. The home-range size depends both on the behavioural characteristics of the monkeys and on the physico-chemical components of the habitat. It is evident from the data that groups with larger home ranges occupy rich environment (forests), have low densities, and relatively low dimorphism. Thus, the monkeys in these groups may be relatively healthier than in other groups.

Group composition and optimal group size

It is significant that of all the habitat groups, temple monkeys have the maximum number of adult males per group (Table 2). Although the number of males per group in the other habitats is almost the same, the number of adult females varies considerably. Temple groups have the most immature rhesus, while village groups have the least, probably because temple groups enjoy more religious protection than do their counterparts in other habitats.

Of the 372 groups sighted, 93 groups were between 21 to 30 animals in size with few groups totalling between 71 to 90 individuals. Average rhesus group size varies from 10 to 50 monkeys, irrespective of the habitat they occupy. There is evidence that with an increase in group size, the number of groups decreased. Larger-sized groups inhabit temples, pond and village + pond biomes.

Although only 14 temple groups in different habitats were sighted during the current surveys, they account for 13.8% of the entire rhesus population censused. Despite the numerous disadvantages for survival of monkeys in urban habitats they were found to occur preponderantly therein. In fact, the maximum number of animals sighted (23.8%) was in urban biomes, followed by forests (16.7%) and village + pond (15.7%).

Table 2. *Average group composition of rhesus in different habitats*

Habitat	Group composition[a]							Groups observed	Mean group size	Range
	AM	AF	SAM	ADF	JUV	INF	NB			
Temple	13.64	47.00	3.64	4.64	37.43	20.21	3.57	14	130.29	5–859
Urban	3.29	9.20	1.16	1.66	8.76	5.98	0.65	103	30.33	3–93
Village	2.69	6.02	0.83	0.74	5.74	4.14	0.21	42	20.38	3–59
Village + pond	4.30	10.46	1.28	1.60	9.88	6.73	0.33	60	34.60	4–121
Pond	3.00	11.96	1.71	1.71	12.35	7.92	0.21	28	38.89	6–118
Roadside	3.70	12.39	1.58	2.24	12.06	8.82	0.45	33	41.24	4–124
Canalside	3.32	8.63	1.09	1.18	9.09	5.59	1.50	22	30.32	5–106
Forest	2.81	8.77	1.18	1.60	10.20	5.64	1.20	70	31.50	4–83

[a] AM = adult male; F = adult female; SAM = subadult male; ADF = adolescent female; JUV = juvenile; INF = infant; NB = newborn

The average group size at each site for rhesus inhabiting temple precincts ranges from 78 to 242; in contrast, the majority of groups from other habitats are between 20 and 30 in size (range 5–242), with groups of 30–40 animals less common. Generally, the village, roadside and canal bank groups consist of fewer monkeys.

Feeding

The data suggest that rhesus display spatial and temporal variability in population size as a consequence of variation in the physical environment. This is particularly evident in temple groups where predator pressure by humans is almost absent. This suggests that both physical and biological processes shape rhesus populations. Even the selection and location of sleeping sites differs markedly between habitats.

Differences between the sexes in the utilisation of feeding levels and feeding sites, in the time expended on exploiting various resources, in the types of food consumed, and in overall activity budgets were noted. Adult male rhesus spent proportionately less time feeding and more time inactive than females or immatures. In general, the sexes have diets of different compositions and in the proportion of time spent feeding on a particular food resource. Differences between age classes were also found; infants and juveniles spent more time feeding on terminal twigs than did mature rhesus of the same group. In general, members of a group fed on the same food items at the same time. It is primarily the differential exploitation of food resources by males and females that acquires adaptive significance.

The time spent in foraging was consistent over time, and was proportional to metabolic rate and thus to basic energy requirements (Kleiber, 1975). Dominant rhesus often fed in regions of trees rich or abundant in ripe fruit, while low-ranking rhesus were often seen feeding in less rich regions. Similar trends have been observed in rhesus elsewhere (Seth & Seth, 1984) and in Toque monkeys (Dittus, 1977). Social dominance appears to regulate feeding/social behaviour, suggesting that population regulation occurs mainly within the naturally occurring rhesus groups within their home range.

Feeding group size differs significantly between habitats, and foraging rhesus monkeys must be alert in order to survive. Since forest monkeys can feed at will, they do not experience the same levels of hunger reached by their urban/village counterparts from whom food is generally withheld.

Time budgeting

The data reveal considerable day-to-day variability as well as individual variability. Climate and season affect the timing of the onset of daily activities, and there are seasonal differences in food items consumed. The daily feeding patterns for forest monkeys are fairly regular with periods of active feeding in the early morning and in the evening. Forest rhesus spend relatively more time feeding than do rhesus in other habitats, based on preliminary analyses (Table 3). When habitats and range utilization patterns are viewed in the light of dietary proclivities, forest rhesus differ from the common impression of temple-dwelling free-ranging groups. There are, of course, differences in the percentage of time devoted to the basic routine activities between monkeys in different habitats (Table 3). Rhesus do exhibit a pattern linking a set of behaviours in different habitats. Further data collection on time budgeting for subsistence activities is in progress, to facilitate a clear understanding of the adaptive significance of subsistence patterns and activity budgets.

A multivariate quantitative analysis of the relations among social variables observed in rhesus populations was performed using data from their behavioural repertoire. The resulting correlation coefficient matrix is shown in Table 4. There is a general lack of high correlation with the 'available food' variable. The most interesting result from the correlation analysis is the positive correlation between three social variables – grooming, play, and aggression. The correlation coefficients are slightly lower but positive for these variables with resting.

Discussion

These data suggest a distinct social order in the free-ranging populations of rhesus monkeys. Behavioural differences can be attri-

Table 3. *Percentage of time devoted to basic routine activities*

	Temple	Urban	Pond	Road-side	Canal-side	Forest
Feeding	26.69	16.21	20.12	23.25	16.36	40.08
Locomotion	18.67	17.19	19.08	20.62	16.92	26.21
Resting	34.85	46.16	37.64	39.36	62.32	27.70
Grooming	11.01	14.49	8.87	8.96	2.06	1.79
Playing	5.85	4.06	12.26	6.86	1.30	2.88
Other	2.73	1.89	2.03	1.01	1.04	1.34

buted to differences between habitats. The variety of activity patterns exhibited by rhesus on the basis of habitat gradients, namely the eight-point scale in tropical and arid zones clearly demonstrates the complexity of the dynamics affecting population attributes. Activity levels change with time, indicating a dynamic process of habitat selection.

That the social behaviour of rhesus monkeys is regulated is evidence that behaviour is influenced by natural selection to respond to changing conditions in a manner yielding the greatest possible benefit for individual survival and reproductive fitness (Darwinian fitness). For instance, a threat during foraging would stop the threatened individual from feeding or prevent it even from approaching the feeding area of the dominant individual. The ratio of threats received to those given roughly reflects differential access to food and water resources between age and sex classes. Adult males had the greatest relative access and infant females the least. Empirical data show that free-ranging rhesus monkeys are extremely selective in their exploitation of available resources. The abundance, predictability and distribution of food, water, and roosting sites determined to a significant extent the home range, this being a function of metabolic propensities conceivably influenced by dietary biases (Blumenberg, 1981; Seth & Seth, 1985).

Ecological variation contributes significantly to the biology of primates. In addition, the frequency of contact between humans and monkeys, the degree of overlap in space, the degree of commensality, and the amount of time primates spend in agricultural areas are all of biological significance.

Demographic consequences may stem from several types of habitat changes. Rapid urbanisation and industrialisation is not only changing the location of rhesus home ranges but is also forcing them into

Table 4. *Correlation coefficient matrix for social behaviour variables*

	Locomotion	Resting	Sleeping	Grooming	Play	Aggression
Feeding	−0.18	0.01	−0.36	−0.25	−0.11	−0.02
Locomotion		−0.28	−0.52	−0.03	0.15	−0.41
Resting			−0.19	0.22	0.24	0.36
Sleeping				−0.14	−0.32	0.02
Grooming					0.59	0.45
Play						0.10

smaller home ranges, thereby increasing their density (Seth & Seth, 1984). Such ecological disturbance results in distortions of ranging patterns as large areas of habitat are rendered unsuitable for occupation by rhesus with the destruction of trees, undergrowth, and arboreal pathways. This transforms behaviour as well as ranging patterns, at least temporarily if not permanently. Thus habitat transformation could ultimately cause isolation of rhesus populations into small pockets with potentially harmful genetic consequences.

Acknowledgements

The authors wish to thank the research team of the main programme at the Department of Anthropology, University of Delhi for their excellent technical assistance. This investigation received financial support from the Department of Science & Technology, Government of India.

References

Blumenberg, B. (1981) Observations on the palaeoecology, population structure, and body weight of some Tertiary hominoids. *J. Hum. Evol.*, **10**, 543–64

Dittus, W. P. J. (1977) The social regulation of population density and age-sex distribution in the toque monkey. *Behaviour*, **63**, 281–322

Kaul, O. N. (1977) Vegetation and ecology of Indian Himalayas. *Ecologie et Géologie de l'Himalaya*, (*Coll. Int. CNRS*), no. 268, 149–61

Kleiber, M. (1975) *The fire of life: an introduction to animal energetics*. New York: R. E. Krieger Publishing Co.

Koford, C. B. (1965) Population dynamics of rhesus monkeys on Cayo Santiago. In *Primate Behaviour*, ed. I. DeVore. New York: Holt, Rinehart & Winston

Puri, G. S. (1960) *Indian Forest Ecology*. New Delhi: Oxford Book & Stationery Co.

Ripley, S. D. (1965) *The Land and Wildlife of Tropical Asia*. Netherlands: Time-Life International

Seth, P. K. & Seth, S. (1983) Population dynamics of free ranging rhesus monkeys in different ecological conditions in India. *Am. J. Primatol.*, **5**, 61–7

Seth, P. K. & Seth, S. (1984) Population trends in naturally occurring rhesus monkey populations in India. *Int. J. Primatol.*, **5**, 380

Seth, P. K. & Seth, S. (1985) Ecology and feeding behaviour of the free ranging rhesus monkeys in India. *Indian Anthropol.* **15**, 51–62

Seth, P. K. & Seth, S. (1986) *The Primates*. Delhi: Northern Book Centre

Seth, P. K., Seth, S. & Shukla, A. K. (1983) Sociobiology of free ranging rhesus monkeys. In *Perspectives in Primate Biology*, ed. P. K. Seth. New Delhi: Today & Tomorrow's Printers & Publishers

III.2

Time budgets and activity patterns in free-ranging rhesus monkeys

I. MALIK

Introduction

The way in which primates use time and organize activity patterns is an important aspect of behavioural ecology. In both field (Chalmers, 1968) and laboratory (Zimmerman, Wise & Strobal, 1973) studies it has been noted that the frequency of the majority of social interactions, especially aggression, may be related to food availability and distribution. Changes in food availability also affect day range (DeVore & Hall, 1965; Crook, 1966; Altmann & Altmann, 1970). As foraging is second in order of importance in the daily activity pattern (Washburn, 1971) it has been postulated that time spent on other social interactions increases when time spent foraging decreases (Hall, 1963; Crook, 1970; Rowell, 1972). Teas (1978) found that feeding became the most time-consuming activity in fall as food availability decreased, whereas it was the second most predominant activity in summer. Play is observed when the most immediate needs are met, i.e. food (Bekoff, 1972) and decreases when food decreases (Altmann, 1959; Loizos, 1967). Therefore play can be classified as non-primary activity (Loizos, 1967). This is borne out by Loy's (1970) observation of rhesus on Cayo Sentiayo and Baldwin & Baldwin's (1972) data on squirrel monkeys.

This paper presents data based on a long-term study on the activity patterns and percentage of time spent on various activities in different seasons of an expanding population of rhesus monkeys in India; namely the rhesus monkeys at Tughlaqabad on the southern outskirts of New Delhi.

Tughlaqabad was first referred to in the population survey of rhesus monkeys in northern India by Southwick, Beg & Siddiqi (1965). No long-term study has been conducted prior to this study.

Study area

Tughlaqabad is an ancient city site with a fourteenth-century fort and is situated on the southern edge of New Delhi at latitude 30° 25′ north and longitude 78° 76′ east. The home ranges of rhesus monkey groups under study extend throughout the fort and surrounding areas, covering approximately 5 km^2 (2.5 × 2.0 km). The fort rises about 15–30 m (50–90 feet) above the surrounding plain. The outer walls of the fort form a polygon with a circumference of nearly 5 km. The flat and fertile area surrounding the fort contains croplands, pasture, two forested areas, and encroaching suburban development.

Fig. 1. Map of the Tughlaqabad area showing forest plantations, agricultural fields, canal, and roads surrounding the fort and tomb. The interior of the fort is not depicted accurately but some structures are shown to indicate a complex pattern of walls, turrets, ramparts, steps, shrubs, and small trees which form a diverse primate habitat.

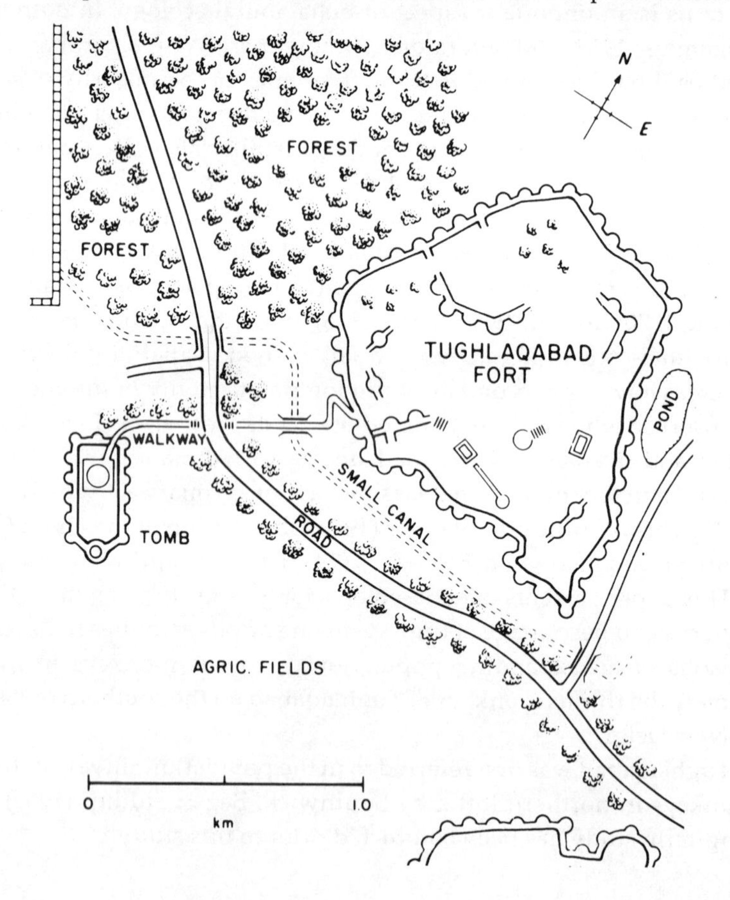

Trees lining a road in the southern area are used by the rhesus monkeys for sleeping at night and resting during the day.

The fort constitutes a quarter of the total area, two forest plantations occupy another quarter, and the surrounding open areas of cultivation and pasture constitute the remaining half (Fig. 1).

Tughlaqabad has a sub-tropical climate with marked seasonal changes. During the months of May and June, daytime temperatures often reach 40–45 °C; in the months of December and January, daytime temperatures fall to 7–9 °C. Monsoon occurs from the end of June or early July until mid-September, with an annual average of 567 mm of rain. Winter and spring rains occur sporadically and are usually light.

The natural vegetation inside the fort is xerophytic, generally grasses and arid forbs and shrubs. Outside the fort, vegetation is more mesophytic, and better groundwater supports trees and crops, primarily wheat and pulses. The main trees present are Indian jujube (*Zizyphus jujuba*), neem or margosa (*Azadirachta indica*), sheesham or sissoo (*Dalberigia sissoo*), oak (*Quercus incana*), acacia (*Acacia arabica*), pipal (*Ficus religiosa*) and date palm (*Phoenix dactylifera*). Other than people, the dominant fauna includes rhesus monkeys, cattle, buffalo, donkeys, goats, dogs, jackals, mongooses, lizards, and a great variety of birds, both migratory and resident. Peacocks, partridges, pigeons, crows, sparrows, vultures, mynahs and kites are all common.

Field methods

Field work was carried out from March of 1980 to August, 1983. The data were collected during 5800 contact hours, covering three breeding seasons. Scan techniques (Smith, 1968) were used to collect data on activity patterns and time budget.

Results

The daily activity cycle varied with the solar cycle (Table 1). In the warmer months (mid March–October), the sun rose between 05.22 and 06.32 and monkeys were up between 04.45/04.50 and 06.00. During the winter months (November–mid March), sunrise was between 06.31 and 07.15 and the waking up time was roughly between 06.00 and 06.45. Thus they invariably awoke at least half an hour before sunrise.

During winter mornings, monkeys chattered, descended from the trees, and sat huddled up in groups (up to 18 in a huddle) till 07.00–08.30 h, when the sun was fully risen. At full sunrise monkeys slowly and randomly moved out of their 'huddles'. 'Huddles' would only temporarily break before this, if food was offered by passers by. Some rhesus (sentinels) from within these huddles gradually moved to places from where their surroundings could be surveyed; others socialised (groomed or played), punctuated by slow feeding and locomotion. From 09.00 until 12.00 h, social activities and resting were interspersed with feeding and foraging. The warmest part of the day, 12.00–15.00 h, was spent resting and grooming in the open while infants and juveniles indulged in active play in between bouts of resting. Between 15.00 and 17.30 h was a time of intense feeding. The

Table 1. *Diurnal activity patterns in each season*

	Activity	
Time	Summer (mid Mar.–Oct.) 18.7–40.5°C Sunrise 05.22–06.32 Sunset 18.29–19.22	Winter (Nov.–mid Mar.) 7–17°C 06.31–07.15 17.35–18.28
04.45	Arise	
05.00	Chatter, descend Eat, groom	
06.00		Arise
06.45		Chatter, descend Eat, groom
07.00	Travel, feed	
08.30	Intense feed, play Groom, farm raid	Travel, slow feed, groom, play
09.00		Travel, intense feed, Groom, play
10.30		
11.00	Rest, intense play Intense groom	
12.00		Rest, intense play Lazy feed, intense groom
13.00		
14.00		
15.00		Travel, intense feed, groom Play, farm raiding
16.00	Travel, intense feed Play	
17.00		
17.30		Groom, play
18.05		Sleep
18.15	Eat, groom, play	Sleep
19.00	Sleep	
19.50	Sleep	

rhesus monkeys were most likely to raid the cultivated fields at this time. The monkeys were back on their sleeping sites, or at least near them, within a half hour of sunset, between 18.00 and 19.00 h.

The difference in inhibitory effects of temperature could be clearly seen between seasons. In summer, daylength increased, morning temperatures were higher, foraging began earlier, and there were no huddles. Siesta time increased in duration (11.00–16.00 h) in the cooler shady recesses in the tomb. The day was quite predictable, in so far as behavioural sequences and time allotted to the specific behaviours were concerned, but changed in terms of location within the territory.

Annual daily activity budgets

At Tughlaqabad the overall percentage of time spent in different activities was: rest 30.0%; locomotion 18.6%; grooming 11.5%; fed by humans 9.5%; play 9.3%; foraging 8.0%; and drinking 3.4% (Table 2). During summers, time spent on resting (4.40 h/day) and drinking water (0.56 h/day) was almost twice the time spent on these activities during winters (2.79 h/day and 0.24 h/day, respectively), whereas the order was reversed in the case of locomotion (2.64 h/day in winters and 1.81 h/day in summers). There were only slight changes in the time spent on activities like grooming, being fed by humans, play and foraging in the two seasons.

Table 2. *Average percentage of time and number of waking hours per day spent on various activities in winter and summer*

	Winter		Summer		Annual	
Activities	% Time	Waking hours per day	% Time	Waking hours per day	% Time	Waking hours per day
Rest	23.3	2.79	36.7	4.40	30.0	3.60
Locomotion	22.0	2.64	15.1	1.81	18.6	2.22
Groom	13.0	1.55	10.1	1.20	11.5	1.38
Fed by humans	9.0	1.08	10.0	1.20	9.5	1.14
Play	9.8	1.17	8.8	1.05	9.3	1.11
Eating natural vegetation	7.3	0.88	8.5	1.02	8.0	0.96
Drink	2.0	0.24	4.7	0.56	3.4	0.42
Others	13.7		6.1			

Feeding

The Tughlaqabad area provides a wide range of two types of food: primarily, that provided by humans and, secondly, natural foods. Food provided by humans is consistent, so consistent as to be almost ritualistic. As the days in summer become longer, people have a few more daylight hours to feed the monkeys. Thus the monkeys spent 10% and 9% of their time during summers and winters, respectively, on food given by humans. The variation in natural food comes in months (1) when the crops have been sown and the trees bear fruits, and (2) when crops have been harvested and the trees bear no fruit (May and November, respectively). In the first instance, when the monkeys did not have enough food provided by man, they relied upon natural foods. In the second instance, when natural foods were low in abundance, they spent more time foraging, i.e. 11% in the month of May as compared to at least 6.5% in the months of March, September and December. Thus the dietary pattern was variable at different times of the year. The peak feeding time was in the morning when the animals either were fed by the humans or had waited long enough to be fed by humans and, if not fed, resorted to natural vegetation. The other period of equally intense feeding was in the evening.

Drinking and water requirements

During winters most water requirements seem to be met by consuming moist leaves and juicy fruits. Thus time spent on drinking was only 2%. In summer, with insufficient water available from leaves, juicy fruits or ditches and ponds, they spent more time (4.7%) looking for drinking water. Little direct drinking behaviour was witnessed in the early part of March. However, towards the end of March and from April onwards, direct drinking from ditches and ponds was observed and the majority of the animals drank (up to three or four times a day).

Grooming

Weather was a contributing factor to the changes in grooming pattern. During winter, grooming was observed only after the monkeys had fed and the sun was sufficiently high to provide enough warmth for them to bask and groom. In the summer season, grooming was seen even on waking. This did not take priority over feeding but was sufficiently frequent to be conspicuous.

Grooming was observed right until the animals slept, in both the seasons. The peak time for grooming was in the afternoons. It differed in the two seasons as a result of difference in the timing of resting periods, from 12.00–15.00 h in winters to 11.00–16.00 h in summers. Secondly, grooming was slightly more frequent in winter because less time was spent sleeping during the afternoon rest period. Thus the percentage of time spent on grooming was 13.0% (1.55 h/day) on average in winter and 10.1% (1.20 h/day) on average in summer.

Locomotion

Rhesus have ample protection from the wind in the fort, and there are trees for shade and a tomb with a cool interior. Thus the locomotor behaviour of the rhesus monkeys is fairly predictable in all seasons. On cool, cloudy days they were observed locomoting for pleasure, covering 1 km in over an hour at a speed of about 100 m/min and covering the extremes of their territory. Bouts of grooming, feeding and relaxation were observed in between locomotion. Monkeys locomoted less on hot summer days. Locomotion in summer was only done in the mornings and evenings, that too for specific purposes of food and water acquisition. Afternoons were spent in cool, shady places. In winters, time spent on locomotion was 22% (2.64 h/day) as compared to 15.1% (1.81 h/day) on average in summers.

Discussion

Components of the ecosystem affect the majority of rhesus activities (Shukla, Seth & Seth, 1982). The rhesus monkeys of Tughlaqabad spent 80% of their waking hours on the ground, i.e. in the fort, playground, cultivated fields and the tomb. Only about 17% of the daytime was spent feeding. The time spent foraging on a particular day depended upon the availability of food from visitors. When food was provided (by visitors) in abundance, the monkeys spent more time on other activities. When there were too few visitors to supply enough food, the time spent on other activities decreased and time spent foraging increased. Southwick (1962) found that macaques (of the temple population in Aligarh) spent approximately 10% of their waking hours in feeding. Southwick *et al.* (1982) found feeding to be the most important behaviour in the daily activity pattern of rhesus monkeys of Nepal, accounting for 27% of time. Altmann (1962) on the other hand, in his study of rhesus macaques, reported that a maximum of 80% of their time was spent on foraging, which is far greater than 17% for the Tughlaqabad monkeys.

The Tughlaqabad monkeys spent 30.0% of their time resting; more in summers and less in winters. Shukla (1979) reported that the two Maroth groups spent 9.3% and 16.1% of their time on resting; Southwick *et al.* (1982) in their study of rhesus monkeys of Nepal reported that rest took up 8% of the monkeys' time, and play, 3%. Oppenheimer (in press), in his study on the Hanuman langur (*Presbytis entellus*), found play to take up about 9% of their time which compares with 9.3% of time spent in play for the Tughlaqabad monkeys. Koford (1963) reported that the time spent on locomotion was 15% and Oppenheimer (in press) reported that locomotion consumed 10% of the daytime; these compare with 18.6% for the rhesus of Tughlaqabad, while in Nepal monkeys spent 25% of their time in locomotion.

Since the Tughlaqabad monkeys spent relatively little time foraging, more time was available for other activities. Perhaps as a result of provisioning, the study animals were in good health and spent more time locomoting in the home range.

It is significant that the time spent on locomotion was higher in winter than in summer. In winter, animals looked for warm sunny places in the fort where the walls would protect them from the cold wind; secondly, without the oppressive heat of the summers to bind them to cool shady places, the animals were free to roam as they pleased. At this time of the year it was not unusual for them to be raiding farms, at times when in summers they would normally have retired for the 'siesta'. Wind and rainfall are the climatological factors which influence the locomotory behaviour of the Tughlaqabad monkeys. Wind has an inhibitory effect: on every windy day there was a marked decrease in activity. As soon as they sensed a wind storm, they moved up into the fort, where they sat huddled up in the crevices. Wild swaying of the branches affected their feeling of security; banging of the branches and whistling of the wind through the trees producing auditory cues of a high pitch seemed to increase their feeling of insecurity, hence they sought refuge in the fort.

The animals under study spent only 20% of their waking hours on trees, whereas Jay (1965) in a study of langurs in North India, Koford (1963) in his study of an inland population of rhesus monkeys and Oppenheimer (in press) in his study of langurs noted that the animals spent 30–50%, 75–85% and 90%, respectively, of their waking hours on trees. The Tughlaqabad rhesus spent 15.1% of their time on locomotion in summers and 22% in winters. The significance of comparison with the above data is that the study areas exhibit a warm climate, hence only the time spent on locomotion during the summers at Tughlaqabad coincides with the data given by Koford and Oppenheimer; the time spent locomoting during winters was much higher at Tughlaqabad, almost as long as for monkeys in Nepal. It is indicative of the inhibitory effect the warm weather has on the ability of the monkeys to locomote.

Oppenheimer (in press) gives 5% of the waking time to grooming. Lindburg (1971) gives an account of approximately 7.28 bouts in a day. The time taken up by grooming in Tughlaqabad was on average 11.5% (31 bouts/day) which is much higher than that reported by both Oppenheimer and Lindburg, but is nearer to the 15% of time spent on grooming reported by Southwick *et al.* (1982) in their study of Nepal monkeys.

Field studies of non-human primates indicate that there is a considerable variation in the extent to which different species engage in social grooming (Bekoff, 1972). Marler (1965) reported that the species with rigid and clearly defined dominance relations tend to show the highest levels of grooming. The rhesus monkeys under study fit this concept very well.

Conclusions

The components of the ecosystem that most affect the behaviour of rhesus monkeys at Tughlaqabad are:

1. Protection provided to rhesus by their physical habitat and by their religious connotations for local humans. The percentage of time spent on the ground was four times that spent on trees.
2. Abundance of food. Rhesus spent around 78% of their time resting and on activities like locomotion, grooming, and play.
3. Weather. This affected the percentage of time spent in locomotion, which was 1.5 times more in winters than in summers, as well as affecting the location of different behavioural sequences of their daily activity pattern in the territory.

References

Altmann, S. A. (1959) Field observations on a howling monkey society. *J. Mammalol.*, **40**, 317–30

Altmann, S. A. (1962) A field study of the sociobiology of rhesus monkey. *Ann. N.Y. Acad. Sci.*, **102**, 338–435

Altmann, S. A. & Altmann, J. (1970) *Baboon Ecology*. Basel: Karger & Chicago: University of Chicago Press

Baldwin, J. D. & Baldwin, J. (1972) The ecology and behaviour of squirrel monkeys (*Saimiri oerstedi*) in a natural forest in Western Panama. *Folia Primatol.*, **18**, 161–84

Bekoff, M. (1972) The development of social interaction, play and meta communication in mammals: an ethological perspective. *Q. Rev. Biol.*, **47**, 412–34

Chalmers, N. R. (1968) The social behaviour of free living mangabeys in Uganda. *Folia Primatol.*, **8**, 263–81

Clutton-Brock, T. H. & Harvey, P. H. (1977) Primate ecology and social organisation. *J. Zool., Lond.*, **183**, 1–39

Crook, J. H. (1970) The socio-ecology of primates. In *Social Behaviour in Birds and Mammals*, ed. J. H. Crook, pp. 103–66. New York: Academic Press

DeVore, I. & Hall, K. R. L. (1965) Baboon ecology. In *Primate Behaviour: Field Studies of Monkeys and Apes*, ed. I. DeVore, pp. 20–52. New York: Holt, Rinehart & Winston

Crook, J. H. (1966) Gelada baboon herd structure and movement: a comparative report. *Symp. Zool. Soc. Lond.*, **18**, 237–58

Hall, K. R. L. (1963) Observational learning in monkeys and apes. *Br. J. Psychol.*, **54**, 201–26

Jay, P. (1965) The common langur of North India. In *Primate Behaviour*, ed. I. DeVore, pp. 197–249. New York: Holt, Rinehart & Winston

Koford, C. B. (1963) Group relations in an island colony of rhesus monkeys. In *Primate Social Behaviour*, ed. C. H. Southwick, pp. 136–52. Princeton, N.J.: Van Nostrand

Lindburg, D. G. (1971) The rhesus monkey in North India: an ecological and behavioral study. In *Primate Behavior*, vol. 2, ed. L. A. Rosenblum, pp. 1–106. New York: Academic Press

Loizos, C. (1967) Play behaviour in higher primates: a review. In *Primate Ethology*, ed. D. Morris, pp. 179–219. Chicago: Aldine

Loy, J. (1970) Peri-menstrual behaviour among rhesus monkeys. *Folia Primatol.*, **13**, 286–97

Marler, P. (1965) Communication in monkeys and apes. In *Primate Behaviour*, ed. I. DeVore, pp. 544–84. New York: Holt, Rinehart & Winston

Oppenheimer, J. R. (in press) Effects of intra and interspecific competition and habitual structure on use of time and space by Hanuman langur (*Presbytis entellus*). Ph.D. thesis, Johns Hopkins University, Baltimore, Maryland

Rowell, T. E. (1972) Female reproductive cycles and social behaviour in primates. *Adv. Study Behav.*, **4**, 69–105

Shukla, A. K. (1979) Activity patterns in the rhesus monkeys (*Macaca mulatta*). Research Proceedings Dept of Anthropology, Univ. of Delhi

Shukla, A. K., Seth, P. K. & Seth, S. (1982) The ecology of free ranging rhesus monkeys (*Macaca mulatta*) in an arid forest of India. In *Symposium on National Primate Programme*. Delhi: Primatological Society of India (Abstract)

Smith, C. C. (1968) The adaptive nature of social organization in the genus of tree squirrels *Tamiasciurus*. *Ecol. Monogr.*, **38**, 31–63

Southwick, C. H. (1962) Patterns of inter-group social behaviour in primates with special reference to rhesus and howling monkeys. *Ann. N.Y. Acad. Sci.*, **102**, 436–54

Southwick, C. H., Beg, M. A. & Siddiqi, M. R. (1965) Rhesus monkeys in North India. In *Primate Behavior*, ed. I. DeVore, pp. 111–59. New York: Holt

Southwick, C. H., Teas, J., Richis, T. & Taylor, H. (1982) Ecology and behaviour of rhesus monkeys (*Macaca mulatta*) in Nepal. *Nat. Geogr. Soc. Res. Rep.*, **14**, 619–30

Teas, J. H. (1978) Ecology and behaviour of rhesus monkeys in Kathmandu, Nepal. Ph.D. dissertation, Johns Hopkins University

Washburn, S. L. (1971) On understanding man. *Rehovot*, **6**, 22–31

Zimmerman, R. R., Wise, L. A. & Strobel, D. A. (1973) Dominance measurements of low and high protein reared rhesus mascaques. *Behav. Biol.*, **9**, 77–84

III.3

The ecology of the hoolock gibbon, *Hylobates hoolock*, in Tripura, India

R. P. MUKHERJEE

Introduction

A census of non-human primate populations in the Indian state of Tripura was started in November 1976 to study the distribution, abundance and present status of different primate species. The behaviour of selected species in certain areas was also examined. The present paper reports on a study of the ecology and behaviour of the hoolock gibbon of Tripura.

In the past, Tripura (known as Tipperah) was part of the former province of Bengal. The hilly portion of Tipperah now lies within India, while the majority of the plains fall in Bangladesh. The state of Tripura (23°50′N, 91°15′E) in northeastern India has a total area of 10477 km^2 with a human population of 147 people per km^2. It borders Bangladesh on the north and southwest, and Assam and Mizoram in the northeast. Tripura is divided into three districts: north, south and west.

Despite its small size, Tripura harbours a large number of primate species, ranging from tree shrews to gibbons (see Mukherjee, 1977, 1982*a*,*b*). The total area of tropical wet evergreen and semi-evergreen forest is just under half the state. The forests consist of mixed tree and bamboo species with open scrub jungle and grasslands. The remaining area is agricultural, with the main crops being paddy, jute, cotton, and tea.

The climate is tropical, hot and moist, with three seasons: summer from March to May has occasional rains; the rainy season starts in June and ends in October; and winter lasts from November to February. Average annual rainfall is 1582 mm and the average relative humidity varies between 68 and 71% (see Mukherjee, 1982*b*, for further details).

Methods

Since the study began in 1976, a number of field trips of 21–40 days in duration were conducted during different seasons. The latest field season was in March–April (summer) 1983. Two groups of hoolock gibbons, one at Abhoya in the south and another at Khasi Bari in the north, were selected for detailed study. Hoolock gibbons are diurnal, and found in Tripura between 80 and 400 m elevation. They are clearly visible in the forest, but difficult to locate and follow when they retreat into dense foliage or up to greater canopy heights. The study groups at Abhoya and Khasi Bari were less shy of humans than were other groups in the interior of forests elsewhere. The activities of these groups could be observed at close distances without disturbing the gibbons.

The forests inhabited by the gibbons are composed of tall trees of *Mangifera, Dipterocarpus, Lagerstoemia, Shorea, Albizzia, Artocarpus, Ficus* and other species. The terrain is generally not rugged. The Abhoya area also includes thick growths of bamboo. The forests, providing year-round food and shelter for the gibbons, are selectively logged at intervals. Cultivation extends up to the forest edge, particularly in Khasi Bari. Both forests are frequently visited by villagers, which may have resulted in the tolerance of the gibbons to human observers.

The major activities of the groups were recorded continuously for 10 min in each hour, and rare interactions (grooming, aggression, vocalizations, travelling and sexual behaviour) were recorded as and when they occurred.

Results

Distribution and abundance of hoolock gibbons in Tripura

The present study is concerned with the distribution, abundance, and ecological aspects of behaviour of the hoolock gibbon in Tripura. Although this species, also known as the white-browed gibbon, is found from southeast of the Brahmaputra river in India to the Chindwin and Salween rivers in Burma, with reports of sightings in western Burma and southern China, nothing is known about this species in other areas of its distribution.

The hoolock gibbon is a small ape with long limbs and long thick fur. Adults are sexually dimorphic in colour, males are black while females are brown. Both sexes of young are black. Adult males and young females have a silver-white band above the eyebrows and a brown, white or pale grey bear-like tuft in the groin. The black coat colour of infant females fades to a pale yellowish brown at puberty. They are wholly arboreal and move by brachiating through the trees.

The gibbons are more abundant in the north district than elsewhere in Tripura. The population estimates and social group composition of the hoolock gibbons are compared with those of the other larger primates in Tripura in Table 1. Relative to the other primates, the hoolock gibbon is among the rarest of the species, with only the pig-tailed macaque (*Macaca nemestrina*) found less frequently. Rhesus monkeys are the most common primate. As shown in Table 1, most of the gibbon groups observed in Tripura consisted of three members, typically one adult male, one adult female and one offspring. The same pattern of group structure was observed during censuses in Manipur. Two groups with four members (two offspring) were recorded at Tripura. Two groups, one in Tripura and the other in Manipur, had only two members – an adult male and female.

Using an estimated group size of 3.2 individuals and a home-range size of 3–4 km^2 (although Gittins & Tilson (1984) report smaller home ranges of 18–30 ha in Assam and Bangladesh), Mukherjee (1982*a*) estimated 340–435 groups, or 1000–1400 hoolock gibbons in Tripura. Based on the most recent survey, with a group size of 3.0, the total population is likely to be lower.

Feeding behaviour

Since the hoolock gibbons of Tripura are exclusively arboreal, they explore for food in a hanging posture. They may remain in one area of the home range for several days before moving, depending on food availability. They move primarily through the upper and middle canopies during feeding, at between 18 and 22 m in height. Occasionally they descend to the lower levels. All group members move as a unit from tree to tree while searching for food. Both hands are used equally during feeding; they have never been seen using lips or teeth for this purpose.

Their period of daily activity varies from season to season and on average they remain active for 8–10 h each day. About half this time is spent feeding. Hoolock gibbons become active well after dawn, when they descend from their sleeping positions. From the time of descent to 11.00 h, 50–70% of time is spent feeding. Between 11.00 and 14.00 h, time spent feeding is between 25% and 40%, and there is a midday rest period. After 14.00 h until settling into their sleeping trees, only 10–20% of time is spent feeding. They retire to sleeping trees well

Table 1. *Social structure of different non-human primates of Tripura*

Species	Total no. of groups seen	Total no. of groups counted	Total no. of monkeys	Average group size	No. of adult males	No. of adult females	No. of juveniles	No. of infants
Hylobates hoolock (hoolock gibbon)	10	9	27	3.00	9	9	7	2
Macaca mulatta (rhesus macaque)	45	33	1020	30.91	131	476	219	194
Macaca arctoides (stump-tailed macaque)	5	5	62	12.40	6	34	15	7
Macaca nemestrina (pig-tailed macaque)	1	1	9	9.00	1	4	3	1
Presbytis phayrei (phayre's leaf monkey)	49	49	523	10.67	90	224	114	95
Presbytis pileatus (capped langur)	25	18	154	8.55	30	69	34	21

before sunset (1.5–2 h before), generally much earlier than other primates. Different sleeping trees are selected each night.

The food of the hoolock gibbons includes leaves, fruits, buds, flowers and animal protein (Table 2). About 50% of daily feeding time is spent eating leaves, and 40% is spent feeding on fruits. The remaining 10% is accounted for by buds, flowers, and animal protein. During seasons when fruits are less abundant, the consumption of leaves may rise to 60% of feeding time, and fruit intake is reduced to 30%. The leaves of trees are consumed the most, while those of climbers are rare in the diet. A species list of foods can be found in Mukherjee (1982*b*).

Ranging behaviour

Feeding and ranging behaviour are highly correlated in the hoolock gibbons at Tripura. The distance covered and the time spent travelling per day varies seasonally (Table 3). When food is abundant

Table 2. *Food items of different non-human primates of Tripura*

Species	Leaf (L)	Flower (FL)	Fruit (FR)	Seed (S)	Insect (I)	Bird's egg (BE)	Main food items
Hylobates hoolock	+	+	+		+	+	FR, L
Macaca mulatta	+	+	+	+	+		S, FR
Macaca arctoides	+	+	+	+			S, FR
Macaca nemestrina	+	+	+	+			S, FR
Presbytis phayrei	+	+	+				L, FR
Presbytis pileatus	+	+	+				L, FR

Table 3. *Range of movement of hoolock gibbons during different seasons*

Season	Percentage of time spent in feeding	Day range of movement (m)
Winter	65	400–800
Summer	50	700–1000
Rainy[a]	40	300–500

[a] Data related to the beginning of the rainy season

within a small area of the home range, less time is spent travelling and a short distance is covered during each day. They spend more time in a single tree or group of trees when foods are clumped. Feeding time and ranging distances are particularly reduced in the morning hours when exploiting clumped foods. The average day range is 600 m with a range of 300–1000 m. After the gibbons have spent a number of days feeding in the same site, they then spend more time travelling and cover longer distances when they shift to a new area.

Behaviour within the group

Group members are highly cohesive, but rates of interaction are low. Either the adult male or the adult female may lead the group during travel, but the adult male generally leads the group over long distances. During feeding, travel, and resting, group members remain within 10–20 m of each other. The adult male occupies the central position in the group and other members stay close to him. Grooming and play are rare interactions and are of short duration when they occur.

Interactions with other primates

The hoolock gibbon tolerates the presence of most animals, including other non-human primates, on feeding trees. At least two species of leaf monkey, *Presbytis pileatus* and *P. phayrei*, squirrels and many bird species approach hoolock gibbons in feeding trees with few signs of agonistic behaviour.

Vocalizations, which are common among hoolock gibbons, play an important role in communication between different groups and in maintaining territories. Among the groups in Khasi Bari, vocalisations are uncommon. Population densities are low in this area. Neither sex vocalizes on a daily basis, and when they do call, they sing only for a short period. This suggests that the frequency of vocalization depends upon the presence of other groups in an area. If there is more than one group in the same area, vocalizations are daily occurrences and of long duration.

Conservation problems

Hoolock gibbons are found in many eastern states of India but their present status in most of these areas is unknown. It is likely that their numbers have been greatly reduced. In Tripura, clearing of forests for cultivation, the discontinuity of forests, and shifting cultivation are the major factors affecting gibbon populations. Similarly,

Chivers (1977*a*) indicated that forest clearing for cultivation, hunting for food and the export of gibbons are the major problems for their conservation in Malaysia.

If these problems are not redressed, the population of the hoolock gibbon in Tripura will decline further. The forest ecosystems, and especially those within reserves, must be preserved and indeed expanded if a viable population of gibbons is to be maintained.

Discussion

Hoolock gibbons are found in the Indian states of Assam, Meghalaya, Tripura, Mizoram, Manipur, Nagaland, and in a small pocket of the southwestern part of Arunachal Pradesh adjoining Burma (Fig. 1). Chivers (1977*b*) reported that while the hoolock gibbon is restricted to the forests of Assam, they also occur in some parts of Bangladesh, South China and Burma. Of the seven states in eastern India with hoolock gibbons, only in Tripura have their distribution, abundance, status and ecological aspects of behaviour been studied.

Fig. 1. Map showing the distribution of hoolock gibbons in North-eastern India

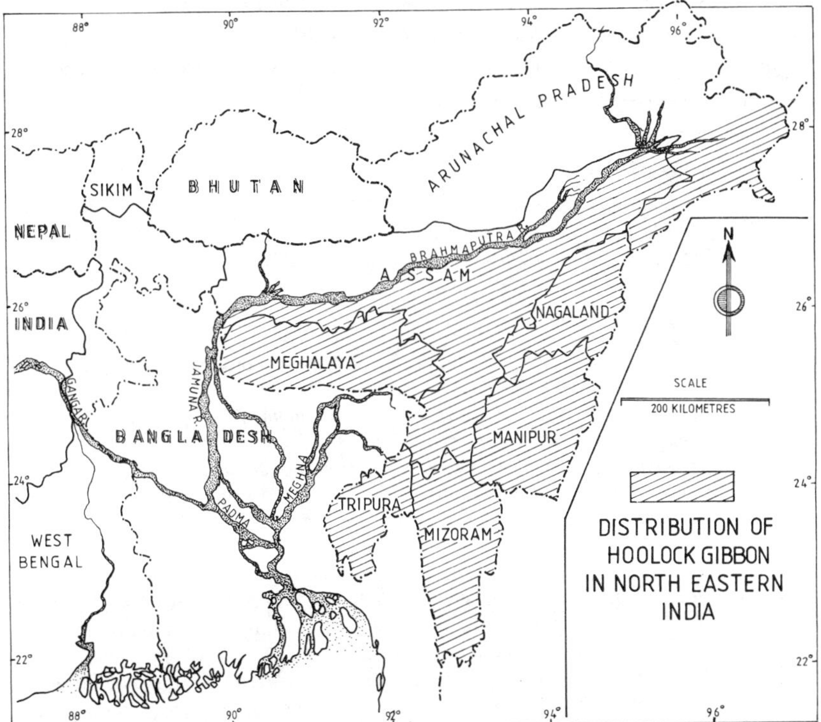

Nothing is known of their present status in other areas. Past records show that they were once more abundant in Burma than in other parts of their range, but recent surveys indicate that their range is greatly reduced in that country as well as in India.

Within Tripura, the hoolock gibbon is more abundant in the north district. In the present surveys, groups of 2–4 individuals, with an average group size of 3, were observed. Blanford (1888–91) recorded hoolock gibbon 'flocks' of 50–100 individuals, while McCann (1933) reported that the largest group was 7. Agrawal & Bhattacharya (1977) found 2–5 individuals in different parties at Ambasa and Atharamurra in Tripura. Tilson (1979) recorded group sizes of 2–6 in Assam. Chivers (1977*b*) estimated 80000 hoolock gibbons in Assam only, which is probably an overestimate. Mukherjee (1982*b*) estimated a total population of 1000–1400 hoolock gibbons in Tripura. The conservation of these apes is of major urgency in the remaining parts of their range.

Hoolock gibbons in Tripura have shorter day ranges than the 1.5 km reported for the lar and agile gibbons (Chivers, 1980). Hoolock gibbons leave their sleeping trees later in the morning and retire to them earlier in the evening than do other primates. They eat primarily leaves and fruits, the proportions of which vary seasonally. Seasonal patterns of food abundance affect day ranges and range use. The gibbons appear to monitor areas of local abundance and exploit them intensively during the different seasons.

Acknowledgements

My thanks are due to the Director, Zoological Survey of India, Calcutta, for providing facilities of work and to the staff of the Forest Department of Tripura for their cooperation and help in the field.

References

Agrawal, V. C. & Bhattacharya, T. P. (1977) Report on a collection of mammals from Tripura. *Rec. Zool. Surv. India*, **73**, 135–57

Blanford, W. T. (1888–91) *The fauna of British India including Burma and Ceylon: Mammalia*. London: Taylor and Francis

Chivers, D. J. (1977*a*) The ecology of gibbons: some preliminary considerations based on observations in the Malay Peninsula. In *Use of Non-Human Primates in Biomedical Research*, ed. M. R. N. Prasad & T. C. Anand Kumar. New Delhi: India National Science Academy

Chivers, D. J. (1977*b*) In *The Lesser Apes*. In *Primate Conservation*, ed. Prince Rainier & G. H. Bourne, pp. 539–98. New York: Academic Press

Chivers, D. J. (1980) Primate ecology, rain-forest ecosystems and land development. *Trop. Ecol. Dev.*, 341–51

Gittins, S. P. & Tilson, R. L. (1984) Notes on the ecology and behaviour of the hoolock gibbon. In *The Lesser Apes: Evolutionary and Behavioural Biology*, ed. H. Presuchoft, D. J. Chivers, W. Y. Brockelman and N. Creel, pp. 258–66. Edinburgh: Edinburgh University Press

McCann, C. (1983) Notes on the coloration and habits of the white-browed gibbon or hoolock (*Hylobates hoolock* Harl). *J. Bombay Nat. Hist. Soc.*, **36**, 395–405

Mukherjee, R. P. (1977) Rhesus and other monkeys of Tripura. *Newsl. Zool. Surv. India*, **3**(3), 111

Mukherjee, R. P. (1982*a*) Survey of non-human primates of Tripura, India. *J. Zool. Soc. India*, **34**(1,2), 70–81

Mukherjee, R. P. (1982*b*) Phayre's leaf monkey (*Presbytis phayrei* Blyth, 1847) of Tripura. *J. Bombay Nat. Hist. Soc.*, **79**(1), 47–56

Tilson, R. I. (1979) Behaviour of hoolock gibbon (*Hylobates hoolock*) during different seasons in Assam, India. *J. Bombay Nat. Hist. Soc.*, **76**(1), 1–16

III.4

Social organization of two sympatric galagos at Gedi, Kenya

L. T. NASH

Introduction

A field study examined two sympatric species of galagos in East Africa for differences in habitat usage and behavior. The species are *Galago zanzibaricus*, a small-bodied species weighing 140–160 g, and *G. garnettii*, a larger-bodied species weighing 800–900 g. The larger species has been considered conspecific with *G. crassicaudatus* but differs from that species in several reproductive and morphological features (Olson, 1979; Eaglen & Simons, 1980). *G. crassicaudatus* and *G. senegalensis moholi* are similar species which have been studied in allopatry in South Africa (Clark, 1978; Bearder & Martin, 1980).

At the coastal study site, these species differ in diet, habitat usage, and reproductive patterns (Nash, 1983*a*,*b*). While both eat insects and fruit, the larger *G. garnettii* incorporates more fruit in the diet than does *G. zanzibaricus*. Though both species breed seasonally and mainly have singleton births, the smaller *G. zanzibaricus* has two successive pregnancies per year. This paper reports on additional similarities and differences between these two species in their social organization as it is revealed by the overlap of individual animals' ranges.

Methods

The study area is included within the ruins of a thirteenth- to fifteenth-century Arab walled town which was excavated in the 1950s. It is now fully protected as the Gedi National Monument. The site encompasses about 44 ha of lowland rainforest. This field study lasted from February, 1979, to November, 1980. Individual animals' ranges were established by trap–retrap and radio tracking methods. The site was divided into 50 m squares by a grid of paths. All trap sites were located within a 25 m square quarter of the 50 m quadrats demarcated by the paths. All the adult *G. garnettii* were trapped more than once (mean number of trappings = 7). Of the adult *G. zanzibaricus* trapped, 69% were trapped more than once (with a mean of 4.9 trappings each. Some animals, however, were only trapped in one location.

Radio tracking equipment was available for the final 7 months of the field study. The subject's location was noted every 10 min during whole- or half-night focal animal radio tracking sessions. These were employed for 178 h on five male and four female *G. zanzibaricus* and for 130 h on two male and three female *G. garnettii*. For *G. zanzibaricus*, it was possible to estimate which 25 m square quarter of the 50 m quadrat the subject was in. For *G. garnettii*, which spent time on the edges of the gridded area, it was impossible to establish the quarter quadrat about 50% of the time. Their ranging data are specified only at the 50 m square level of precision. All *G. zanzibaricus* sleeping sites and 91% of *G. garnettii* sleeping sites were located within known 25 m quarter quadrats.

In both species, age was judged by weight and dentition. For males, maturity was also assessed by the size of the testes and the presence of penile spines. Females' parity was judged on the basis of nipple length and observations of a palpable fetus and/or infant carrying. 'Multipares' are females judged parous at first trapping. 'Primipares' are females initially judged to be nulliparous who produced at least one infant during the 2-year study period.

Fig. 1. Ranges of adult female *G. zanzibaricus* as shown by trapping data only. Circled numbers represent animals trapped at only one site. The square border encompasses the study area. Position of ranges in all figures are shown relative to this same border.

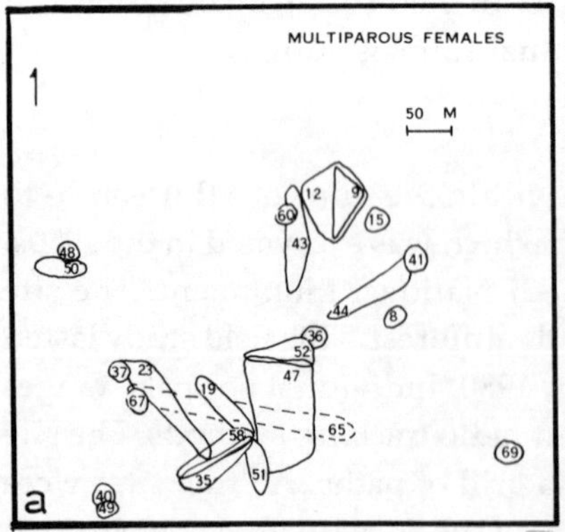

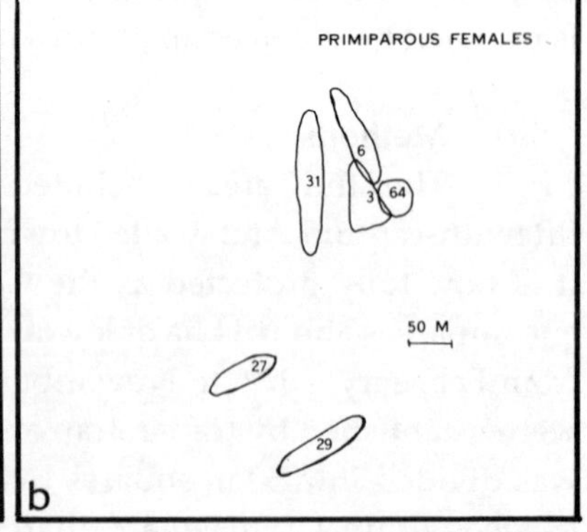

Results

G. zanzibaricus

Females Forty different females were trapped. This included 24 multipares, 6 primipares, and 9 infant, juvenile or young adult nullipares. The spatial distribution of trapsites was consistent with relatively non-overlapping territories for most multipares (Fig. 1a). At most, two multipares showed the same general range area. The primipares all showed ranges overlapping those of multipares (Fig. 1b). Thus 'female clusters' of a multipare and younger females were formed. One of these primipares (F3) shared a sleeping tree hole with another (F64) during the time the latter was radio tracked. Both had infants at this time.

Two multiparous females (F12, F43) and one primiparous female (F64) were radio tracked just after they had given birth (Fig. 2a). While F64 shared her range with F3 (see above), the two multipares did not share their range with other multipares and had nearly non-overlapping ranges. Range sizes were 1–1.25 ha.

Males This discussion excludes juvenile males who shared their mothers' ranges. There were two juvenile males in 1979 that by 1980 had made major range shifts away from their natal range.

Males which were judged sexually mature when first trapped were divided into two weight classes, since there was a major gap in weights between 140 g and 145 g ($N = 11$ of males weighing 145 g or more; $N = 8$ of males weighing 140 g or less). A social distinction has

Fig. 2. Ranges of *G. zanzibaricus* animals from radio tracking data. (See Fig. 1 for border convention.)

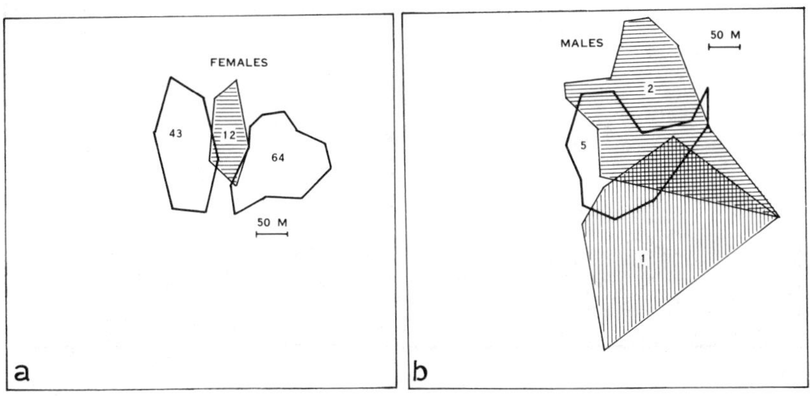

been found between large and small adult males in South African *G. senegalensis moholi* (Bearder & Martin, 1980). At Gedi, the two classes did not differ in mean range size (about 2.2 ha), but the larger adult males were less variable in range size (Figs 3a and 3b).

Both large and small adult males overlapped extensively with 0–2 other males (of either size class). There was no tendency for overlap to be between one large and one small male. Males also had some range overlap with several more males on 'boundaries'. Based on tracking males M1, M2, and M5, there may have been areas on which males converged.

Male/female range overlap Females had extensive range overlap with as many as three adult males (of either size class), and some overlap with up to three more males. However, it was more usual for any one female's range to overlap extensively with only one male's range. Large males seemed to have extensive overlap with fewer females than did small males. The same is true for 'female clusters'. Like range size, the variability in number of females or 'female clusters' overlapped was larger for small males than for large males.

Social organization of G. zanzibaricus Consistent with other studies of small galagos, females probably live in territories shared only with matrilineal relatives. At most, two multiparous females seemed to share a range. These could have been mother/daughter pairs or adult sisters. Additionally, immature females and young nulliparous adult

Fig. 3. Ranges of adult male *G. zanzibaricus* as shown by trapping data only. (See Fig. 1 for conventions.)

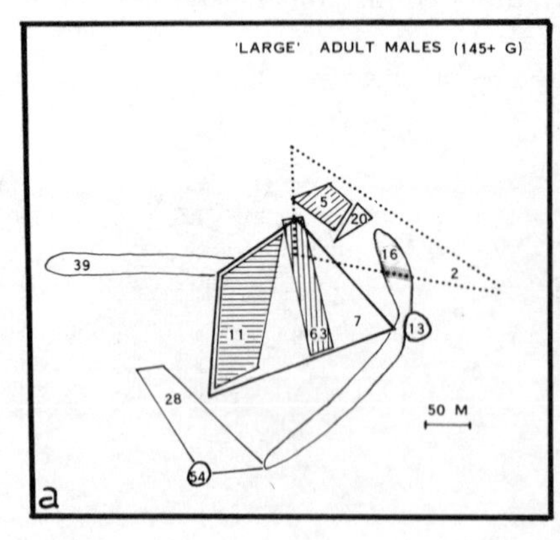

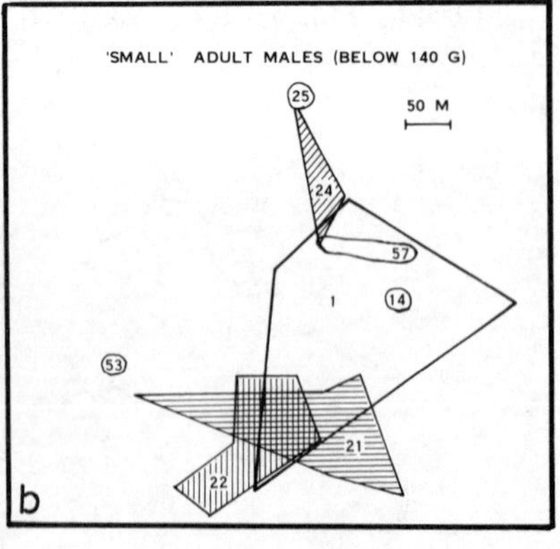

Fig. 4. Ranges of *G. garnettii* individuals. (See Fig. 1 for border conventions.)

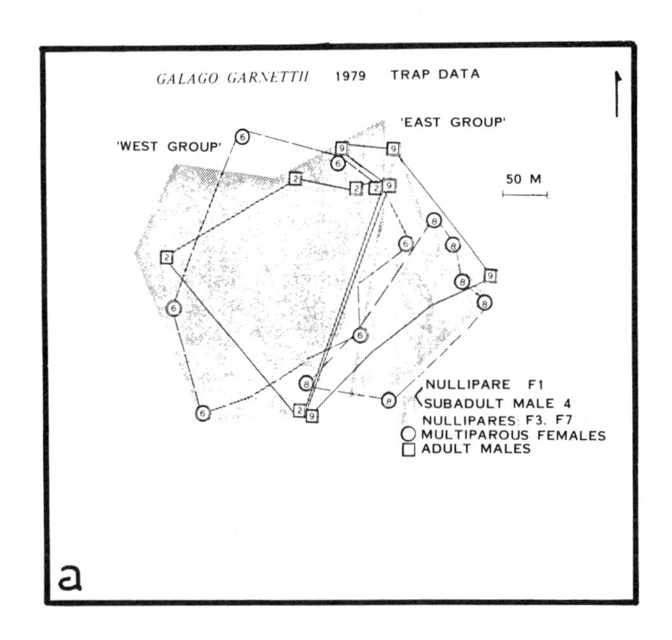

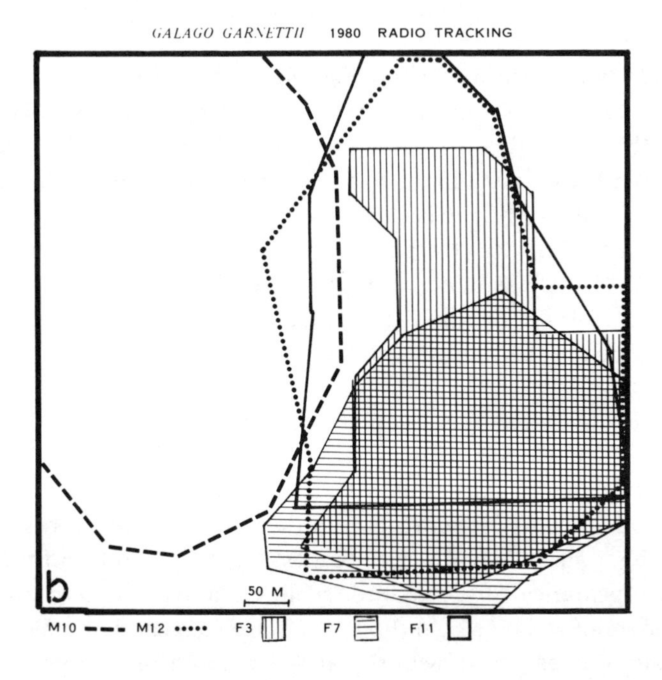

females might share the range of parous adult females in these 'female clusters'.

Young males leave their mothers' ranges at puberty. Adult males show range overlap, but two possible 'strategies' may exist. Some males have large ranges and contact several 'female clusters'. Other males, perhaps those who are older (and hence larger), may have smaller ranges, or at least less variably sized ranges, and have quite consistent sleeping group associations with a given female or 'cluster'. There are no data to indicate if either possible 'strategy' is more reproductively successful. In *G. senegalensis moholi* from South Africa, Bearder & Martin (1980) suggest that the larger 'central A' males might be received preferentially by the females with whom they consistently associate.

G. garnettii

Ranging behavior In 1979, trapping data on this larger species suggested the existence of two 'family groups', each consisting of a fully adult male, a single multiparous female, and subadults (Fig. 4a). In the 'East Group', relative sizes of the then nulliparous females, F3 and F7, suggested that the former was the elder by a year (since breeding is seasonal). Similarly, in the 'West Group', subadult male M4 was older than nulliparous female F1. A juvenile was also seen with multipare F6 in late 1979.

By 1980, when animals were tracked, a similar pattern persisted (Fig. 4b). In the 'East Group', M9 was not trapped and M12 had apparently migrated in from elsewhere. Primipares F3 and F7 had had infants, while the year-old nullipare F11 had not. F8 was still trapped in her 1979 range, but was not tracked. In the 'West Group', only subadult male M10 was tracked, but M2 and primipare F1 were seen with him. M10 is assumed to be the juvenile seen with F6 in 1979. These two 'West Group' males (M2, M10) were occasionally seen in the overlap zone in association with 'East Group' females.

Social organization In *G. garnettii*, a female or group of related females share a territory. Their ranges are smaller than those of males. At this site, only one fully adult male at a time associated with a female group. Subadult male offspring might also be present in this range. The younger animals had larger ranges than fully adult individuals. The disappearance of a subadult male from West Group and the replacement of M2 by M9 in the East Group suggest that males leave their natal ranges. West Group males had more range overlap and contact with the females of East Group than did the West Group females. The range of the youngest East Group female (F11) was the largest of the females in her group. This suggests that she might disperse, since her range might have been 'crowded' with three other resident, reproductively active, females.

G. crassicaudatus umbrosis is the only other large galago which has been studied in the field. It is similar to *G. garnettii* in having related female clusters sharing ranges. However, births are usually twins rather than singletons and range overlap is more extensive between and within the sexes. Clark (1985) notes as many as six females (four adults and two nulliparous sisters) and four males (two adults and two subadult brothers) showing extensive range overlap. Males in this species, like *G. garnettii*, apparently have larger ranges than females and disperse more regularly from their natal range than do females (Clark, 1978). In *G. garnettii*, both males and females overlap ranges with fewer like and opposite sex animals than in *G. c. umbrosis*. In the probably less 'patchy' environment of Gedi, female groups show larger ranges and less range overlap than in South Africa.

Acknowledgements

This project was partially funded by the Center for Field Research, the Wenner–Gren Foundation for Anthropological Research, and National Institute for Mental Health grant 1 RO3-MH353736-01. I thank the Government of Kenya and the Institute of Primate Research, National Museums of Kenya, for permission to work at Gedi, and IPR for assistance. The staff at Gedi helped in many ways and the participation of M. P. Nash was crucial.

References

Bearder, S. K. & Martin, R. D. (1980) The social organization of a nocturnal primate revealed by radio tracking. In *A Handbook on Biotelemetry and Radio Tracking*, ed. C. J. Amlaner, Jr & D. W. Macdonald, pp. 633–48. London & New York: Pergamon Press

Clark, A. B. (1978) Sex ratio and local resource competition in a prosimian primate. *Science*, **201**, 163–5

Clark, A. B. (1985) Sociality in a nocturnal 'solitary' prosimian: *Galago crassicaudatus*. *Int. J. Primatol.*, **6**, 581–600

Eaglen, R. H. & Simons, E. L. (1980) Notes on the breeding biology of thick-tailed and silvery galagos in captivity. *J. Mammol.*, **61**, 534–7

Nash, L. T. (1983*a*) Differential habitat utilization in two species of sympatric *Galago* in Kenya. *Am. J. Phys. Anthropol.*, **60**, 231

Nash, L. T. (1983*b*) Reproductive patterns in galagos in relation to climatic variability. *Am. J. Primatol.*, **5**, 181–96

Olson, T. (1979) Studies on aspects of the morphology and systematics of the genus *Otolemur* Coquerel, 1859 (Primates: Galagidae). Ph.D. dissertation, University of London

III.5

An experimental study of territory maintenance in captive titi monkeys (*Callicebus moloch*)

C. R. MENZEL

Introduction

In the wild, titi monkeys are monogamous and live in small, well-defined home ranges (Mason, 1966, 1968; Robinson, 1977; Kinzey, 1981). Many individual characteristics contributing to monogamy in this species have been identified in studies of captive animals (Mason, 1974, 1975; Cubicciotti, 1978). It has proved more difficult to explore the organization of territoriality in *Callicebus* under captive conditions. In previous studies, established family groups have been introduced to a common living area, such as an outdoor enclosure or small forested island, but the groups typically failed to maintain a spatial separation for more than about a week (Mason & Epple, 1969; Mason, 1971). In no previous instance have two groups established a stable 'boundary' area where ritualized confrontations occurred in the manner reported for titis in the wild.

During the past 4 years I have studied two family groups of titi monkeys that have maintained separate home ranges inside a large outdoor enclosure at the California Primate Research Center. To my knowledge this is the first instance in which any primate has shown well-developed territoriality in captivity, and it provides a unique opportunity to describe the social use of space and intergroup relations in titi monkeys. It also presents an occasion to examine experimentally some of the immediate causes of territoriality in this species. Here I will present normative data to illustrate the typical spacing of social groups in the field cage, and then offer a brief review of major findings from an experimental study conducted in the field cage with the same individuals.

Materials and methods

Subjects

The subjects were seven *Callicebus moloch* that lived in family groups within a 2-acre field cage. Group 1 comprised an adult heterosexual pair and a young adult female (the offspring of the male). Group 2 comprised an adult heterosexual pair, a young adult female (the offspring of the male) and a juvenile female (the offspring of both pairmates). The males were wild caught and laboratory habituated, and the others were born in the laboratory. Six of the animals had been introduced to the field cage by the late Daniel Cubicciotti, who documented the initial formation of territories by these groups. All animals had lived in the field cage for at least 6 months prior to the present study and were well habituated to humans. Except during experimental sessions the animals had unrestricted 24-h access to the field cage. They were fed ample portions of chow, fruit, eggs and vitamins once daily at approximately 14.00 h, and could supplement their diet with a wide range of natural foods available within the cage.

Environment

The field cage (91.4 × 88.4 × 9.1 m high) was 'insulated' from the outside environment by walls and a ceiling constructed of poultry mesh. Inside the cage were over 200 domestic fruit trees, vines, and other nonfruiting trees. They were planted in approximately equal numbers in 11 groves, which were spaced in an approximately uniform distribution throughout the field cage. The cage also enclosed artificial structures including an elevated wooden runway system, 10 rectangular wooden perching grids and 10 sub-enclosures (measuring 1.8 × 3.7 × 2.4 m high and enclosing a heated hutch box).

Normative observations

The objective of these observations was to describe the use of space by individuals and family groups under everyday conditions. Data presented here were collected on 54 separate days over a 12-week period (May–August 1982). A maximum of three samples was collected per day, one in each of the following time blocks: 06.00 to 11.00 h, 1.00 to 15.00 h, and 15.00 to 18.00 h. The total numbers of scans collected at these times were 54, 21 and 17, respectively. To collect a sample the observer entered the field cage, stood in full view of the animals at a distance of 7.6–9.1 m, and recorded, using pencil and paper, (1) the location of each individual to the nearest 3.1 m, employing marked runway sections and other prominent structures as reference points, and (2) the distance between all pairs of animals, employing the following mutually exclusive distance categories: contact, <3.1 m, <22.9 m, or >22.9 m.

Results

The location scores collected for the two adult males are shown in Fig. 1. The males occupied clearly separate portions of the field cage, with very little overlap in their home ranges. The location of each male was also representative of the male's remaining family group members. Thus, family members were found in contact with one another on 11% of intervals and within 3.1 m of one another on 67% of available intervals. The corresponding values for nongroup members were 0%

Fig. 1. Use of space by two adult male titis. Circles = male from Group 1; triangles = male from Group 2; heavy line = perimeter of field cage; thin line = runway system; R = release cages; H = hutch boxes used in experiment (see text). Trees and other ecological details omitted for clarity.

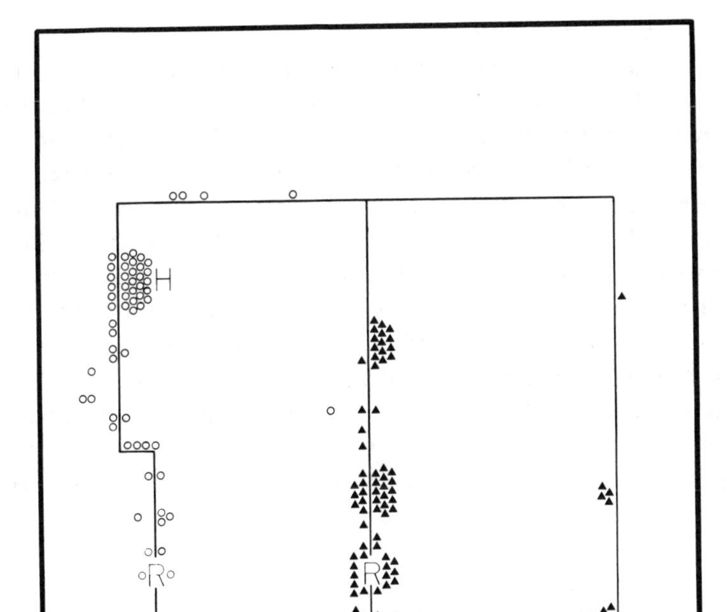

and 0% of intervals. Group members followed one another and travelled together throughout their common home range. In the evening they retired to a common sleeping site and spent the night in contact. The social and spatial separation of groups revealed by these data was characteristic of the months immediately preceding and following the study and, moreover, was the typical pattern shown during the 4 years that I observed the animals.

Other normative data collected showed that groups did not remain as far apart as possible in the field cage but converged toward one another in boundary regions and engaged in postural displays and loud vocalizations, similar to those described for the field. Thus, one or more animals from the neighboring groups were found within 22.9 m of one another on 11% of available intervals. Intrusions into the neighboring home range occurred periodically, and these often caused the neighbors to approach, and to direct postural and vocal displays toward the encroaching individuals. Sometimes the residents physically chased the intruders out of their home range. If unopposed, however, the intruders often withdrew to their own home range after several minutes.

Experimental study

An experimental study was conducted to explore more systematically the way that territories were maintained by these captive titi monkeys. The specific aims were to determine: (1) whether animals tended to avoid close proximity to the neighbors' home range; and (2) whether they also tended to avoid close proximity to the neighboring animals themselves. Expectations were that animals would travel more slowly toward the neighboring home range than away from it (to obtain a food incentive), and that they would also travel to food on fewer opportunities; that the mere visual presence of the neighbors and the added potential for direct interaction would both serve as deterrents to travel, and that both would produce increments in the latency to obtain food; and that the frequency of vocalizations and postural displays would increase as the potential for direct contact between the groups increased.

Subjects and test apparatus

Subjects were six of the animals used for normative observations. The young adult female in Group 1 was omitted from the study because she could not be trained to enter the release cage of the test apparatus. (She was confined out of sight to a hutch box during testing.) An identical test apparatus was located in each home range. The release cage (see Fig. 1) measured 0.8 × 1.2 × 1.8 m high and was equipped on its south side with a small (40 × 30 cm) door and on its north side with a larger door (80 × 80 cm) that could be raised and lowered by a rope-and-pulley system. (Outside of test hours both doors were left open, allowing the animals to pass through the release cage.) The larger door of the release cage opened onto the existing runway system, which extended 2.9 m north. At the end of the existing runway was a test runway of equivalent height (0.9 m). The test runway was perpendicular to the existing runway and consisted of removable horizontal units. The test runway, when completely assembled, extended 13 m (from the existing runway) directly toward the neighboring group's home range. It also extended 13 m in the opposite direction (away from the neighboring home range). At each end of the test runway was a platform (1.2 × 0.9 × 0.9 m high). Miniature marshmallows (sugar candies) were used as a food incentive. The marshmallows were placed inside clear plastic cups (10.2 cm high, with the open end facing upward), and the cups were placed on top of the platform. The marshmallows were thus easily accessible to any animal who reached the platform. The distance separating the two platforms that were between the neighboring home ranges was 4.6 m.

Design

The experiment consisted of four phases; only the first three will be described here (see Fig. 2). In Phase 1 each group was tested individually under two conditions, movement toward or away from the territory of the neighboring group (which was confined out of sight during testing). In Phase 2 each group was tested individually under the same conditions as in Phase 1, but now the neighboring group could be seen inside its home range, confined to its release cage. In Phase 3 both groups were tested simultaneously under the same conditions as in Phases 1 and 2, and thus the groups were either allowed to move directly toward or directly away from one another. During Phases 1–3 the animals received a total of 36, 24 and 24 5-min trials per condition; conditions were presented in a balanced order within each phase. A test session consisted of six trials, and each group was given one session per day, 6 days per week. Before formal testing began, all animals were given 30 training trials, designed to familiarize each animal with the test procedures. During training the animals travelled parallel to the neighboring home range.

Procedures

Test sessions were conducted between 08.00 and 12.00 h. The temperature ranged from 14 to approximately 32°C. To begin a session, all animals not scheduled for testing were confined to a hutch box, located at least 30.5 m from the test apparatus of the subject group(s); see Fig. 1. The animals scheduled for testing were then offered a preferred liquid (orange-flavored Tang) inside their release cage. After all group members had entered the cage, the observer confined the animals by raising the door. Next, the appropriate runway sections

Fig. 2. Experimental design for Phases 1–3. The subject group (on the left) moves toward and away from the neighboring home range (on the right), with increasing potential for direct social interaction.

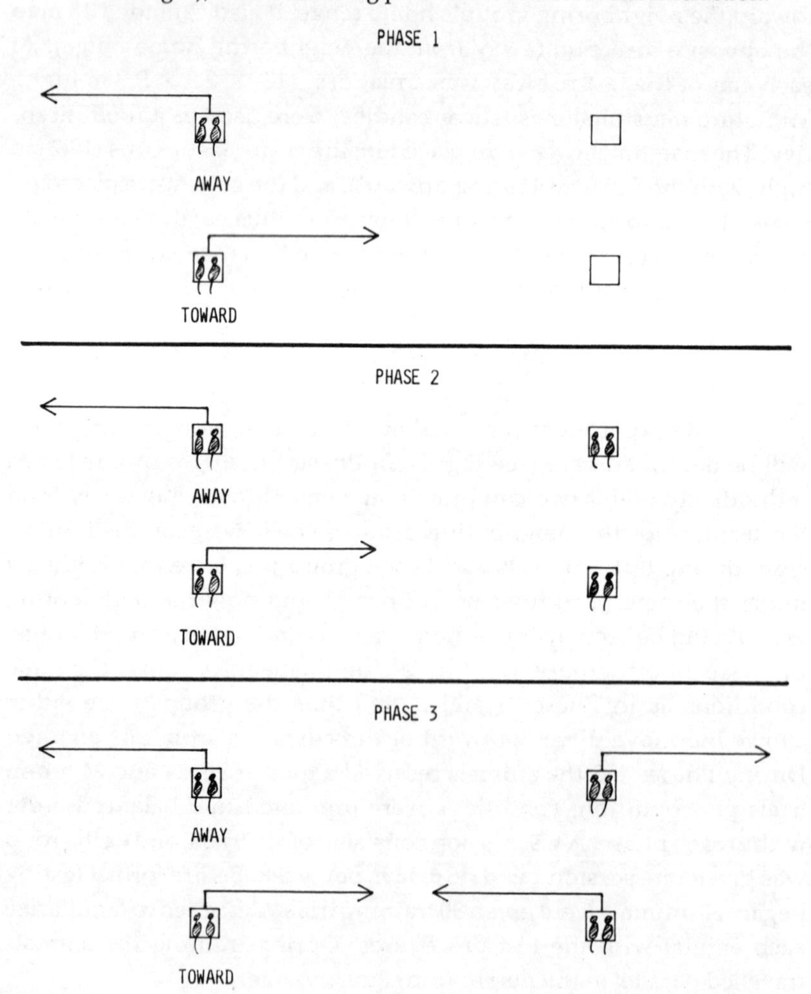

were assembled. These included either the sections leading toward the neighboring home range, or those heading in the opposite direction. Two food cups per animal were placed on top of the platform, and two half-pieces of a miniature marshmallow were placed inside each food cup.

One observer sat behind each release cage. About 15 s after both were seated, one observer started a tape recorder, which provided an audible 5 s time signal for the coordination of experimental procedures and for recording data. The trial started when each observer lowered the door of his release cage, allowing the entire group to exit. The trial ended 30 s after two animals in the group reached the feeding platform, or after 5 min, whichever came first. If at the end of the trial, however, the animals had initiated additional unusual activities, such as exploration of the neighboring test apparatus, direct interaction with the neighbors, or protracted vocal and postural displays, the length of the trial was extended for up to 5 additional minutes. During the trial the observer recorded (for each animal in his subject group) the time at which the animal reached the platform, and the presence or absence within each 5 s interval of certain stereotyped postural displays (arch, tail lash) and vocalizations (part call, full call). When the trial ended the animals were returned to the release cage (usually, offering Tang inside the cage was sufficient to accomplish this), and then preparations were begun for the next trial.

The data were analyzed by t-tests and analyses of variance for related samples.

Results

The results presented here are restricted to the scores from the four adult pairmates. Vocalizations and postural displays provided the best evidence that adults discriminated among the various phases and test conditions (see Fig. 3). The mean number of displays per adult (vocalizations and postural displays combined) showed a significant increase across phases, from 0.01 to 0.76 different displays per trial (maximum possible score = 4.0 per trial, $p < 0.01$). The adults showed few if any displays during Phase 1 when neighbors were absent, but they showed a significant increase in displays during Phase 2 when the opposite group could be seen confined within its home range ($p < 0.05$). The adults showed an additional and substantial increase in the frequency of displays from Phase 2 to Phase 3, when they could see the opposite group moving within its home range (Phase 2 *vs* 3, toward direction only, $p < 0.05$).

There was also a significant main effect of travel direction on displays ($p < 0.05$). During Phase 2, the adults performed 0.16 different displays per trial when travelling toward the confined neighboring group, and only 0.05 per trial during travel away from the neighbors. All adults tested followed this pattern in Phase 2, but the difference in travel directions was not statistically significant. In Phase 3, the adults gave displays at a significantly higher frequency when travelling toward rather than away from the neighbors (1.40 *vs* 0.11 different displays per trial, $p < 0.02$). Furthermore, adults gave full calls (the louder and more protracted type of vocalization) on an average of 3 out of the 24 trials on which the two groups were able to move toward one another. None of the adults performed full calls during any other phase or condition. Finally, the proportion of displays performed (within each phase) during movement toward the neighboring home range increased steadily from 0% in Phase 1 to 92% in Phase 3 (phase × direction interaction: $p < 0.005$).

All of the adults tested reached the feeding platform on fewer trials when travelling toward the neighboring home range than when moving away from it: 46% *vs* 54% of total trials, respectively (all phases combined, $p < 0.05$). The mean percentage of successful trials

Fig. 3. Combined frequency for postural and vocal displays. T = travel toward; A = travel away (from the neighboring home range).

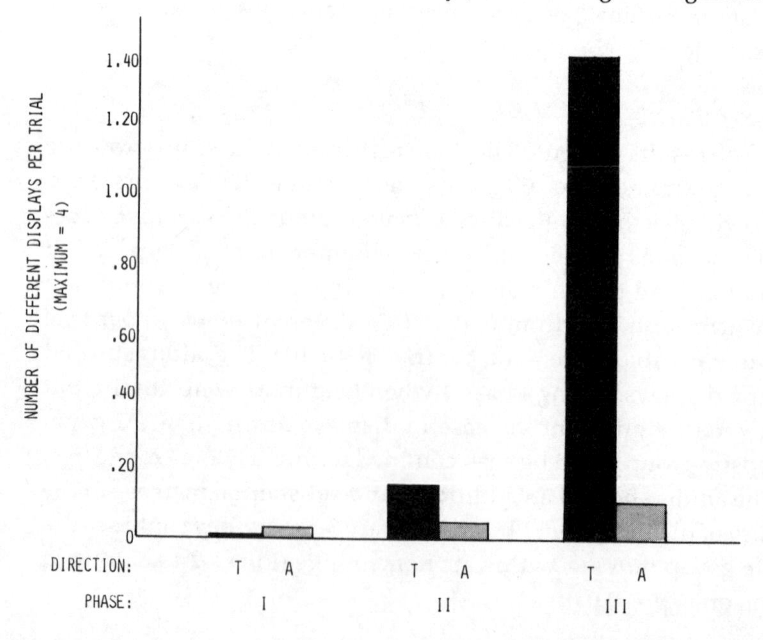

per adult conformed to this pattern in each phase, but for no single phase was the difference statistically significant (toward direction: 42%, 49%, 50%, *vs* away direction: 51%, 54%, 58%; Phases 1–3, respectively).

The animals also took longer to reach the feeding platform when they were travelling toward the neighboring home range. Thus, in Phases 1–3 the latency for the first adult in the group to reach the platform was significantly higher when moving toward rather than away from the neighboring home range (109 *vs* 69 s, uncompleted trials included as 296 s). Moreover, the adults reached the platform that was located closer to the neighboring home range more quickly during the last half of Phase 2 – in which the neighbors were merely seen and heard – as compared with the first half of Phase 3 – in which the neighbors were approaching the adults directly (67 *vs* 105 s, uncompleted trials included as 296 s). When travelling away from the neighboring home range, however, the adults showed only a slight, nonsignificant increase (from Phase 2 to Phase 3) in their latency to reach the platform (55 *vs* 57 s).

Discussion

In nature, family groups of dusky titi monkeys occupy and defend small, intensively used home ranges that show little overlap with the home ranges of neighboring groups. Although titis show this pattern consistently under natural conditions, individuals are not necessarily 'driven' to acquire and defend an exclusive area. Well-developed territoriality seldom occurs in captivity even under seemingly favorable conditions. A more typical outcome in captivity is for established family groups to mingle freely and travel together, to overlap extensively in their use of space, and to show little chasing or threatening when introduced to a common living area (Mason & Epple, 1969; Mason, 1971). The present study is based on the first known instance of two family groups maintaining stable, spatially distinct home ranges in captivity and provides new information on some of the factors contributing to territoriality in this species.

Individuals did not habituate to the presence of non-group members in spite of the fact that groups were often in clear view of one another during a 4-year period. Many trees were present in the field cage, but these did not form a continuous visual barrier between the two home ranges. Visual contact was essentially a repeated, daily occurrence.

Defense of a limited food supply against the incursion of non-group members apparently played no role in the maintenance of separate home ranges in the present instance. Both groups had more than enough food within their respective areas to meet their energetic and nutritional requirements. They were provisioned daily with large quantities of a standard laboratory diet (more than they consumed) and each had access to available natural food items, such as leaves, buds, flowers, fruits and insects. In these respects the animals exceeded by a wide margin their minimum ecological requirements.

The relations between groups appeared to involve a large element of ambivalence, in which the tendency to approach members of the neighboring group was usually over-ridden by the tendency to avoid or withdraw from them. Individuals from neighboring groups were seldom found in close proximity, particularly when accompanied by other group members. The proximity of a neighbor tended to elicit marked visual interest, and frequently approach, but this was often accompanied by a variety of signs of agitation and aggressive displays. Such displays were particularly common at customary boundary locations.

The experimental findings confirm that a perceived opportunity for direct interaction with a neighboring group increases agitation and arousal. In fact, the data for completed *versus* uncompleted trials suggest that adult titis are not only reluctant to approach an area currently occupied by a neighboring group, but also an area which neighbors customarily occupy. The latency scores provide additional evidence that adult titis are reluctant to encroach upon a neighboring home range. They also suggest that the immediate potential for direct interaction has a more inhibiting effect on travel than does the mere visual (and auditory) presence of neighbors who are stationary and located at a distance within their own home range.

The combined findings of the present experimental study suggest that the tendency to avoid close proximity to members of neighboring groups and areas occupied by neighboring groups contributes to the species-typical pattern of intergroup spacing. Additional behavioral factors contributing to territoriality in *Callicebus moloch* probably include the tendencies of individuals to use a small portion of the total available space and to organize routine daily activities – such as feeding, resting and travel – around familiar social partners and familiar physical structures (Mason, 1966, 1968; Fragaszy, 1979, 1980). Finally, it seems likely that the territorial pattern is reinforced by the frequent confrontations between groups occurring along home range boundaries. During confrontations, pairmates draw together and often give loud vocalizations which (as Robinson, 1977, 1979, has shown) have powerful effects on the recipients, particularly when located along the boundary.

Acknowledgements

I wish to thank Drs William Mason and Peter Rodman for their generous sponsorship and encouragement, Dr Mason for conceptual assistance in the planning and execution of this study, and Ryne Palombit for assistance in conducting the experiments. Financial support was provided by an NSF and a University of California Regents' graduate fellowship, by a University of California graduate research award to the author, and by NIH #RR00169 to the California Primate Research Center.

References

Cubicciotti, D. D. III (1978) Comparative studies of social behavior in *Callicebus* and *Saimiri*: heterosexual jealousy behavior. *Behav. Ecol. Sociobiol.*, **3**, 311–22

Fragaszy, D. M. (1979) Titis and squirrel monkeys in a novel environment. In *Captivity and Behavior*, ed. J. Erwin, T. Maple & G. Mitchell, pp. 172–216. New York: Van Nostrand

Fragaszy, D. M. (1980) Comparative studies of squirrel monkeys (*Saimiri*) and titi monkeys (*Callicebus*) in travel tasks. *Z. Tierpsychol.*, **54**, 1–36

Kinzey, W. G. (1981) The titi monkey, genus *Callicebus*. In *Ecology and Behavior of Neotropical Primates*, Vol. 1, ed. A. F. Coimbra-Filho & R. A. Mittermeier, pp. 241–76. Rio de Janeiro: Academia Brasileira de Ciencias

Mason, W. A. (1966) Social organization of the South American monkey, *Callicebus moloch*: a preliminary report. *Tulane Stud. Zool.*, **13**, 23–5

Mason, W. A. (1968) Use of space by *Callicebus* groups. In *Primates: Studies in Adaptation and Variability*, ed. P. C. Jay, pp. 200–16. New York: Holt, Rinehart & Winston

Mason, W. A. (1971) Field and laboratory studies of social organization in *Saimiri* and *Callicebus*. In *Primate Behavior: Developments in Field and Laboratory Research*, Vol. 2, ed. L. A. Rosenblum, pp. 107–37. New York: Academic Press

Mason, W. A. (1974) Comparative studies of social behavior in *Callicebus* and *Saimiri*: behavior of male–female pairs. *Folia Primatol.*, **22**, 1–8

Mason, W. A. (1975) Comparative studies of social behavior in *Callicebus* and *Saimiri*: strength and specificity of attraction between male–female cagemates. *Folia Primatol.*, **23**, 113–23

Mason, W. A. & Epple, G. (1969) Social organization in experimental groups of *Saimiri* and *Callicebus*. In *Proceedings of the 2nd International Congress on Primatology*, Vol. 1, pp. 59–65. Basel: Karger

Robinson, J. G. (1977) Vocal regulation of spacing in the titi monkey, *Callicebus moloch*. Unpublished Ph.D. Dissertation, University of North Carolina at Chapel Hill

Robinson, J. G. (1979) Vocal regulation of use of space by groups of titi monkeys *Callicebus moloch*. *Behav. Ecol. Sociobiol.*, **5**, 1–15

III.6

Leadership: decision processes of group movement in yellow baboons

G. W. NORTON

Introduction

The mechanisms by which a cohesive social group moves to and utilizes resources are complex and poorly understood (Leuthold, 1977). For example, individual members of a baboon troop must in some way reach a consensus on the time, place and route of movement, whereby the spatial integrity of the troop is maintained, and the necessary resources are found, monitored and utilized in an effective manner. The issue of how such a consensus is reached has generally fallen under the rubric of 'leadership' (Stolba, 1979). The problems of leadership have rarely been discussed and even more rarely studied on a quantitative basis. This paper contains a preliminary discussion of such a study conducted on yellow baboons in Mikumi National Park, Tanzania, between 1980 and 1984. It will address some of the concepts subsumed under the rubric of leadership, the theoretical implications of these concepts and problems of definition and methodology.

Definitions

The term leadership has been used to refer to different types of social phenomena. For example, it has been used to refer both to several aspects of high social status (de Waal, 1975; Wilson, 1975; Hinde, 1983) and to animals who hold the forward or first, hence leading, positions within a moving group (Richard, 1985). There may be a relation between social rank or relative spatial position and group movement, but there is no *a priori* reason to assume that movement decisions are made by animals of high rank or a forward position (Etkin, 1964). Leadership should refer to behaviors which are necessary for decisions about the time, route or place of group movement

(Leuthold, 1977). Leadership behavior may be exclusive to specific individuals, to the interactions of specific individuals or common to all group members. Behavior resulting in the change of a group's movement behavior need not be exclusive to movement decision processes. Such behavior may serve a variety of functions and result in movement decisions when it occurs in certain environmental or social circumstances. A definition should make no assumptions about the characteristics of individuals who effect movement decisions, or about the behaviors employed other than that they can be shown to influence group movement. In this paper leadership refers to the decision processes of group movement.

Hypotheses

Questions of leadership are concerned with the integration and co-ordination of individual behaviors, what Dunbar (1983) calls the supra-individual features of social relationships. The questions of primary importance are: How do the actions of individuals, or subgroups of individuals, within a social group, influence the time, route and place of movement? Do certain individuals or types of individuals have a greater influence on the decisions of movement and location which occur at the group level? How does the pattern of such decisions vary with spatio-temporal differences in environment? Do such decisions differentially effect different individuals or types of individuals in a group? Finding answers to these questions will allow explanation of how a large group of individuals moves cohesively from one place to another.

Movement pattern of groups or aggregations may be determined by proximate conditions of the environment. An example of such environmentally determined group movement might be the movements of fish schools (Shaw, 1962; Wilson, 1975). Whichever individual happens to be in front effectively moves a school through proximate temperature, light and nutrient gradients. 'Leaders' in this case are random with respect to individuals. Individual movement behaviors within a fish school may not determine school position but rather may maintain group cohesion. Complex age and sex-structured social groupings present a different problem in assessing group movement processes. In complex social groups, movement may be determined by a process of communication and consensus between group members, or by the behaviors of specific individuals within the group. Movement choices may not be dependent on the immediate environmental milieu. Decisions might be reached using prior knowledge of home range and its potential resources.

Baboons forage in a cohesive stable group, over a large definable home range (Altmann & Altmann, 1970). The resources available within this home range vary in quality, and are frequently patchy and widely dispersed (Post, 1984). Baboons often appear to make choices or decisions about locations and resources which are outside the immediate sensory environment (Kummer, 1971). The ability to recall the nature and location of resources distant in time and space, is an important one for animals using resources as varied and widely distributed as those of baboons. It has been suggested that baboons possess a cognitive or mental map of their home range and the resources in it (Sigg & Stolba, 1981). Such a map would greatly facilitate effective and safe foraging.

Hypotheses on the ways by which decisions on group movement can be made are listed below in order of increasing complexity. These general hypotheses are not mutually exclusive as the mechanisms of choice in a group may vary with social and environmental constraints.

1. The actions of individuals within a group have no effect on the movement patterns of the group. If so, movement decisions would probably be almost totally controlled by the immediate environmental conditions. For group movement to occur individuals must be responding to environmental cues in similar and consistent ways. Motivational state and monitoring accuracy must be similar for all individuals. Movement in response to the behavior of other group members would be limited to that necessary to maintain social proximities and group cohesion. This type of movement choice might be most plausible in environmentally determined aggregations rather than in complex and structured social groupings (Brown, 1975).
2. Individuals or subgroups of individuals affect the time, direction or place of group movement but each individual (and all possible groupings) has an equal probability of doing so in any decision context. For example, the first animal to move or suggest movement might stimulate movement in all group members but the individual who moves first is totally random for any given case. This type of decision process would also probably be under the influence of environmental information and constraints similar to those of the first hypotheses. Differential information processing by individuals can occur

but reception and communication of this information would still be random with respect to individuals.

3. Decisions and choices are made on the basis of consensus between group members. No individual, or type of individual leads. Individuals indicate choices and the group movement behavior changes whenever a sufficient number of individual suggestions are in agreement. This assumes that a large percentage of the group can and will participate in the choice process. Such a consensus process has been called voting (Stolba, 1979) and may be seen as an example of graded signaling (Wilson, 1975). In this hypothesis the concept of a cognitive map becomes plausible. It is possible that individuals can both recall and suggest movement to places that are outside the immediate sensory milieu. However, individual suggestions may only indicate satiation on a resource or choices between proximate alternatives.
4. The behavior of specific individuals or types of individuals effect or determine group movement decisions. This does not preclude decisions based upon consensus or the interactions between individuals. It does imply that such interactions will occur within a specific and recognizable subgroup of the total group. Members of such a subgroup could vary in the way their behavior influences movement choices as well as in the strength of that influence. If certain individuals determine the group choices, this need not imply the functioning of a cognitive map but the use of such a map could increase the number of choices possible. The most effective functioning of a cognitive map with the greatest benefit to individuals may occur in groups where clear and rapid decisions can be made by a few group members. Such decision-makers might be those with the most efficient map abilities. In groups in which the operation of a cognitive map seems likely, the existence of specific leaders seems most plausible.

Methodology

The form of group movement decisions may vary with the choices to be made and it was necessary to collect data on the behavior and spatial relations of a group which encompassed the various conditions constraining decisions. One must demonstrate that a decision has occurred and that certain individuals or social-behavioral processes caused a definable choice to be made. In species where it is possible that decisions are reached about locations outside the immediate sensory milieu, it is necessary to demonstrate the functioning of a cognitive map. Decision processes could be examined by: the description of pattern and context for a group's movement, foraging and ranging behavior; the prediction of a group's future location or movement by its present action; and the correlation of individual behavior with group choices.

That a decision has been made is demonstrable and may be examined by looking at changes on a few dimensions of group behavior – changes in movement activity, in location, and in group direction or behavior. Three important movement contexts should be specified: stopped (no progressive movement), moving and resultant stop. Transitions between these three states are points at which group decisions are likely to occur. One should seek to define the start and finish points of movement and the transition between them. Decisions about location and group movement may occur during the progress of a movement, making it crucial also to look at the patterns of movement, independent of endpoints.

The central isssue remains how group choices result from the behavior of individuals within the group. Four methods by which the effect of individual behavior on troop movement might be examined are:

1. Examination of changes in group movement patterns with changes in group composition (Altmann, 1980; Rasmussen, 1983). This assumes that a correlation between the frequency of certain classes of animals and the pattern of range use/ movement might suggest a causal relation between those individuals and the movement behaviors of the troop. This method may be useful in suggesting possible leaders, but will provide little information on how they effect movement. It is difficult to assess if a relation is causal or a collateral result of other factors.
2. Quantification of individual behavior and spatial pattern under different movement conditions (Rhine & Westland, 1981; Collins, 1984). Recording the position or location of individuals relative to other troop members during a movement progression is an example of this technique. This has demonstrated that certain types of individuals consistently assume similar positions within a moving troop. Spatial organization may or may not be related to movement choices. It may support the contention that leadership occurs, but it

does not demonstrate a causal relation. Nor does it demonstrate the important behaviors which result in the observed spatial organization.

3. Experimentation, either in a controlled captive situation or by controlled interference with a natural population (Menzel, 1971; Bachmann & Kummer, 1980). The importance of different individuals to movement decisions might be assessed by selective removal and/or introduction. The usefulness of this technique for wild populations would depend on background data about ranging, foraging and movement behavior in the study group and about the behavior of the individuals manipulated.
4. Observation and recording of individual behaviors in conjunction with ongoing group decision processes (Stolba, 1979). This approach is also primarily correlational; it measures the behavior of individuals under a variety of possible movement and decision contexts and relates this behavior to changes in group movement and location. Perhaps the most important attribute of this method is that it permits prediction. One should be able to predict both how movement behavior may change from one state to another (stopped/moving) and in what direction changes will occur. If cognitive maps are functioning, the future location of a group can be predicted. Stolba described decision processes and the functioning of a cognitive map in hamadryas baboons using this approach (Stolba, 1979; Sigg & Stolba, 1981).

Description of this study

The quantitative data for this study consisted of three types of systematically collected information. These were: (1) specification of troop movement patterns and behavioral activities, which noted such variables as troop speed, location, direction, width, spatial structure, ongoing behavior and shape; (2) records of the time and context in which changes in troop movement activities or location occurred; and (3) focal animal observation, which recorded individual behavior in conjunction with more fine-grained information on the ongoing troop movement, locations and activity. These data were collected on three troops of differing size and structure, and subjects were all adult animals.

It is possible here to discuss qualitatively the patterns observed to illustrate complexities of the question and to suggest likely results, but further analysis may require modification of the results suggested. This discussion will be limited to the smallest study troop which consisted of 16 animals with a maximum of 6 adults – 1 male and 5 females. The small size of this troop made it amenable to observations of the effect of a single individual's behavior on the movement and location choices of the troop.

The tendency for certain types of individuals to hold predictable positions within the spatial structure of a troop was observed (Rhine & Westland, 1981). The relation of this positioning to movement decisions was not clear. A pattern repeatedly observed in the small troop was illuminating. The single adult male was regularly observed in the first position when the group was moving. Prior to onset of movement many members of the troop could be observed showing orienting behavior which was defined by body position, directional looking or scanning, and short movements in consistent directions. Frequently the male was observed to move first, or in conjunction with the movement of particular females. This suggests that his movement initiated that of the group. The male often oriented or closely attended to the behaviors of adult females. His movement usually resulted in troop movement on those occasions where certain adult female orientations agreed with the male initiation. This suggested that the adult male was serving the function of an initiator for the time of movement and that some or all females were acting as deciders (Kummer, 1968).

Despite the male's forward position, the direction or place of movement was not necessarily being determined by the male. This was observed when direction or goal may have changed or been misinterpreted by the male; the male moving in a forward position would become isolated and would scan for the troop location and direction, returning, usually, to the front of the troop.

In all troops, orienting behavior (as described above) was observable and measurable during daily activities and in a variety of contexts. A direction and shape could often be assigned to a troop when a majority of individual orientations were in agreement, even when there was no change in location, nor ongoing movement. The orientations of individuals may represent movement suggestions and the resultant troop orientations may represent consensus decisions. It is possible that agreement by key individuals to suggestions was necessary for a decision or consensus to occur. This was supported by the observation that direction of orientation by the troop could change rapidly. Orientation did not appear to be a good indicator of when movement was likely to occur.

In the small troop, orienting behavior was most conspicuous in adult females. Orienting and its apparent influence on group decisions varied between females. Two females conspicuously oriented more than other individuals. They would move to the troop's edge, climb into a raised position and orient body and gazes in a particular direction while also monitoring group activities. The directions and apparent endpoints or goals indicated by these females were frequently taken by the troop. One of the two females was the oldest with the largest matriline and of middle to low social status, and the other was the highest-ranking female. Until her death, the older female was the most frequent and influential of the two in the indication of direction or place of troop movement. If the younger dominant female joined the older one this increased the probability of movement in the direction they were indicating. The time of movement in a direction was highly variable. While they were apparently indicating direction or place of movement, they were not necessarily influencing the time of movement. All adult animals were observed to display orientation and appeared to influence troop choices through it. It appeared that the agreement of the two most influential females and often of the adult male was necessary for other individual's suggestions to influence group decisions.

The distinctions between time, place and route of movement was evident when a decision was apparently made but not yet implemented. There were a number of observations when, given the time and context of a group's activities, the endpoint of movement was predictable, e.g. cases where a goal was in proximity to the troop. For example, in late evening it was often possible to observe a troop feeding near a known sleeping grove which would be used that night. The troop while perhaps orienting to the grove had not yet made the decision to approach nor as to which of several possible approach routes to use. The place of future movement was determined but not route or time. Prior to movement to such goals, orienting behavior and attending to other group members was conspicuous. Usually purposeful movement toward the established goal was initiated by the movement of a few animals or a single animal, often an adult male. The directional movement of one animal towards the goal was followed almost instantly by movement of all troop members. The classic file progression frequently was the resultant pattern of movement. This suggests that the time of move was being determined by consensus. If a large number of adults were indicating the goal by orientation, short moves and the cessation of other behaviors, movement was quickly implemented as soon as an adult male moved. Thus males determined the onset of movement and appeared to determine the route by which the goal was approached, given that adult females and others had already indicated the goal and a readiness to move.

In summary, all the systems for social group decisions suggested in the above general hypotheses appeared to operate in the study troops; the system used for a decision depending on the context of that decision. Movement decisions could occur by all troop members responding to environmental cues in essentially similar or compatible ways. A decision may be effected by whomever makes it first, and such decisions may occur when part of the necessary choice has already been effected, for example, when a goal is already determined. Some decisions will occur as a result of consensus between a percentage of adult troop members. This form of choice may apply to decisions on time or speed of movement. Specific individuals can and do influence the choices made by a group. Some individuals or combination of individuals are more influential than others. Different individuals can determine different components of a decision. A consensus of the most influential animals may often be sufficient for a decision to occur. The influence of important decision-makers can be subtle and indirect. Any individual might suggest or indicate choices. For that suggestion to be successful, agreement by decision-makers may be necessary. They may not respond with conspicuous indications of their own. Conversely the indications of influential animals will have a high probability of success, regardless of the agreement or alternative suggestions of others.

The described patterns are derived primarily from observations on a particularly small troop. Similar patterns apparently occurred more complexly in the larger study troops. The pseudopodal form of movement suggested by Kummer (1971) may be operating in larger troops and may be the outcome of alternative suggestions made by individuals with a varying degree of influence on the decision process.

Discussion

The questions remain as to why certain individuals are influential and others less so, and what is the importance of an understanding of group decision processes to questions about social structure, resource use and individual success.

That the ability of an individual to forage optimally is constrained both by the distribution of resources and by previous choices about space and time, and that a set of basic movement/resource use decisions must be made by individuals, has often been stressed (Pyke, 1983; Post, 1984). It is more rarely stressed that an individual's ability to make such decisions may be constrained by its membership in a social group (Murton, Isasicson & Westwood, 1966). Resources available to an individual and its ability to utilize them are constrained not only by its ability to make decisions but also by its ability to implement decisions within a social context. The advantages and disadvantages of group living will depend on the nature and distribution of resources and the extent to which individuals differ in their resource needs. Unless all individuals within a cohesive social group have similar decisive abilities about resources and similar resource needs, it seems unlikely that all individuals will be capable of distributing their time and spatial location optimally among resources while remaining within the social group. The plausibility of all individuals having an equal influence on group movement seems low, since compromise between a number of different choices may result in a choice useful to none. In groups where individual needs and choices are likely to be in agreement, it is plausible though not necessary, to suggest equal individual influence on group decisions. Commonality of decision processes between individuals might occur in social groups which are specialized to utilize a few abundant resources which may be distributed in large, discrete, easily locatable, or monitorable patches.

Individuals are also constrained in resource use by proximate social competition. The ability to influence group movement decisions has the potential of being a form of competition if by such influence an individual can maximize its own optimal or effective resource use and minimize the effective resource use of group members with which it is in competition. Conversely, the ability of an individual to influence group choices may have beneficial or even altruistic consequences for other group members. Some individuals may be better able to make the best choices between alternatives or to locate resources useful to other group members. Such an individual may make choices to meet its own needs but may coincidentally benefit other individuals. An individual may acquire the ability to make decisions useful to other animals, to insure its own effective foraging, to gain social benefits, or to provide benefits to its relatives. Subordinate animals may develop the ability to locate, remember or indicate choices advantageous to a large number of animals in order to insure that some benefits are available to itself or its relatives (Murton *et al.*, 1966). Parents might make choices less than optimal to themselves but important for offspring. Some individuals may attempt to influence decisions for potential social benefits, such as access to females by males, rather than for immediate benefits to resource foraging. Individuals may ultimately face the choice of accepting group decisions they cannot influence or of leaving the group to optimize their own foraging strategy.

Thus the study of the ways individuals make and implement decisions about group movement processes can be revealing of the causes, costs and benefits of sociality.

Acknowledgements

This research was supported by grants to Dr Ramon J. Rhine from the National Institute of Mental Health, the Harry Frank Guggenheim Foundation and support grants from the University of California, Riverside. This project was originally planned and inspired by Dr Ramon J. Rhine who has been a source of ideas, aid and insight throughout its progress. The Serengeti Wildlife Research Institute, the Tanzania National Scientific Research Council and the Tanzania National Parks enabled this project to be undertaken and completed and, in addition, provided significant encouragement and support. Many individuals of the Mikumi National Park staff provided important logistical support and companionship as did numerous other researchers working in Mikumi. Dr P. C. Lee provided a great deal of advice and encouragement without which this paper could not have been completed and an anonymous referee provided many insightful comments which helped shape its final form. Special thanks are owed to Charles Kidung'ho who provided help in the field throughout the duration of the research.

References

Altmann, J. (1980) *Baboon Mothers and Infants*. Cambridge, Mass.: Harvard University Press

Altmann, S. A. & Altmann, J. (1970) *Baboon Ecology: African Field Research*. Chicago: University of Chicago Press

Bachmann, C. & Kummer, H. (1980) Male assessment of female choice in hamadryas baboons. *Behav. Ecol. Sociobiol.* **6**, 315–21

Brown, J. L. (1975) *The Evolution of Behavior*. New York: Norton

Collins, D. A. (1984) Spatial pattern in a troop of yellow baboons (*Papio cynocephalus*) in Tanzania. *Anim. Behav.*, **32**, 536–53

Dunbar, R. I. M. (1983) Structure of gelada baboons reproductive units. IV. Integration at group level. *Z. Tierpsychol.*, **63**, 265–82

Etkin, W. (1964) Co-operation and competition in social behavior. In *Social Behavior and Organization Among Vertebrates*, ed. W. Etkin. Chicago: University of Chicago Press

Hinde, R. A. (1983) Description of and proximate factors influencing social structure. 10.1 Description. In *Primate Social Relationships*, ed. R. A. Hinde. Oxford: Blackwell

Kummer, H. (1968) *Social Organization of Hamadryas Baboons*. Basel: Karger

Kummer, H. (1971) *Primate Societies: Group Techniques of Ecological Adaptation*. Chicago: Aldine

Leuthold, W. (1977) *African Ungulates: A Comparative Review of their Ethology and Behavioral Ecology*. Berlin: Springer-Verlag

Menzel, E. W. (1971) Communication about the environment in a group of young chimpanzees. *Folia Primatol.*, **15**, 220–32

Murton, R. K., Isasicson, A. J. & Westwood, N. J. (1966) The relationships between wood-pigeons and their clover supply and the mechanism of population control. *J. Appl. Ecol.*, **3**, 55–96

Post, D. G. (1984) Is optimization the optimal approach to primate foraging? In *Adaptations for Foraging in Nonhuman Primates*, ed. P. S. Rodman & J. G. Cant. New York: Columbia University Press

Pyke, G. H. (1983) Animal movement: an optimal foraging approach. In *The Ecology of Animal Movement*, ed. I. Swingland & P. J. Greenwood. Oxford: Clarendon

Rasmussen, D. R. (1983) Correlates of patterns of range use of a troop of yellow baboons, (*Papio cynocephalus*). II. Spatial structure, cover density, food-gathering and individual behavior patterns. *Anim. Behav.*, **31**, 834–56

Rhine, R. J. & Westland, B. W. (1981) Adult male positions in baboon progressions: order and chaos revisited. *Folia Primatol.*, **35**(2), 77–116

Richard, A. (1985) *Primates in Nature.* New York: W. H. Freeman

Shaw, E. (1962) The schooling of fishes. *Sci. Am.*, **206**, 128–138.

Sigg, H. & Stolba, A. (1981) Home range and daily march in a Hamadryas baboon troop. *Folia Primatol.*, **36**, 40–75

Spencer, C. (1975) Interband relations, leadership behaviour and the initiation of human-oriented behaviour in bands of semi-wild free ranging *Macaca fascicularis. Malay. Nat. J.*, **29**, 83–9

Stolba, A. (1979) Entschedungsfindung in Verbanden von *Papio Hamadryas.* Ph.D. Thesis, University of Zurich

Waal, F. B. M. de (1975) The wounded leader. A spontaneous temporary change in the structure of agonistic relations among captive Java-monkeys (*Macaca fascicularis*). *Neth. J. Zool.*, **25**, 529–49

Wilson, E. O. (1975) *Sociobiology: The New Synthesis.* Cambridge, Mass.: Harvard University Press

Part IV

Ecology and Mating Systems of New World Primates

Editors' introduction

Despite much work on the New World primates in captivity, and especially on their reproductive behaviour, long-term field studies of the ecology and social organisation of these species have been lacking. In this section, the reproductive strategies of nine primate species studied in Manu National Park, Peru, are related to specific ecological variables such as the size of food patches and their productivity, to inter-individual and inter-group competition for those resources, and to the problems of avoiding predation. Comparisons with the ecology and mating systems of the Old World primates are made in order to understand the differences in ecology between species and to derive some general rules about how these differences influence social behaviour.

Since many of the New World primates live in areas of tropical rainforest that are threatened with destruction, emphasis is placed on the application of the findings presented here to the conservation and management of wild populations. In addition, suggestions for the management of captive colonies are made, both for assisting with conservation and with general practices of captive propagation.

The range of social systems found in primates throughout the world is represented by the different species of New World primates covered in this section. Wright (Chapter IV.1) begins by discussing why some of the smaller species are monogamous and differentiates between the species in their abilities to exploit large or small food patches and to exclude other animals from a food tree. Janson (Chapter IV.2) relates the maintenance of the size of small multi-male groups among different species of capuchin monkeys to the costs and benefits of inter-individual competition for food. McFarland (Chapter IV.3) compares the quality, distribution and size of food resources between chimpanzees and spider monkeys in order to explain the similarities (and differences) in their social systems. Goldizen and Terborgh (Chapter IV.4) discuss resource distribution and cooperation among tamarins in relation to the evolution of polyandry.

In the final paper, Terborgh (Chapter IV.5) summarises the social systems of the New World primates, and provides an overview of the different ecological variables, food distribution and abundance, competition and predation, that may have influenced the evolution of the different social systems.

IV.1

Ecological correlates of monogamy in *Aotus* and *Callicebus*

P. C. WRIGHT

Introduction

In mammals there is strong selection pressure for males to breed polygynously whenever possible, since by doing so individuals will increase their reproductive success (Alexander, 1974; Wilson, 1975; Clutton-Brock, 1978). Because of the constraints of lactation, monogamy is rare in mammals and found in only 3% of mammalian species (Kleiman, 1977). Yet monogamy has evolved in about 14% of all primate species and occurs in 8 out of 11 primate families (Rutberg, 1983). What are the advantages to primates of a social system in which one adult pair breeds exclusively with one another over several breeding seasons? In birds monogamy has been correlated with distribution of food (Brown, 1964; Orians, 1969) or with situations where male assistance is necessary for successful raising of offspring (Lack, 1968; Trivers, 1972). In order to assess the importance of these two factors in the evolution of monogamy in primates, I studied the feeding behavior and parental care in two species of strictly monogamous New World monkeys (*Aotus* and *Callicebus*) and have compared my results with information on a sympatric polygynous New World monkey (*Saimiri*) of similar body size.

Methods

The socioecology of *Callicebus moloch* and *Aotus trivirgatus* was studied for a total of 15 months at the Cocha Cashu Biological Research Station in the Manu National Park, Madre de Dios, Peru. Two focal groups of *C. moloch* and one of *A. trivirgatus* were each followed for 5 complete days (or nights) each month. General activity was recorded at 5-minute intervals for each group member of the *C. moloch* group. Opportunistic notes were taken on playing, weaning, conflicts, infant carrying, nursing, insect foraging, intergroup fighting and calling. Position of group members in progressions while traveling, and entering and exiting food trees was noted. Crown diameters of all fruit trees were measured. Although it is difficult to observe the activity of each group member in an *Aotus* family, with the aid of an ITT pocket image intensifier, a Varo Noctron IV scope, moonlight and my ears, I could record a general group activity every 5 minutes throughout the night. Data on infant development could be recorded and was compared with infant development and parental care data taken on captive *Aotus* (Wright, 1984). Crown diameters and distances between fruit trees were recorded for *Aotus* throughout the study.

For comparison with the two monogamous species, data were used from a study of *Saimiri sciureus*, studied by a team headed by John Terborgh in 1976 and 1977. These squirrel monkey groups overlapped the territories of all the *Aotus* and *Callicebus* groups that I studied. Squirrel monkeys seemed appropriate for this comparison since they are approximately the same size as *Aotus* and *Callicebus*. Several squirrel monkey groups were followed all day for 20 day samples for a total of 60 days including both wet and dry seasons. Scan sampling techniques were used at 10- or 15-minute intervals to determine activity and more detailed *ad libitum* notes taken on other behaviors. Crown diameters of all fruit trees and distances between each tree fed in were recorded.

Botanical censuses were conducted to determine density of trees of different crown diameters and phenology of fruiting was determined both by fruit traps and by biweekly census of 130 trees of 48 species (Wright, 1985).

Results

Ranging patterns and size and distribution of resources

It has been suggested that monogamy would evolve as a mating system if resources were evenly distributed in space, predictable over time, found in small patches and defendable (Orians, 1969; Brown, 1964). In contrast, a polygynous animal would use larger, rarer resources (Verner & Wilson, 1966; Horn, 1968; Orians, 1969; Wittenberger, 1976). Do these correlations between resource size and distribution hold true for primates in South America?

Comparison between polygynous and monogamous species Aotus and *Callicebus* – Group size of both species ranges from 2 to 5 individuals and composition consists of an adult pair and 1–3 sequential offspring.

One infant is born each year and the subadult emigrates in his third year. Small territories are used exclusively by one family group and defended from intruders of the same species (Mason, 1966; Robinson, 1979; Wright, 1985). There is no strict breeding season, but there is a birth peak at the beginning of the rainy season.
Saimiri – Group size at the Manu National Park is about 35 individuals. There are approximately an equal number of males and females. One offspring is usually born to each mother annually, and it is not known if or when subadults emigrate from the group. The home range of one group overlaps extensively with that of several other squirrel monkey groups and is not defended against conspecifics. The breeding season occurs in May, and all females in one group come into estrus within 2 weeks of one another. There is fierce male–male competition for access to females during the breeding season.

Home range size and daily path length *Aotus* and *Callicebus* – The territory size of both species at Cocha Cashu is small (6–12 ha). The mean nightly path length of the focal *Aotus* group was 708 m (S.D. = 243, N = 55). The mean daily path length of *Callicebus* Group I was 552 m (S.D. = 105.8, N = 25) and that of *Callicebus* Group II was 670 m (S.D. = 192.9, N = 64). In both species group members traveled and fed in close proximity and group spread averaged 10 m. All areas of the territory were monitored within 2–4 days (or nights).

Saimiri – The home range of *Saimiri* is large (250 ha) (Terborgh, 1983). The mean daily path length of the squirrel monkeys was 2100 m (S.D. = 255.8, N = 18). These monkeys, which are similar in size to the night and titi monkeys, traveled three times as far each day. The group could be spread over a 100-m area at one time. Not all areas of the home range were visited in a 2–4 day time span.

Distances between consecutively used fruit trees *Aotus* and *Callicebus* – The mean inter-fruit tree distance for *Callicebus* was 57 m (confidence interval, $p = 0.001$: ±12.2 m) (Fig. 1). The difference between these two species was not significant (t test for samples with different variance, $t = 0.29$, $p > 0.05$, d.f. = 120) (Fig. 1).
Saimiri – The inter-fruit tree distance for *Saimiri* was 206 m (confidence interval, $p = 0.01$: ±69.8 m) (Fig. 1). The distance between fruit trees used consecutively by squirrel monkeys is significantly different from the distance between fruit trees used consecutively by night or titi monkeys. (*Aotus–Saimiri*, $t = 5.39$, $p < 0.001$, d.f. > 120; *Callicebus–Saimiri*, $t = 5.859$, $p < 0.001$, d.f. > 120.)

The squirrel monkeys traveled farther between fruit trees than either *Aotus* or *Callicebus*.

Size and availability of resources Aotus and *Callicebus* – In order to determine the size of resources used by the monkeys, the percentage of feeding minutes spent in fruit trees in three size classes over an annual cycle was calculated. *C. moloch* spent nearly 80% of its feeding minutes in small crowns (10 m in diameter) and only 5% in large crowns (21–40 m). *Aotus* spent over 50% of its time feeding in medium crowns (11–20 m), and 33% of feeding minutes in large crowns (21–40 m (Fig. 2).

Forty-five percent of the species of fruit trees used by *Callicebus* ripened fruits over a 2-month period and 82% of *Callicebus* fruit-feeding minutes were on trees which ripened a few fruits over a 2–5-month period.

Saimiri – In contrast, the squirrel monkeys spent 51% of their feeding minutes in fruit trees with large crown diameters (21–40 m) (Fig. 2), especially fig trees (Terborgh, 1983). The remainder of the feeding minutes were divided between small and medium-sized crowns.

Fig. 1. Distance between fruit trees used by *Aotus*, *Callicebus moloch* and *Saimiri* in the Manu National Park, Peru. Vertical bars are confidence intervals for $p = 0.001$. The difference between *Aotus* and *Callicebus* was not significant (t test for samples with different variance, $t = 0.29$, $p > 0.05$, d.f. = 120), but the differences in fruit tree distance seen in the *Aotus–Saimiri* comparison ($t = 5.39$, $p < 0.001$, d.f. > 120) are significant.

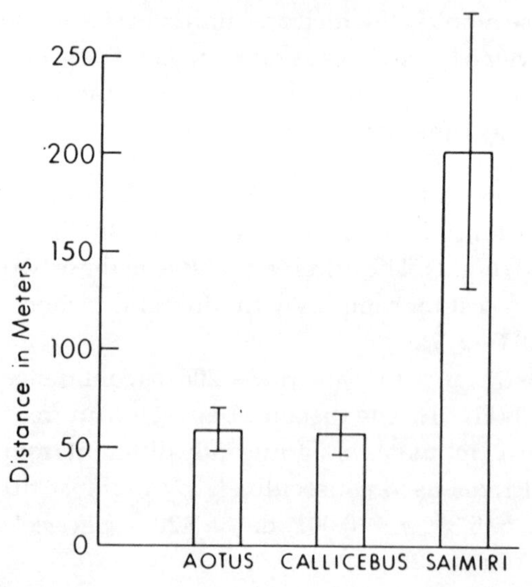

According to botanical censusing and phenology of selected trees (Wright, 1985), there are more small-crowned fruit trees (<10 m) in each hectare than large-crowned trees, and small-crowned trees have small amounts of fruit which ripen a few at a time throughout several months. Large-crowned trees (for example, figs) have abundant fruit, but the fruits on each tree ripen over a 2-week period. Trees with crown diameters of over 20 m are rare.

Male assistance

Aotus and *Callicebus* – The males in both monogamous species invest heavily in their offspring (Dixson & Fleming, 1981; Wright, 1981, 1984; Fragaszy, Schwartz & Shimosaka, 1982). The father, rather than the mother, carries the infant almost exclusively in the first 2 months. At 3 weeks, the infant is on the father's back 80–90% of the time, returning to the mother only for nursing bouts. As the infant gains more independence the father carries it less. By age 3 months,

Fig. 2. The percentage of feeding minutes spent in fruit trees in three size classes by *Aotus*, *Callicebus* and *Saimiri* in the Manu National Park, Peru.

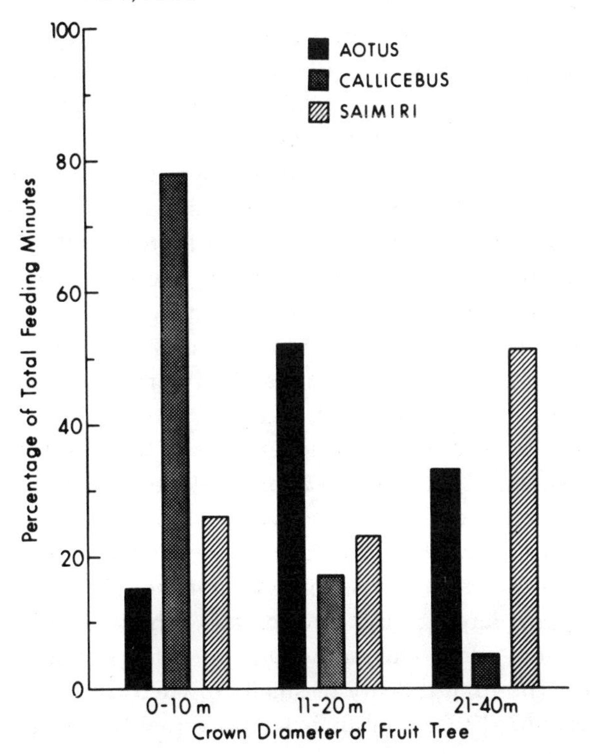

the infant (now over 40% of its father's weight) is carried 38% of the time and is independent for 60%. At the age of 4 months, the infant only rides on the father during long leaps between trees, or in times of danger. Throughout the first year, it is the father who plays with, shares food with, and guards the infant. The mother is near her infant only for nursing bouts (Wright, 1984).

The lactating mother leads the group into fruit trees and eats fruit at a faster rate than other group members. The lactating mother eats three times as many insects as the father who is burdened with an active infant (Wright, 1984, 1985).

Saimiri – Squirrel monkey males do not carry, guard, or teach infants new foods. Infant squirrel monkeys play with peers rather than adults (Terborgh, 1983). Subadult females sometimes relieve mothers from infant carrying (Baldwin & Baldwin, 1981). Predator detection is provided by more eyes and ears of large group size, and predator protection may be enhanced when the predator is confused by 35 animals leaping in all directions.

Discussion

Resource use and mating system

The data presented here support the prediction that the monogamous primate species would rely on small, predictable, evenly spaced resources. But these data do not explain why one species selects small crowns, while another prefers large crowns.

Large groups cannot feed efficiently in small crowns because there are not enough fruits available to make traveling costs economical. A squirrel monkey group travels farther to reach the big trees, but once the group has arrived, there is food for all members. These large trees attract large monkeys (*Cebus apella, C. albifrons* and *Ateles paniscus*) which chase smaller monkeys. Because of the large number of squirrel monkeys in each group, chases are ineffective. But interference competition does exclude titi monkeys from these large trees (Wright, 1984, 1985). Titi monkeys then feed on vine leaves, and the few fruits found in small crowns. The night monkeys avoid the large monkeys by being nocturnal, and feed unmolested in crowns of all sizes.

The small-crowned trees on which the titis rely are often ignored by monkeys with large group sizes because there are not enough fruits in small patches to support large groups and females cannot afford to tolerate other females (an additional mother and her offspring nearly doubles group size). The high density of small-crowned trees in all areas of the forest results in little variation in territory quality. This fact is a reason why additional females would not readily accept the costs of polygyny (Wittenberger & Tilson, 1980). The male then has the choice of mating with one female, or visiting the separate territories of several females. There may be several reasons why the latter choice is not acceptable: (1) widely dispersed females coming into estrus once a year would not be easy to locate; (2) ranging widely may be energetically expensive; and (3) unprotected infants may not survive.

This reliance on small, predictable patches is advantageous in certain circumstances. In harsh habitats, at high altitudes or in subtropical South American forest patches, there are few large trees with fruit eaten by monkeys. Here the monkeys with small group size do well, while monkeys with large group size (except *Alouatta* which rely on leaves) are absent. In addition, successional and damaged areas of forest lack large trees and small monkeys with small body size can subsist there.

Do these economic factors hold true for monogamous primates outside South America? There is some evidence that gibbons rely on small-crowned fruit trees, but siamangs feed mainly on figs and leaves (which could be related to large body size) (Gittins & Raemaekers, 1980).

Male assistance

The obvious reason why *Aotus* and *Callicebus* males invest so heavily in infants, while *Saimiri* males do not, is that the monogamous fathers are assured of their paternity. But not all monogamous primate fathers invest as heavily in offspring. For example, some gibbon (Tenaza, 1975) and indri males (Pollock, 1979) do not carry infants.

The differences between species may be related to size. All the New World monkeys with male assistance are small (1 kg or under) while gibbons and indris are much larger (8–10 kg). Seven species of hawks and eagles eat monkeys in the Amazonian rain forest (Terborgh, 1983), and large monkeys chase small monkeys (Wright, 1984). The only recourse for small monkeys with small group size is to hide or flee. In larger species males can attack predators to protect offspring, while females with infants escape. The much smaller titi or night monkey male would not be effective attacking a raptor or capuchin monkey, and if fleeing from danger is the best protective strategy, then it may be more effective for the male (not stressed by the energetic costs of lactation) to make a quick retreat with an extra weight on his back. If the survival of the offspring is at stake, it may be to the advantage of both mother and father for the father to carry the infant (Wright, 1984).

Summary

In the New World primates *Aotus* and *Callicebus*, monogamy is correlated with (1) distribution of food into small, predictable, uniform patches and (2) the need for male assistance to raise offspring successfully.

Acknowledgements

This research was supported by NSF grant BNS 81-15368, NSF grant BNS 77-24921, a grant from the World Wildlife Fund–US Primate Program and the Wenner–Gren Foundation for Anthropological Research Grant 4282. I am also grateful to the Peruvian Ministry of Agriculture, and the General Directorate of Forestry and Fauna of Peru for permission to work in the Manu National Park. Special thanks to Patrick Daniels, Cirilo Lujan Munares and Peg Stern for field assistance and to John Terborgh, Charles Janson, Anne Wilson Goldizen, Warren Kinzey, John Oates, John Fleagle, Robin Foster, Jorg Ganzhorn, David Sivertsen and Louise Emmons for support and expert advice. I am grateful to John Terborgh for permission to use unpublished data on the ecology of the squirrel monkey in Manu National Park.

References

Alexander, R. D. (1974) The evolution of social behavior. *Ann. Rev. Ecol. Syst.*, **5**, 325–83

Baldwin, J. D. & Baldwin, J. I. (1981) The squirrel monkeys, Genus *Saimiri*. In *Ecology and Behavior of Neotropical Primates*, Vol. 1, ed. A. F. Coimbra-Filho & R. A. Mittermeier. Rio de Janeiro: Brasilian Academy of Science

Brown, J. L. (1964) The evolution of diversity in avian territorial systems. *Wilson Bull.*, **76**, 160–9

Clutton-Brock, T. H. (1978) Mammals, resources and reproductive strategies. *Nature*, **273**, 191–5

Dixson, A. F. & Fleming, D. (1981) Parental behavior and infant development in owl monkeys (*Aotus trivirgatus griseimembra*). *J. Zool. (London)*, **194**, 25–39

Fragaszy, D. M., Schwartz, S. & Shimosaka, D. (1982) Longitudinal observations of care and development of infant titi monkeys (*Callicebus moloch*). *Am. J. Primatol.*, **2**, 191–200

Gittins, S. P. & Raemaekers, J. J. (1980) Siamang, lar and agile gibbons. In *Malayan Forest Primates: Ten Years Study in Tropical Rainforest*, ed. D. J. Chivers. New York: Plenum Press

Horn, H. S. (1968) The adaptive significance of colonial nesting in the Brewer's blackbird (*Euphagus cyanocephalus*). *Ecology*, **49**(4), 682–94

Kleiman, D. G. (1977) Monogamy in mammals. *Q. Rev. Biol.*, **52**, 39–68

Lack, D. (1968) *Ecological Adaptations for Breeding in Birds*. London: Methuen

Mason, W. A. (1966) Social organization of the South American monkey, *Callicebus moloch*: a preliminary report. *Tulane Stud. Zool.*, **13**, 23–8

Orians, G. H. (1969) On the evolution of mating systems in birds and mammals. *Am. Naturalist*, **103**, 589–603

Pollock, J. I. (1979) Female dominance in *Indri indri*. *Folia Primatol.*, **31**, 143–64

Robinson, J. G. (1979) Vocal regulation of use of space by groups of titi monkeys, *Callicebus moloch*. *Behav. Ecol. Sociobiol.*, **5**, 1–15

Rutberg, A. T. (1983) The evolution of monogamy in primates. *J. Theoret. Biol.*, **104**, 93–112

Tenaza, R. R. (1975) Territory and monogamy among Kloss' gibbon (*Hylobates klossii*) in Sibertu Island, Indonesia. *Folia Primatol.*, **24**, 68–80
Terborgh, J. W. (1983) *Five New World Primates: A Study in Comparative Ecology*. Princeton, N.J.: Princeton University Press
Trivers, R. L. (1972) Parental investment and sexual selection. In *Sexual Selection and the Descent of Man*, ed. B. Campbell, pp. 136–79. Chicago: Aldine
Verner, J. & Willson, M. F. (1966) The influence of habitats on mating systems of North American passerine birds. *Ecology*, **47**, 143–7
Wilson, E. O. (1975) *Sociobiology: The New Synthesis*. Cambridge, Mass.: Harvard University Press
Wittenberger, J. (1976) The ecological factors selecting for polygyny in altricial birds. *Am. Naturalist*, **110**, 779–99
Wittenberger, J. F. & Tilson, R. L. (1980) The evolution of monogamy: hypotheses and evidence. *Ann. Rev. Ecol. Syst.*, **11**, 197–232
Wright, P. C. (1981) The night monkeys, genus *Aotus*. In *Ecology and Behavior of Neotropical Primates*. Vol. 1, ed A. F. Coimbra-Filho & R. A. Mittermeier, pp. 211–40. Rio de Janeiro: Brazilian Academy of Science
Wright, P. C. (1984) Biparental Care in *Aotus trivirgatus* and *Callicebus moloch*. In *Female Primates: Studies by Women Primatologists*, ed. M. E. Small. New York: Alan R. Liss
Wright, P. C. (1985) Costs and benefits of nocturnality for *Aotus* (the night monkey). Unpublished Ph.D. dissertation, City University of New York, New York

IV.2

The mating system as a determinant of social evolution in capuchin monkeys (*Cebus*)

C. H. JANSON

Introduction

Most studies of primate social behavior can be divided into those examining *external* forces affecting group structure (e.g. predation, food distribution and abundance), and those focusing on *internal* factors that influence social behavior (kinship, aggression, reproductive behavior). Although external factors have been shown in numerous instances to correlate with gross aspects of social organization (e.g. social group size: Clutton-Brock, 1974; Hladik, 1975; Terborgh, 1983; feeding group size; Klein & Klein, 1975; Leighton & Leighton, 1982; day range: Clutton-Brock & Harvey, 1977; Struhsaker & Leland, 1979; van Schaik *et al.*, 1983; and spatial structure: Rhine & Owens, 1972), only a few authors have shown how external forces might also influence more detailed aspects of social structure (individual spatial position: Robinson, 1981; female–female bonding: Wrangham, 1980; mother–infant relations: Altmann, 1980; female mate choice: Janson, 1984). An underlying assumption of many of these socioecological studies is that ecologically similar species should have similar social structures.

The two species examined here, the brown capuchin (*Cebus apella*) and the white-fronted capuchin (*Cebus albifrons*) present a challenge to traditional views of socioecology. Although very similar ecologically (Terborgh, 1983; Terborgh & Janson, 1983), they differ markedly in social organization (Table 1). Thus, I either had to reject ecology as a basis for explaining these differences in social structure, or I had to explain how a *small* ecological difference between species can be amplified into *major* social changes. In answering the latter challenge, I became convinced that the mating system is the crucial variable needed to understand social differences between the two capuchins.

Materials and methods

Study site

I studied the two capuchin species in moist tropical forest at the Cocha Cashu Biological Station of the Manu National Park in Peru. The Cocha Cashu Biological Station comprises a 3 km^2 trail system in the floodplain of the Manu River at 400 m elevation. There is strong seasonal variation in fruit production (the capuchins' major food), with the lowest production in the dry season when capuchins switch to other foods (Terborgh, 1983; Janson, 1984). The flora and fauna of the area are diverse: over 1200 plant species are recorded from the study area (R. Foster, unpublished), and the over 500 bird species and 120 mammal species include major predators on capuchins (harpy eagle, Guianan crested eagle, ocelot) and many potential competitors for fruit and insects, including eight other common primate species. For more details, see Terborgh (1983).

Table 1. *Comparison of ecology and social structure of capuchins*[a]

	Brown capuchin	White-fronted capuchin
Ecology		
1. Mean fruit diameter	1.9 cm	1.6 cm
2. No. fruit trees visited/day	10.4	8.5
3. No. of fruit species	96	68
4. Percent of feeding spent in trees of crown diameter:		
0–10 m	**63**	**40**
10–20 m	25	19.5
20–50 m	**12**	**40.5**
5. Frequency of encounters with conspecific groups	once/3.7 days	once/7.6 days
Social structure		
1. Aggression:		
a. Relative rate of aggression[b] between:		
Ad. ♂–Ad. ♂	1.77	1.52
Ad. ♀–Ad. ♀	**1.11**	**2.00**
Ad. ♂–Ad. ♀	**0.55**	**1.41**
Adult–Juv.	1.05	0.41
Juv.–Juv.	1.80	0.0 (infants only)
b. Defense of juveniles by dominant males	Rapid; nearly always occurs when juv. squeals	Almost never occurs, even when juv. squeals

Table 1. *Continued*

	Brown capuchin	White-fronted capuchin
2. Mating system:		
a. ♀ solicitation	Continuous for 4–5 days, toward dominant ♂	Intermittent, <2 h at a time
b. Behavior of dominant ♂ to estrous ♀	Mildly aggressive, rare approaches, no sniffing of her urine	Follows ♀, sniffs her urine, tries to mount her even if she is not proceptive
c. ♂–♂ aggression	Rare, mostly on next-to-last day of a ♀'s estrus	Rare, throughout the estrus
3. Spatial association[c]:		
Dom. ♂–Ad. ♂♂	**0.27**	**1.72**
Other Ad. ♂–Ad. ♂♂	**0.85**	**1.61**
Dom. ♂–Juv.	**1.42**	**0.77**
4. Cooperation in defense:		
a. Aerial predators	Dominant ♂ only one to remain visible, gives special loud bark	Dominant male usually hides at first, no special bark
b. Ground predators	Most group members except subordinate ♂♂ approach and mob, use distinct call	Dominant and other males approach together, give silent threat display
c. Other group, not at food tree	Little aggression, subordinate ♂♂ approach other group individually	Very aggressive, dominant ♂ & other adult ♂♂ confront other group's ♂♂
d. Other group at food tree	Dominant ♂ leads attack, followed by adult ♀♀, and rarely subordinate ♂♂	Adults of both sexes run in at same time

[a] Statistics are omitted from numerical comparisons for simplicity, but the most significant contrasts between the two species are printed in bold face.
[b] The relative rate of aggression is the ratio of observed to expected number of aggression interactions between two age–sex classes. The expected number of interactions is obtained by assuming that each individual will initiate aggression toward all subordinate group members (but not to dominants) at the same rate.
[c] The relative spatial association is the proportion of time two individuals or age–sex classes were observed within 10 m of each other, divided by the time expected if neighbors were chosen from each age–sex class in proportion to the group's actual composition. All comparisons shown are significant. Associations for other age–sex classes did not differ more than 7% between brown and white-fronted capuchins.

Field protocol

My field associates and I spent over 4000 h watching four groups of the brown capuchin (BC) and three groups of the white-fronted capuchin (WFC). The groups spanned the range of common group sizes (BC: 3–14 individuals; WFC: 10–17) (e.g. Terborgh, 1983). We followed the monkeys continuously from dawn to dusk, marking every fruit tree they visited, and taking instantaneous samples at 5-minute intervals of the foraging and social behavior of focal animals (Altmann, 1974). *Ad libitum* (Altmann, 1974) notes were used for infrequent conspicuous behaviors of the focal animal or other group members. Although only one observer was recording from a given study group at most times, I tried to assure inter-observer agreement by checking their foraging and behavioral descriptions in the field.

Results

Ecological differences between the two capuchin species

The non-fruit portions of the diet are very similar. Both species spend about half their time foraging for invertebrates in trees and palm crowns (Terborgh & Janson, 1983), and use the same foraging substrates 80% of the time (Terborgh, 1983). Neither species consumes much leaf material (<7% per year), but the BC depends on the pith of *Scheelia* palm frond petioles when all other foods are scarce, whereas WFCs very rarely eat this resource.

At first glance, it is also hard to detect much difference between the two species in their use of fruit (Table 1). The BC feeds on slightly larger and harder fruits and seeds than the WFC, but the differences between them are small compared to the range observed among all the primates in the study area (Terborgh, 1983). The actual species lists of fruits are fairly similar, with 58 of the 68 recorded food species of the WFC also eaten by the BC (Table 1). More importantly, the food species which the two capuchins share are the most important components of their diets, accounting for over 90% of the feeding time of each species (Janson, unpublished).

However, there is a significant difference in how the two capuchins use their shared fruit species (Table 1). WFCs feed in trees with a mean crown diameter of 17.1 m, significantly larger than the mean crown diameter of 12.3 m used by BCs. This contrast is even more striking when viewed as the amount of time spent feeding in these different-sized trees (Table 1). WFCs derive nearly half their diet from trees with crowns larger than 20 m diameter, whereas 63% of the BC diet comes from trees 0–10 m in diameter.

The difference between the two capuchins in size of fruit tree in which they feed affects the pattern of within-group competition (Fig. 1). In the small trees typically used by BCs, aggression is frequent and dominants can easily monopolize feeding opportunities, achieving feeding rates up to 300% greater than subordinates (Fig. 1A). By contrast, in large fruit trees, like figs, aggression is rare and dominants of either capuchin species feed no more than 30% better than subordinate group members (Fig. 1B). Thus, for a majority of their feeding time, WFCs feed in trees in which dominants *cannot* easily affect the feeding success of subordinates.

Aggression

The two capuchin species differ markedly in the frequency and form of aggression within groups (Table 1). BCs are often aggressive, especially in fruit trees, where over 10 fights/h per group may occur. Interactions are usually brief and dyadic, and the outcome of a given interaction is easily predicted by an animal's dominance status – subordinates, even in coalitions, displace dominants in less than 1% of all aggressive acts (Janson, 1985). WFCs show much less aggression in general, and rarely fight in fruit trees. Chases and displacements are not as useful in defining dominance status in WFCs because dominants rarely actively displace subordinates, but instead receive *spontaneous* submissive behaviors (silent threats, noisy 'kekkers') from subordinates.

The distribution of aggression also differs noticeably between BCs and WFCs (Table 1). In BCs, most aggression occurs among adult and subadult males, and females receive relatively little aggression (relative to their proportion of the group composition). By contrast, in WFCs, aggression is relatively more common among adult females than among adult males, and females receive much more aggression from males than they do in the BC. In both species, juveniles are not frequently involved in aggression directly, but in the BC, juveniles often cause aggression as the dominant male chases away another group member that tries to displace a juvenile. In the WFC, such defense of juveniles by the dominant male against other adults has been seen only once in over 1200 observation hours.

Mating system

The mating system of the BC is described in detail elsewhere (Janson, 1984). It is unusual among primates in that males rarely contest access to estrous females, yet all females consistently prefer to

Fig. 1. Relation between feeding success and aggressive success within capuchin groups. Feeding success is measured as the number of fruits ingested by each group member per time period, relative to the group's mean rate. Aggressive success is the proportion of all fights won by an individual, out of all fights in which it engaged. A. Brown capuchins in small-crowned *Scheelia* palms. Aggression is frequent (10.4 fights/group per hour) and feeding success is strongly affected by aggressive success (linear regression, $y = 1.91x + 0.31$, $r = 0.91$, $p < 0.001$). B. Brown capuchins (dots and solid line) and white-fronted capuchins (crosses and dashed line) in large-crowned fig trees. Aggression is uncommon, even for brown capuchins (1.3 fights/h), and feeding success is not strongly related to aggressive success (for brown capuchins, $y = 0.30x + 0.85$, $r = 0.73$, $p < 0.05$, one-tailed; for white-fronted capuchins, $y = -0.02x + 0.98$, $r = -0.08$, $p > 0.05$, one-tailed).

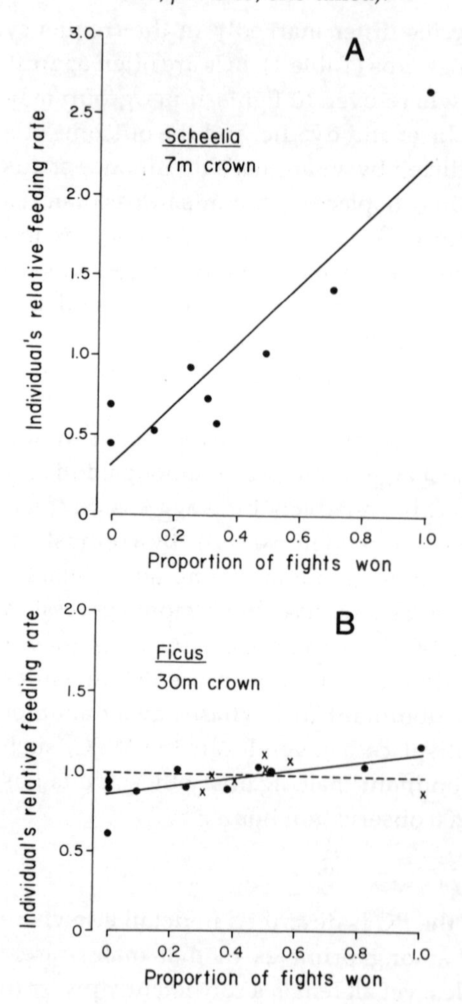

mate with only the dominant male. The dominant male shows little interest in an estrous female, yet her constant active solicitation of him for 4–5 days of her 5–6-day estrus assures that he is the only male to copulate with her in the middle of estrus, when laboratory data suggest that she is most likely to ovulate (Wright & Bush, 1977).

Our mating data are much less complete for WFCs (only four estrous occasions partially observed), yet there are several striking differences from BCs (Table 1). Unlike the long-term consortship typical of BCs, the estrous female WFC switches between bouts of normal foraging and brief periods (never more than 2 h) of following and soliciting males. The dominant male WFC takes an active interest in female estrous condition, often sniffing her urine and attempting to mount her even when she does not cooperate, behaviors never seen in BCs. Our sample is too small to judge the relative mating success of dominant *versus* subordinate male WFCs, but in the socially similar white-faced capuchin (*Cebus capucinus*), the dominant male did *not* achieve the largest share of matings (Oppenheimer, 1968), unlike the 3-to-1 advantage of the dominant male BC (Janson, 1984).

Spatial structure

Spatial associations of males and juveniles are strikingly different in the two capuchins. In the BC, foraging males strongly avoid each other, whereas WFC adult males associate together more strongly than any other age–sex class except juveniles. Young BCs also stay close to each other, but they do so because of a shared attraction to the groups' dominant male (Janson & Wright, 1980), whereas WFC juveniles show no strong association with any group male. In both species, foraging adult females typically avoid other adults except the dominant male.

Group cooperation

In both species, the dominant male is usually the first individual to encounter and threaten foreign groups in fruit trees. In the BC, the dominant male may receive help from adult females and occasionally subadult natal males, but rarely from subordinate immigrant males. By contrast, most adult and subadult males in a WFC group join the dominant male in threatening and chasing another group, whereas adult females only sometimes help.

In defense against predators, the dominant male BC is the only individual that remains exposed and continues giving loud alarm barks after an attack by an aerial predator. By contrast, the dominant male WFC rarely exposes himself to either kind of predator alone, but instead waits for the other adult and subadult males, who then all threaten the predator together.

Discussion

How can the relatively slight ecological differences between the two capuchin species be reconciled with the striking contrasts in their social structures (Table 1)? The crucial ecological difference between the two capuchin species is not fruit size, food species, or foraging mode and effort, but rather the differential use of small *versus* large food trees (Fig. 1). I propose here a model based on these results, which shows how food patch size can influence the mating system, and differences in mating systems lead to quite divergent social systems (Fig. 2).

To start with, the size of fruit tree determines how easily dominants can affect the feeding success of subordinates (Fig. 2, step 1). In small trees, a dominant male can restrict access to those individuals he tolerates (Janson, 1984), whereas in large trees he can rarely exclude a given individual for more than a few minutes (Janson, 1985).

How easily a dominant male can defend food patches may influence his mating strategy (Fig. 2, step 2). Because the BC depends on small food trees in the dry season when fruit is scarce, the dominant male may be able to offer a female substantial benefits of having mated with him, by tolerating either her or her offspring at food trees. Indeed, the dominant male BC is *not* tolerant of group juveniles that he could not have sired (Janson, 1984). By contrast, the WFCs' stronger dependence on large-crowned food trees may not allow the dominant male to affect

Fig. 2. Model of the evolution of social structure in capuchins. Arrows show the suggested direction of influence of social or ecological factors on each other. For explanation see Discussion.

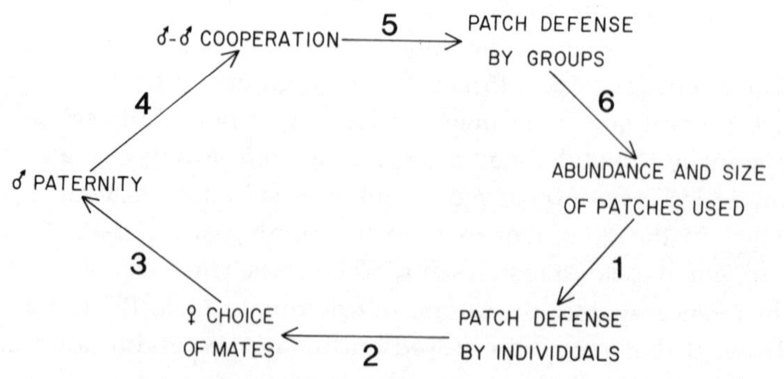

substantially a female's or juvenile's feeding success. Because he cannot offer a female much benefit by his tolerance at food trees, the dominant male may not be the most strongly preferred mating partner in WFCs.

The mating system should affect the kinship structure of the group, showing up in predictable patterns of aggression and affiliation toward group juveniles (Fig. 2, step 3). In BCs, the dominant male has an apparent strong assurance of paternity, and tolerates and helps juveniles, which in turn prefer his company to that of any other group member (Janson & Wright, 1980). Subordinate male BCs, with a low apparent probability and assurance of paternity, are not very tolerant of juveniles. In WFCs, the males may all share an equal, but necessarily lower, assurance of paternity in any given infant. Thus, they should not, and in fact do not, treat particular juveniles with special tolerance or aggression. In response, juveniles do not seek out the dominant male or any other male as a preferred neighbor during foraging, but form their own spatial subgroup.

The contrast in male behavior toward juveniles in the two capuchin species may be the basis for differences in male–male cooperation (Fig. 2, step 4). To the dominant male BC, all subordinate males are competitors for the food that his offspring need to achieve high survivorship and fecundity (e.g. Altmann, 1980). Thus, he is very aggressive toward subordinate males and they in turn have no reason to aid him in cooperative tasks. By contrast, in the WFC, all males may have an equal but low genetic representation in the group's offspring. Thus, no one male should be willing to risk much for the sake of the group, but all males should and do normally cooperate in tasks where their mutual help reduces the risk to each and benefits the entire group.

The degree of cooperation among males can be an important element of a group's success in defending food trees against other groups of the same or other species (Fig. 2, step 5). Usually the winner of these contests among primates is the larger species, or, within a species, the larger group. Yet, WFCs regularly displace BCs from food trees although group sizes may be equal and even though WFCs are smaller and less robust than BCs. The secret to WFC success appears to be coordinated defense. Two or more individuals, usually males, together chase the dominant male BC. Further support that cooperation is important is the fact that BCs can displace some WFC groups, but only those that contain a single adult male. Without the possibility of male–male cooperation, WFCs lose to the physically larger BCs.

Finally, the ability to defend food trees successfully may be important in determining the feeding ecology of a species (Fig. 2, step 6). It may be that at least part of the difference between the two capuchin species in the size of fruit trees used (Table 1) is caused by the BC's inability to forage efficiently in large trees in the presence of the WFC's more coordinated aggression.

Thus, we have come full circle, as it was the use of different-sized fruit trees by the two capuchins that I hypothesized led to their distinctive social systems. That the WFC and not the BC came to specialize on using large food trees may be due to the WFC's less diverse diet, correlated with its less robust build (Kinzey, 1974). Because it cannot take advantage of the greater diversity of food trees available to the BC, the WFC would have more selective pressure on it to be able to use large productive trees with easily swallowed fruit.

This model (Fig. 2) can be substantiated several ways. Some links should be verified in other species, especially the relation between feeding ecology and male mating strategies, and the importance of male–infant interactions on aggression and cooperation between males. The importance of competition *between species* in generating social differences can be tested directly by looking at the ecology and social structure of a species in the presence and absence of major competitors. For instance, the model predicts that BCs, as the inferior competitor at large food trees, should converge on the ecology and social system of WFCs where the latter are absent and appropriate large food patches exist. Indeed, Izawa's (1980) study of BCs in Colombia, where WFCs were absent, suggests they have a social structure more similar to WFCs than to BCs in Peru.

References

Altmann, J. (1974) Observational study of behavior: sampling methods. *Behaviour*, **49**, 227–67

Altmann, J. (1980) *Baboon Mothers and Infants*. Cambridge, Mass.: Harvard University Press

Clutton-Brock, T. H. (1974) Primate social organization and ecology. *Nature*, **250**, 539–42

Clutton-Brock, T. H. & Harvey, P. H. (1977) Primate ecology and social organization. *J. Zool. (London)*, **183**, 1–39

Hladik, C. M. (1975) Ecology, diet, and social patterning in Old and New World primates. In *Socioecology and Psychology of Primates*, ed. R. H. Tuttle, pp. 3–36. The Hague: Mouton

Izawa, K. (1980) Social behavior of the wild black-capped capuchin (*Cebus apella*), *Primates*, **21**, 443–67

Janson, C. H. (1984) Female choice and mating system of the brown capuchin monkey *Cebus apella* (Primates: Cebidae). *Z. Tierpsychol.*, **65**(3), 177–200

Janson, C. H. (1985) Ecological and social consequences of food competition in brown capuchin monkeys. Ph.D. thesis, University of Washington

Janson, C. H. & Wright, P. C. (1980) Parent–offspring relations in the brown capuchin monkey (*Cebus apella*). *Am. J. Phys. Anthropol.*, **52**, 214

Kinzey, W. G. (1974) Ceboid models for the evolution of hominoid dentition. *J. Hum. Evol.*, **3**, 193–203

Klein, L. & Klein, D. (1975) Social and ecological contrasts between four taxa of neotropical primates. In *Socioecology and Psychology of Primates*, ed. R. H. Tuttle, pp. 59–86. The Hague: Mouton

Leighton, M. & Leighton, D. R. (1982) The relationship of size of feeding aggregate to size of food patch: howler monkeys (*Alouatta palliata*) feeding in *Trichilia cipo* fruit trees on Barro Colorado Island. *Biotropica*, **14**, 81–90

Oppenheimer, J. R. (1968) Behavior and ecology of the white-faced monkey, *Cebus capucinus*, on Barro Colorado Island, C.Z. Ph.D. dissertation, University of Illinois, Urbana

Rhine, R. J. & Owens, N. W. (1972) The order of movement of adult male and black infant baboons (*Papio anubis*) entering and leaving a potentially dangerous clearing. *Folia Primatol.*, **18**, 276–83

Robinson, J. G. (1981) Spatial structure in foraging groups of wedge-capped capuchin monkeys *Cebus nigrivittatus*. *Anim. Behav.*, **29**, 1036–56

Schaik, C. P. van, Noordwijk, M. A. van, de Boer, R. J. & Tonkelaar, I. den (1983) The effect of group size on time budgets and social behavior in wild long-tailed macaques (*Macaca fascicularis*). *Behav. Ecol. Sociobiol.*, **13**, 173–81

Struhsaker, T. T. & Leland, L. (1979) Socioecology of five sympatric monkey species in the Kibale forest, Uganda. *Adv. Study Behav.*, **9**, 159–228

Terborgh, J. W. (1983) *Five New World Primates: A Study in Comparative Ecology*. Princeton, N.J.: Princeton University Press

Terborgh, J. W. & Janson, C. H. (1983) Ecology of primates in southeastern Peru. *Natl Geogr. Soc. Res. Reports*, **15**, 655–62

Wrangham, R. W. (1980) An ecological model of female-bonded primate groups. *Behaviour*, **75**, 262–99

Wright, E. M. Jr & Bush, D. E. (1977) The reproductive cycle of the capuchin (*Cebus apella*). *Lab. Anim. Sci.*, **27**, 651–4

IV.3

Ecological determinants of fission–fusion sociality in *Ateles* and *Pan*

M. J. MCFARLAND

Introduction

Early workers studying chimpanzee behavior in the wild were surprised by the apparent lack of structure in their society. Associations between individuals were often extremely short-lived and seemingly random. There was no coherent troop or harem upon which researchers could focus their attention (Reynolds & Reynolds, 1965). Later, longitudinal studies involving large numbers of recognized individuals allowed researchers to elucidate the underlying structure of chimpanzee society (Nishida, 1968; Goodall *et al.*, 1979).

Since the early 1970s, researchers have realized that the social organization of the New World spider monkeys, genus *Ateles*, resembles that found in chimpanzees (Klein, 1972; Cant, 1977). In spider monkeys, as in chimpanzees, it was observed that individuals travel in labile parties or subgroups, the membership of which is drawn from a larger, closed social unit analogous to the chimpanzee community or unit-group. Additional similarities noted included agonistic confrontations between males of different communities and the often solitary nature of the females (Klein, 1972).

In this paper I will address two questions. (1) How similar are the social organizations found in *Ateles* and *Pan*? (2) Are the similarities coincidental or the result of convergent social evolution, where similar selection pressures in the forests of Africa and South America have resulted in the same end product?

Methods

Two communities of habituated spider monkeys (*Ateles paniscus chamek*) were observed for over 1000 contact hours in 1982 and 1983 at the Cocha Cashu Biological Station in the Manu National Park, southeastern Peru. Recognition of individuals was possible through differences in facial and pelage characteristics. Behavioral data were gathered primarily through full-day follows of focal individuals. Detailed ecological and behavioral data were gathered using instan-

Table 1. *Age–sex composition of two communities of* Ateles *at Cocha Cashu*

East	Lake
Adult females	
Darnell	Laurie
Jody	Kathy
Molly	Betty
Lola	Sherry
Michelle	Terry
Nicole	Tina
Emma	Cindy
Maggie	Polly
Tilly	Mabel
Harriet	Nell
WEF	Coney
NFF	
LFF	
Adult males	
Ridge	Rocky
Rusty	Whitey
Lech	Goldy
FFM	Victor
Ernie	Tom
Red	Shavers
Joe	
Subadult females	
Tuna	Barb
Foster	Donna
Ellie	Sally
Subadult males	
Spike	Stumptail
WEM	Billy
Juvenile females	
WEFj	Patty
	Mindy
Juvenile males	
Olly	none
Tuffy	
Total = 28	Total = 24

taneous and *ad lib.* sampling methods (Altmann, 1974). A continuation of this study is currently in progress; results will be reported elsewhere.

Results

Table 1 shows the composition of two communities of spider monkeys in the vicinity of the Cocha Cashu Biological Station. The ratio of adult females to adult males is approximately 2:1. Individuals are classified as juveniles until weaned and as subadults until determined to be reproductively mature (i.e. having offspring or actively cycling). Fig. 1 illustrates the ranges of these two communities as I have been able to delineate them through observation of known individuals. These ranges are minimum estimates and with further observation may reach 150 h.

The pattern of range usage exhibited by males and females differs considerably. In results very similar to those reported by Wrangham & Smuts (1981) for chimpanzees, male spider monkeys were found to range over larger areas of the community range than females. Ranges of an individual male and female from each community are also illustrated in Fig. 1. Males also enter more quadrats per hour over the

Fig. 1. Ranges of two communities of spider monkeys (*Ateles paniscus chamek*) in the vicinity of the Cocha Cashu Biological Station. The range-use patterns of a typical adult male and female in the two communities are indicated.

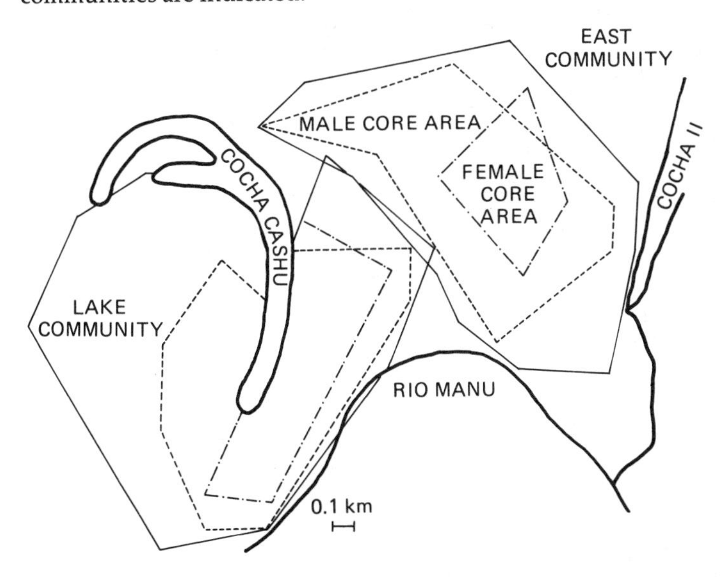

course of the day than females (Fig. 2). This statistic provides an indirect measure of daily path length. A more direct statistical comparison of male and female daily path length is impossible at this time because of the small sample size for males. Fig. 3, however, illustrates the daily ranges of a typical male and female in the early dry season at Cocha Cashu.

Ateles diet composition at Cocha Cashu shows substantial variability even between adjacent communities (Fig. 4). Although the amount of time spent feeding on fruit varies between 45 and 86%, it is always the number-one ranked food type. Further wet season observations will almost surely result in a higher proportion of fruit in the annual diet than illustrated here.

Fig. 2. Number of quadrats entered per hour over the course of the day for male and female *Ateles* at Cocha Cashu.

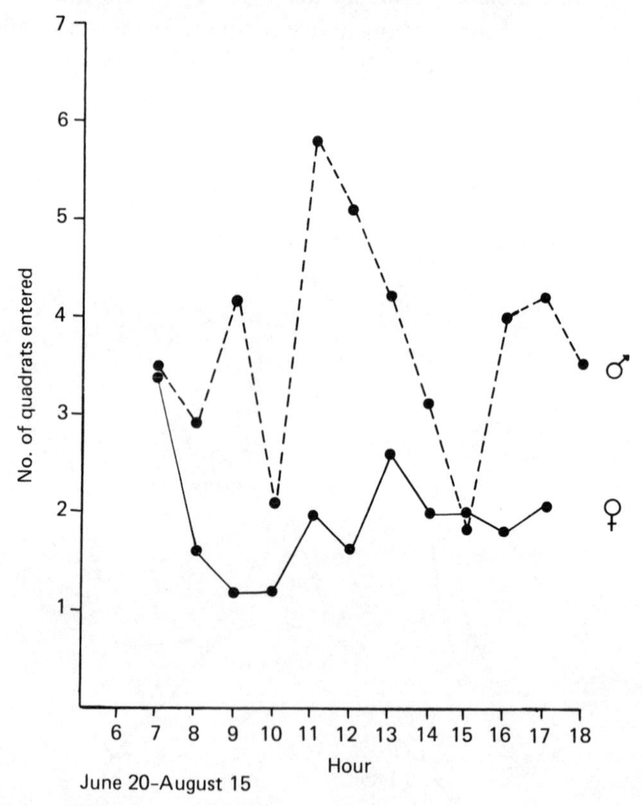

Discussion

Phylogenetic considerations can have little to do with the basic similarity in chimpanzee and spider monkey social organization as their last common ancestor lived over 40 million years ago. More insight may be gained by examining relevant ecological selection pressures, which, in acting upon the individual, serve to mould a species' social organization. For purposes of comparison, this discussion will be confined to what I consider to be the three most important selection pressures: obtaining safety from predators, obtaining access to mates, and maximizing feeding efficiency.

Evidence for strong predation pressure on either chimpanzees or spider monkeys, aside from that imposed by man, is non-existent. Potential primate predators at Cocha Cashu include jaguar, puma, ocelot, margay and a host of raptorial predators including harpy eagle. Recent work with radio-collared jaguar, puma and ocelot (Emmons, unpublished) makes any of these cats seem unlikely predators of *Ateles*. The small and medium-sized raptors, which certainly pose a threat to all of the smaller primates at Cocha Cashu, would be unable to take an adult spider monkey because of its size. Although an adult

Fig. 3. Typical daily path lengths of male and female *Ateles* in the early dry season at Cocha Cashu.

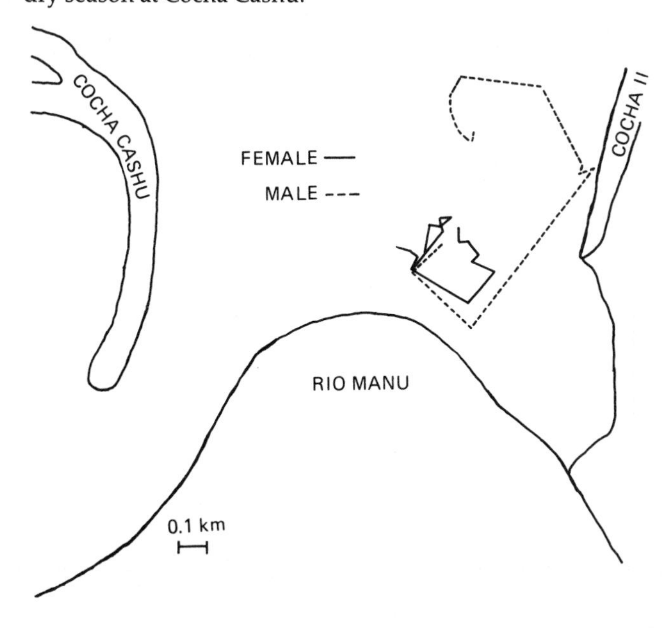

harpy eagle is probably capable of taking all but the largest adult spider monkeys (which may weigh up to 9 kg), evidence for their predation upon *Ateles* is also lacking. In two studies of nesting harpies done in British Guiana, in which a total of 79 prey items were recorded, *Ateles* was not among them (Fowler & Cope, 1964; Rettig, 1978).

With adults weighing 35–40 kg, chimpanzees escape predation as a result of their size even more definitively than does the spider monkey. Although leopards are often cited as potential chimpanzee predators, attempts to verify this through examination of scats has been unsuccessful. In conclusion, I feel it is safe to say that the predation pressure on *Ateles* and *Pan* is low, and, that in determining the overall pattern of fission–fusion sociality, obtaining safety from predators has not been a significant selective force. It could be the case, in fact, that a release from predation pressure is what has allowed once coherent groups to fragment and fission–fusion sociality to evolve.

In examining how individuals obtain access to mates within a fission–fusion society it is instructive to begin by looking at some

Fig. 4. Dry season diet composition of *Ateles* at Cocha Cashu.

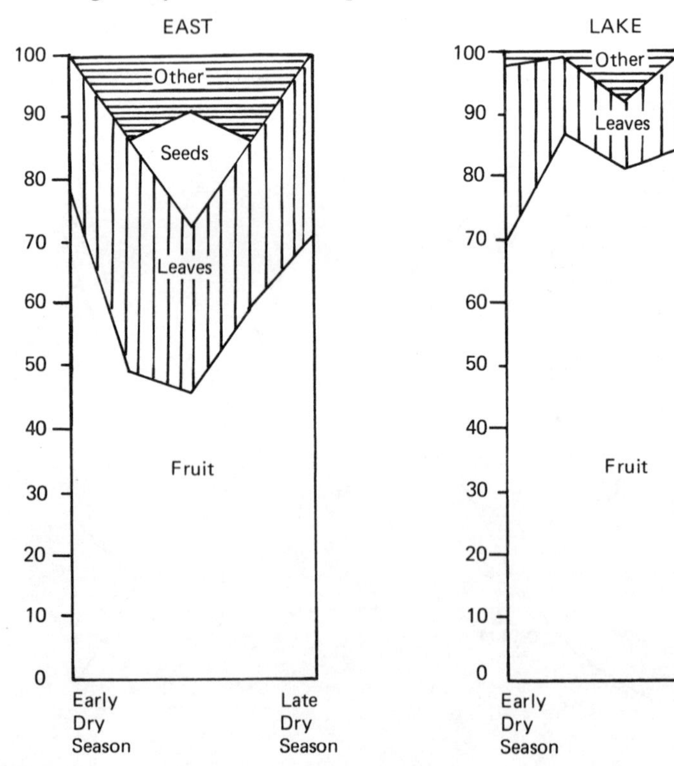

parameters of the female reproductive strategy in *Ateles* and *Pan* (Table 2). In both groups, the onset of reproductive maturity is late and the interbirth interval is long, resulting in a maximum lifetime reproductive output of about four offspring for both female chimpanzees and female spider monkeys. Although both advertise sexual receptivity, one through visual signals and the other through olfactory cues, female spider monkeys appear to be receptive for a much briefer period of time during each cycle. This may place the spider monkey male under particularly intense pressure to check females frequently for sexual readiness.

In both chimpanzees and spider monkeys, estrous females occur at such a low level in the environment and range so widely that the size of a territory one male could economically defend would be extremely unlikely to contain even one receptive female within its boundaries. Furthermore, female groups are not coherent; nomadic groups of females cannot be defended. Consequently, an alternative solution, to which both male chimpanzees and male spider monkeys seem to have arrived, is a type of female defense polygyny, employing male–male cooperation to defend a communal territory (see Fig. 1).

The conclusion that a great deal of male ranging behavior and segregation of the sexes in *Ateles* and *Pan troglodytes* is a result of low levels of female receptivity and consequent male searching behavior in order to obtain access to mates is borne out by some of the recent observations reported for pygmy chimps (*Pan paniscus*). Among bonobos, females are more frequently sexually receptive, and almost certainly as a result, parties are larger, more sexually integrated and display a greater degree of coherence (Kitamura, 1983). We might

Table 2. *Reproductive parameters of female* Ateles *and* Pan

Female reproductive strategy		
	Pan	*Ateles*
Age at maturity (years)	11	4.5–5
Age at first birth (years)	14	5–6
Interbirth interval (years)	5	2–3
Lifespan (years)	40–45	15
Duration of estrous (days)	9–10	2 (?)
Female advertisement	Yes (sexual swellings)	Yes (olfactory cues)

predict that when the data become available, little or no difference will exist between the ranging patterns of males and females.

The two selection pressures discussed so far, obtaining safety from predators and obtaining access to mates, are critical in molding a species' social organization. A pressure which is perhaps even more critical, particularly for females, is obtaining access to food and, indeed, maximizing feeding efficiency. The most obvious cost of foraging in a group, and one which may well be responsible for the often solitary nature of female chimpanzees and spider monkeys, is feeding competition. However, if being in a group carries no advantages for the location or exploitation of food sources, then when predation is low, individuals should prefer to forage solitarily at all times.

Since this is certainly not true of spider monkeys and does not seem to be the case for chimpanzees either, we must look for some possible advantages of group foraging. Being in a group has been hypothesized to augment feeding efficiency in three major ways: through female coalitions and cooperative defense of food sources, regulation of return times and information sharing (Cody, 1971; Ward & Zahavi, 1973; Wrangham, 1980; Clark & Mangel, 1984).

Because of the patchy and ephemeral nature of the fruit resources upon which both chimps and spider monkeys primarily depend, the benefits of information sharing could be quite substantial. I hope to explore further the role of information sharing in spider monkey foraging in a continuing project.

The fact that both chimpanzees and spider monkeys are primarily frugivorous tells us very little, however, about the underlying patterns of resource abundance and dispersion which are more directly tied to foraging tactics. Many interesting comparisons remain to be made concerning large-scale patterns of fruit availability in the forests of tropical Africa and South America. For instance, at all sites where detailed measurements have been made, fruit availability has been found to vary seasonally (Foster, 1982; Terborgh, 1983; Gautier-Hion, unpublished). Several strategies of dealing with this temporal variation in the availability of fruit seem to be held in common by chimpanzees and spider monkeys. Seasonal migration to different habitats, an increased reliance on lower-quality fibrous food sources such as leaves, and an increased exploitation of animal foods have been observed in different populations of both *Pan* and *Ateles* (Nishida, 1968).

As well as coping with the variable distribution of fruit in time, chimpanzees and spider monkeys must be able to cope with the variable distribution of fruit in space. A constant amount of food may be dispersed in a variety of ways. Some fruit sources may be very large but extremely dispersed in time and space, while others may be individually much smaller but relatively clumped. Because both chimpanzees and spider monkeys tend to specialize on one or two species when they are in season, we can use these species-specific patterns of distribution to make qualitative and quantitative predictions concerning the grouping, ranging and foraging strategies of chimpanzees and spider monkeys exploiting these fruit resources.

Conclusions

The fission–fusion social organizations found in spider monkeys and chimpanzees are strikingly similar. In both species, individuals travel in small non-permanent subgroups or parties. In both, the membership of these parties is drawn from a larger and more stable social unit that has come to be known as the community or unit-group among chimpanzee researchers. In both chimpanzees and spider monkeys, males participate in territorial border patrols and, in both groups, females are frequently solitary and exhibit few, if any, affiliative behaviors among themselves. In both, the mating system is best described as promiscuous. These similarities appear to be the result of convergent social evolution in response to similar selection pressures for obtaining safety from predators, obtaining access to mates and maximizing feeding efficiency.

The differences observed between the social organizations of these two genera can be very instructive as well. These differences can be found not only between chimpanzees and spider monkeys, but also between chimpanzees and pygmy chimpanzees, and even among different populations of chimpanzees and spider monkeys. These differences are primarily concerned with the size, coherence and composition of parties, but often involves variation in the number and intensity of affiliative behaviors observed within and between the sexes. These differences may well be the result of fine-grained variation in the temporal and spatial patterns of fruit abundance and dispersion and perhaps minor differences in male and female mating strategies.

Acknowledgements

The author would like to thank the Peruvian government, and in particular the General Directorate of Forestry and Fauna, for permission to work in the Manu National Park, and Princeton University for financial support during the course of the study.

References

Altmann, J. A. (1974) Observational study of behavior: sampling methods. *Behaviour*, **49**, 227–67

Cant, J. G. H. (1977) Ecology, locomotion, and social organization of spider monkeys (*Ateles geoffroyi*). Unpublished Ph.D. dissertation, University of California, Davis

Clark, C. W. & Mangel, M. (1984) Foraging and flocking strategies: information in an uncertain environment. *Am. Naturalist*, **123**, 626–41

Cody, M. L. (1971) Finch flocks in the Mohave desert. *Theoret. Pop. Biol.*, **2**, 142–58

Foster, R. B. (1982) The seasonal rhythm of fruitfall on Barro Colorado Island. In *The Ecology of a Tropical Forest*, ed. E. G. Leigh, A. R. Rand & D. M. Windsor, pp. 151–72. Washington, D.C.: Smithsonian Institution Press

Fowler, J. M. & Cope, J. B. (1964) Notes on the harpy eagle in British Guiana. *Auk*, **81**, 257–73

Goodall, J., Bandora, A., Bergman, E., *et al.* (1979) Intercommunity interactions in the chimpanzee population of the Gombe National Park. In *The Great Apes, Perspectives on Human Evolution*, Vol. V, ed. D. A. Hamburg & E. McCown. California: Benjamin/Cummings

Kitamura, K. (1983) Pygmy chimpanzee association patterns in ranging. *Primates*, **24**(1), 1–12

Klein, L. L. (1972) The ecology and social organization of the spider monkey, *Ateles belzebuth*. Unpublished Ph.D. dissertation, University of California, Berkeley

Nishida, T. (1968) The social group of wild chimpanzees in the Mahali Mountains. *Primates*, **9**, 167–224

Rettig, N. (1978) Breeding behavior of the harpy eagle (*Harpia harpyja*). *Auk*, **95**, 629–43

Reynolds, V. & Reynolds, F. (1965) Chimpanzees of the Budongo forest. In *Primate Behavior*, ed. I. DeVore, pp. 368–424. New York: Holt, Rinehart & Winston

Terborgh, J. (1983) *Five New World Primates: A Study in Comparative Ecology*. Princeton, N.J.: Princeton University Press

Ward, P. & Zahavi, A. (1973) The importance of certain assemblages of birds as 'information centres' for food finding. *Ibis*, **115**, 517–34

Wrangham, R. W. (1980) An ecological model of female bonded primate groups. *Behaviour*, **75** (3/4), 262–300

Wrangham, R. W. & Smuts, B. (1981) Sexual differences in the behavioral ecology of chimpanzees of the Gombe National Park, Tanzania. *J. Reprod. Fertil. Suppl.* **28**, 13–31

IV.4

Cooperative polyandry and helping behavior in saddle-backed tamarins (*Saguinus fuscicollis*)

A. W. GOLDIZEN AND J. TERBORGH

Introduction

Though originally considered to be among the few species of monogamous primates, evidence from recent field studies of several callitrichid species suggests that their social systems are more complex than was previously thought. Studies of *Saguinus oedipus*, *S. mystax*, *Cebuella pygmaea* and *Callithrix humeralifer* have shown that wild groups often contain more than one adult male and, less frequently, more than one adult female (Dawson, 1978; Neyman, 1978; Rylands, 1981; Soini, 1982*a*). In the only previous studies of marked tamarins, changes in group composition and transfers of individuals between groups were both more frequent than would be expected in completely monogamous societies (Dawson, 1978; Neyman, 1978; both with *Saguinus oedipus*).

This paper summarizes 5 years' findings on the social system of the saddle-backed tamarin (*Saguinus fuscicollis*). We have found that these tamarins live in remarkably varied social units. While some groups have had monogamous breeding pairs during some years, some groups have had cooperatively polyandrous mating systems, and others have had even more complex social systems, with multiple breeding females.

Methods

The study was carried out at the Cocha Cashu Biological Station in the Manu National Park of southeastern Peru. Beginning in 1979, groups were trapped approximately once each year. All individuals were weighed, measured, and marked with ballchain collars carrying unique combinations of colored beads. Over 2000 hours of behavioral and ecological observations were made on two completely habituated groups. The other marked groups were censused approximately once per month during periods when observers were at the field station. Further details on methods will be found in Terborgh & Wilson Goldizen (1985).

Results

General ecology

At Cocha Cashu, *S. fuscicollis* groups live in sharply defined territories of 0.5–1 km^2 in area, usually sharing their territory with a group of emperor tamarins (*S. imperator*). The two groups often move together around their territory, feeding in the same fruit trees. Neighboring mixed groups frequently meet at territorial boundaries. These encounters typically result in vocal (and occasionally physical) confrontations of several hours, in which the tamarins direct their aggression at conspecifics in the opposing group.

Both species are omnivorous, eating fruits, nectar, exudates, insects and occasionally small vertebrates. They concentrate most of their feeding on one or two species of fruit or nectar at a time, usually common small trees or vines, which ripen only a small number of fruits or flowers simultaneously, but continue producing over long periods of time (Terborgh, 1983).

Adult composition of groups

S. fuscicollis groups at Cocha Cashu have ranged in size from 2 to 10 individuals. The seven marked groups over all years since initial marking had the following compositions of adults (by troop-months): one female and more than one male (62%); one female and one male (17%); two females and more than one male (12%); two females and one male (6%); males only (3%). The composition of each group has changed over time as deaths, immigrations and emigrations have altered the number of adult members (Fig. 1).

Initial composition of new groups

Five new groups of three different adult compositions have formed in the study area since the study began. Two contained two adult males and one adult female; one included two adult males and two adult females, and two were 'bachelor' groups, each composed of two adult males. After the initial joining, both two-male/one-female groups had young within little more than the time of one gestation period. The two-male/two-female group was initiated by two dispersing nulliparous females from outside the study population who joined with two (of four) adult males from one of the marked groups. It is not known whether this group had young during the next year, but the two females and at least one of the males were still together 2 years after the group formed. One of the bachelor groups was observed for only a few days, but the other persisted for over 5 months (in a territory then without other tamarins).

Fig. 1. Adult composition of seven saddle-backed troops over 5 years. Double lines represent compositions observed at least once a month; single lines represent groups whose adult compositions did not change over periods when observers were absent from the study site; dotted lines represent unknown compositions during prolonged intervals when observers were unable to check the groups, usually during the observers' absences from the site; dashed lines indicate transfers between groups. Symbols: ♂ = adult male; ♀ = adult female; ♂, ♀ = animals that had reached adulthood within their natal groups; (+) indicates replacements of adult males and females; – = troop formation or extinction.

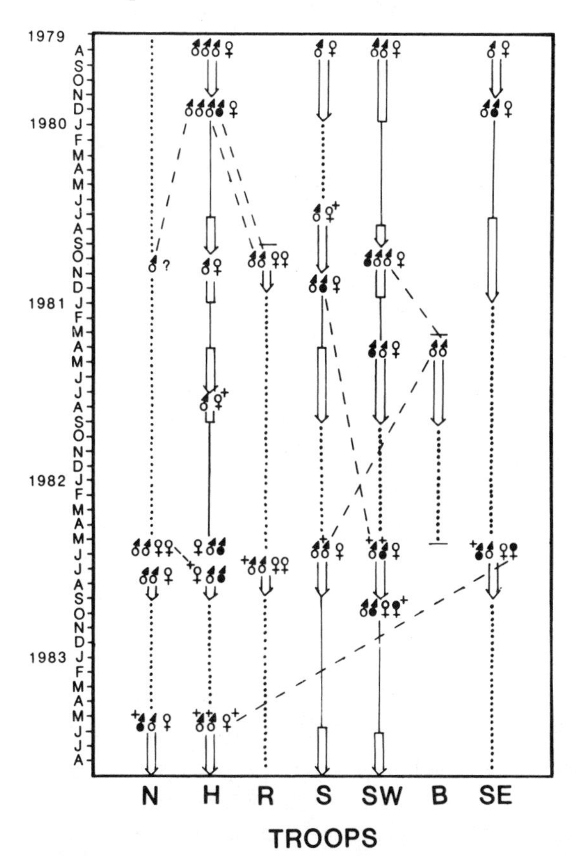

Intergroup transfers

Both sexes transfer between groups. Of 23 documented transfers, 15 were of males and 8 were of females. Two of these dispersal events are of particular interest. In one, an adult male joined a group already containing an adult pair who had bred together in the group raising litters of young for over 2 years. The second case involved a group with two adult females, both of which had had infants less than 6 months before the transfer. The evidence for this is that both females had extended nipples, and there were three juveniles of two different ages (both less than 6 months) in the group. When the female of a neighboring group disappeared, one of the two reproductive females transferred to that group, leaving her juvenile (or juveniles) behind in her old group.

Genetic relationships of group members

The identity of mothers is known for almost all of the infants born into marked groups since the start of the study. Nevertheless, only in 1984 have all of the individuals (seven) in one troop been known since birth. In some of the groups with multiple adult males, pairs of males may have been father and son, or full or half siblings, but in at least one such group, it is virtually certain that the two males were not closely related. One group simultaneously contained a mother and her adult daughter. Both females were observed to mate with the group's one adult male, but the daughter disappeared before observers returned to the study site 6 months later.

Mating behavior

The majority of copulations and conceptions in the study population occurred during April through June or July, a time when observers were rarely at the study site. However, two-male/one-female groups have twice been studied intensively during the mating season. In one group, 11 copulations were seen over 10 days, five by one male and six by the other, and in the second group, 24 copulations were seen over a 2-month period, 14 by one male and 10 by the other. Both times, the two males' copulations were spread quite evenly over the entire period, and the two males often copulated with the female on the same days. In both groups, each male was seen to copulate with the female in full view of the other male, without provoking any signs of aggression from the second male.

Though most of our observations of two-male groups have indicated no obvious differences in the access of the males to the female, an unusual behavior resembling a consort relationship was seen in two such groups, and in a one-male group. During these consortships, one male followed a female for 3–5 days, usually staying within a few meters of her, and copulating with her several times each day. The female appeared to go about her business as usual, playing no active role in maintaining the special relationship. In none of the three cases did any of the other individuals in the group seem interested in the 'consorting' pair. In the two-male group in which 24 copulations were seen, one of the males 'consorted' with the female for 3 consecutive observation days, accounting for 5 of his 14 copulations. During these 3 days he appeared not to allow the second male access to the female. During the rest of the 2-month period the second male was the female's nearest neighbor slightly more often than the first, and there were no signs of either male 'consorting' with the female. The second observed 'consort relationship' occurred between the younger male in a two-male group, and the group's one female. The younger male was only about 22 months old (though probably sexually mature) and still in his natal group. The third case occurred in a group that included one male, an older female and her adult daughter. The male consorted with the daughter, while the mother appeared indifferent. Due to disappearances of the females it is not known whether the second and third cases of 'consorting' resulted in conceptions, and the first case described only occurred 3 months before the writing of this paper.

Infant care

Groups have been studied intensively on five occasions during the period of infant dependence, and in every case all adult group members participated in all aspects of infant care other than lactation, including infant-carrying, 'babysitting' while the rest of the group fed, acting as 'sentinel' (watching for predators while others fed or hunted for insects), and feeding fruit and prey to infants recently weaned or in the process of being weaned. Group members 1 and 2 years old also helped with some aspects of infant care, though more variably.

Babies are carried during all group movements (often both twins on one individual) until they are about 2 months old. Fig. 2 shows the distribution of infant-carrying among the members of two troops. (The H troop had one infant; the SW troop had two.) The H troop had four adult males until the infant was 1.5–2 months old. Then, during a 3-week period, three of the adult males emigrated to other groups. Two yearling members of the group, who had not carried infants at all previously, did a substantial part of the infant-carrying after the males emigrated.

In two groups studied in detail during weaning of the infants and the months following, it was found that all group members, including yearlings, 'donate' food items (insects and fruits) to the infants. Until the infants are at least a few weeks past weaning, older individuals actively share the food items with the infants, either by holding the food while the infant eats it or by allowing the infant to take a food item when the adult could easily have retained it. Only after weaning do infants begin to appear to steal food items from older individuals over protest.

Discussion

Callitrichids have previously been thought to be monogamous (Eisenberg, Muckenhirn & Rudran, 1972; Epple, 1972; Kleiman, 1977), but findings from this study indicate that wild saddle-backed tamarins live in a remarkable variable social milieu. The study population has included monogamous groups, cooperatively polyandrous

Fig. 2. Infant-carrying in two saddle-backed tamarin troops, 1980. Instances of carrying were recorded at intervals ≥15 min, or when one or both infants switched to a new carrier. Between Oct. 20 and Nov. 5, three adult males (numbers 1, 3 and 4) emigrated from the H troop. The composition of the SW troop remained stable over the period.

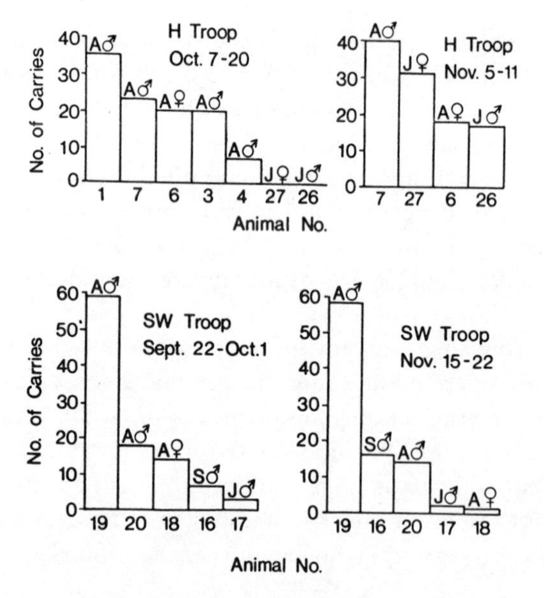

groups and groups with two simultaneously reproductive females. Data on group compositions of other callitrichid species suggest that they may have similarly flexible mating systems (Dawson, 1978; Neyman, 1978; Rylands, 1981; Soini, 1982*a*,*b*).

This study also provides the first data on contributions to infant care by non-parents in wild callitrichids. These data support results of captive studies showing that all group members carry infants and share food with them, with adult males often providing the majority of such care (Box, 1975; Epple, 1975; Ingram, 1977; Brown & Mack, 1978; Hoage, 1978). The extent of non-maternal infant care is substantially greater in callitrichids than in other primate species. Most callitrichid births are of twins, which together constitute from 14 to 25% of the mother's weight (Kleiman, 1977). In view of the large weight at birth and rapid growth of the infants, it is likely that maximum survival of the young is assured when the female devotes herself primarily to food gathering and lactation. Since the individual carrying the young does not feed, more than one carrier, in addition to the female, appears to be essential for successful reproduction. That this is true is supported by the fact that we have never seen a simple pair produce young in the wild.

The presence of supernumerary adults in *S. fuscicollis* groups, and in newly constituted groups in particular, can be readily understood if, in fact, a simple pair without helpers cannot successfully raise infants. Under these circumstances monogamy would occur only when a group included juveniles, subadults or grown offspring who would help with infant care. The only monogamous pairs to have raised young in our marked groups have indeed been accompanied by such younger helpers. Groups without nonreproductive helpers, such as newly formed pairs, typically accept another male as a second breeder and helper. By sharing the probability of fathering the young, and then helping to care for them, cooperating males may enhance their mutual reproductive success. This is a key point that we are as yet unable to support or reject due to insufficient sample size. The presence of occasional groups with two reproductive females emphasizes the variability of the social system. A full understanding of such groups will depend on knowing the degree of relatedness of the females.

Acknowledgements

We wish to thank our field assistants, too numerous to name individually, and express our gratitude to the General Directorate of Forestry and Fauna of the Peruvian Ministry of Agriculture for graciously authorizing our research at the Cocha Cashu Biological Station. This work was supported by NSF grants DEB76-09831, DEB79-04750, BNS79-15079 and BSR-8311782, the Wenner–Gren Foundation for Anthropological Research, the National Geographic Society, Grant no. 220 of the Joseph Henry Grant Fund of the National Academy of Sciences, three University of Michigan Block Grants, the Edwin Edwards Scholarship Fund of the University of Michigan, a Charles Tobach award from the T. C. Schneirla Research Fund, and a Grant-in-Aid of Research from Sigma Xi, the Scientific Research Society.

References

Box, H. O. (1975) A social developmental study of young monkeys (*Callithrix jacchus*) within a captive family group. *Primates*, **16**, 419–35

Brown, K. & Mack, D. S. (1978) Food sharing among captive *Leontopithecus rosalia*. *Folia Primatol.*, **29**, 268–90

Dawson, G. A. (1978) Composition and stability of social groups of the tamarin, *Saguinus oedipus geoffroyi*, in Panama: ecological and behavioral implications. In *The Biology and Conservation of the Callitrichidae*, ed. D. G. Kleiman, pp. 23–38. Washington, D.C.: Smithsonian Institution Press

Eisenberg, J. F., Muckenhirn, N. A. & Rudran, R. (1972) The relation between ecology and social structure in primates. *Science*, **176**, 863–74

Epple, G. (1972) Social behavior of laboratory groups of *S. fuscicollis*. In *Saving the Lion Marmoset, Proceedings of the WAPT Golden Lion Marmoset Conference*, ed. P. D. Bridgewater, pp. 50–8. Wheeling, N.C.: WAPT, Oglebay Park

Epple, G. (1975) Parental behavior in *Saguinus fuscicollis* spp. (Callitrichidae). *Folia Primatol.*, **24**, 221–38

Hoage, R. J. (1978) Parental care in *Leontopithecus rosalia rosalia*: sex and age differences in carrying behavior and the role of prior experience. In *The Biology and Conservation of the Callitrichidae*, ed. D. G. Kleiman, pp. 293–305. Washington, D.C.: Smithsonian Institution Press

Ingram, J. C. (1977) Interactions between parents and infants, and the development of independence in the common marmoset (*Callithrix jacchus*). *Anim. Behav.*, **25**, 811–27

Kleiman, D. C. (1977) Monogamy in mammals. *Q. Rev. Biol.*, **52**, 39–69

Neyman, P. F. (1978) Aspects of the ecology and social organization of free-ranging cotton-top tamarins (*Saguinus oedipus*) and the conservation status of the species. In *The Biology and Conservation of the Callitrichidae*, ed. D. G. Kleiman, pp. 39–72. Washington, D.C.: Smithsonian Institution Press

Rylands, A. B. (1981) Preliminary field observations on the marmoset, *Callithrix humeralifer intermedius* (Hershkovitz 1977) at Dardanelos, Rio Aripuana, Mato Grosso. *Primates*, **22**, 46–59

Soini, P. (1982*a*) *Distribucion Geografica y Ecologia Poblacional de* Saguinus mystax. Report to the Peruvian Ministry of Agriculture

Soini, P. (1982*b*) Ecology and population dynamics of the pygmy marmoset, *Cebuella pygmaea*. *Folia Primatol.*, **39**, 1–21

Terborgh, J. (1983) *Five New World Primates: A Study in Comparative Ecology*. Princeton, N.J.: Princeton University Press

Terborgh, J. & Wilson Goldizen, A. (1985) On the mating system of cooperatively breeding saddle-backed tamarins (*Saguinus fuscicollis*). *Behav. Ecol. Sociobiol.* **16**, 293–9

IV.5

The social systems of New World primates: an adaptionist view

J. TERBORGH

Introduction

The preceding papers in this section have ably illustrated the diversity of social systems to be found among the New World primates. Every known primate social system but the solitary is represented among the platyrrhines, a fact that enhances the appeal of this group for comparative studies.

Ten species of platyrrhines occur together in full ecological sympatry at Cocha Cashu in Peru's Manu National Park. Within this community are one or more species representing each of the known platyrrhine social systems (Table 1). Over the past decade habituated groups of each of the 10 species have been studied in some detail, and observations totalling over 1000 hours have been accumulated for seven of them. The exceptional diversity of species and the wealth of information that is available about most of them provide a data base that is unrivalled for a single locality.

The previous papers have provided some glimpses of the results of this integrated effort. My goal in this chapter is not to add to the detail, but to tie together several threads of argument that have been repeatedly stressed by my colleagues. These recurrent themes are: (1) the pervasive influence of predation on the Cocha Cashu primate community; (2) the effects of intra- and inter-group competition on individual feeding success; and (3) the characteristic use of small or large resource trees by different primate species. As I shall try to make clear, the central factor that unites these themes into a coherent framework is that of group size.

I shall begin by noting first that the typical group size of the 10 Cocha Cashu primates varies over a wide range of values, and secondly, that group size is closely correlated with a species' social system (Table 1).

The smallest groups are found in monogamous species, slightly larger groups are typical of the polyandrous callitrichids, and at the other extreme, large to very large groups virtually always contain multiple adult males. From this broad correlation one is led to ask whether group size is a cause or a consequence of a species' social system. Unfortunately, space limitations preclude a discussion of this important question (something I have done elsewhere, cf. Terborgh, 1983). Instead, I shall launch my argument directly by taking the position that for any species, given the spatio-temporal dispersion of its major food resources, there exists an optimum group size. This is the group size which maximizes the differential between the benefits of sociality and the costs. Once a particular optimum has been established by a species' ecological circumstances, its social system is essentially predetermined. Empirical support for this assertion will be provided in the remainder of this chapter.

I shall begin by discussing what appears to be the overriding benefit of primate sociality, increased safety from predation. I shall then present evidence relating to two of the principal costs of sociality, increased travel and increased within-group competition for access to feeding sites. Next, the presumed variation in these costs and benefits

Table 1. *Some characteristics of Cocha Cashu primates*[a]

Species	Approx. adult weight (kg)	Diet[b]	Group size[c]	Social system
Alouatta seniculus	8.0	F, L, U	5–8	Harem
Ateles paniscus	8.0	F, L, U, Ne	variable	Fission–fusion
Cebus apella	3.0	F, I, Nu, Ne, P, U	8–14	Multi-male
Cebus albifrons	2.8	F, I, Nu, Ne	8–20	Multi-male
Saimiri sciureus	0.9	F, I, Ne, Nu	30–55	Multi-male
Aotus trivirgatus	0.8	F, I, Ne	2–5	Monogamous
Callicebus moloch	0.8	F, L, U	2–5	Monogamous
Saguinus imperator	0.5	F, I, Ne	3–10	Polyandrous
Saguinus fuscicollis	0.4	F, I, Ne, S	3–10	Polyandrous
Cebuella pygmaea	0.1	S, I, Ne	3–10	Polyandrous

[a] Adapted from Terborgh (1983).
[b] Major components of diet listed in order of importance: F = ripe fruit; I = insects and other small prey; L = immature leaves; Ne = nectar; Nu = nuts; P = pith, meristem, etc.; S = sap; U = unripe fruit.
[c] Number of independently locomoting individuals in reproductive groups.

with group size will be represented in a simple graphical model which predicts different optimal group sizes under specified ecological conditions. Finally, by way of testing the model, I shall show how it accounts for some major patterns in the variation of social systems among New World and Old World primates.

Benefits of sociality

Sociality, that is, living in groups, carries with it a variety of concomitant costs and benefits, as judged from the point of view of individual fitness. Greater safety in the event of predator attack is, for example, one of the most widely accepted benefits of sociality. There are other possible benefits as well, such as access to the skills and knowledge of more experienced group members, access to a wider selection of potential mates, greater potential for the discovery of food resources, increased inclusive fitness derived from assisting relatives, and so forth, but in diurnal species under heavy threat of predation, group living offers far greater potential for survival to the age of reproduction than does a solitary existence, and hence enhanced survivorship seems to outweigh any possible advantages derived from these ancillary benefits.

Through the aggregate behavior of their members, groups can effectively foil the efforts of predators in a variety of ways. Simple 'safety in numbers' is the most elementary and universal effect. In the event of an attack, the probability that any particular individual is the target is $1/N$, where N is the number of individuals in the group. The derived benefit increases rapidly with group size at first, and then ever more slowly, but without limit. Much the same can be said of the other effects of groups in thwarting predators, such as enhanced vigilance, group defense, reciprocal warning, benefits derived from the operation of the convoy principle and the confusion of predators by multiple fleeing prey with intersecting pathways (Curio, 1976; Kiltie, 1981; Landeau & Terborgh, in press). No matter which of these mechanisms may prove the most relevant to the situation of any wild primate group, the tendency will be to favor individuals that choose larger groups over smaller groups.

How relevant are these theoretical arguments to the lives of wild primates in general, and to the primates of Cocha Cashu in particular? This is a difficult question to answer directly because successful predation is so rare; at most it can occur once in the lifetime of an individual prey. For the sake of argument, consider a typical *Cebus apella* group of 12 members, including three or four adult females.

These females, on average, will produce about two young per year. In a stable population, births will be balanced by an equivalent mortality of two individuals per year. Now, suppose that half the mortality in this population is due to predation, i.e. one individual per troop per year dies in the talons of an eagle. Given that eagles hunt only by day (discounting any possible nocturnal predation), the death rate due to predation is 1 per 4380 daylight hours per *Cebus apella* troop. How many primate studies have logged over 4000 contact hours? No wonder that few primatologists have ever seen predation!

Although the sample size is small, the evidence from Cocha Cashu suggests that predation is a highly significant mortality factor for most primate species. In 1976–1977 a team of four observers logged over 2700 contact hours with troops of five species. In this time a small hawk (*Accipiter bicolor*) killed a tamarin (*Saguinus fuscicollis*) in front of our eyes, and a healthy adult male capuchin (*Cebus apella*) disappeared one afternoon while its group was foraging in a 2–3 ha area. A few days

Table 2. *Potential primate predators in the raptor community at Cocha Cashu*[a]

Raptor species[b]	Weight (g)[c]	Reported diet[c]	Attacks observed at Cocha Cashu[d]
Harpia harpyja (harpy eagle)	4000–?	Sloths, monkeys, tamanduas, curassows	*Cebus, Alouatta, Saimiri*
Morphnus guianensis (Guianan crested eagle)	?–1750	Monkeys, marsupials	*Cebus, Saimiri, Callicebus*
Spizaetus ornatus (ornate hawk-eagle)	900–1600	Large birds, kinkajous	*Saimiri, Callicebus, Saguinus* spp.
Spizaetus tyrannus (black hawk-eagle)	900–1100	Opossums	None observed
Spizastur melanoleucos (black-and-white hawk eagle)	750–?	Birds, mammals	None observed
Leucopternis schistacea (slate-colored hawk)	?–1000	Snakes, frogs	*Saimiri*
Accipiter bicolor (bicolored hawk)	200–450	Birds	*Saguinus*, squirrel

[a] Adapted from Terborgh (1983).
[b] Listed in descending order of size.
[c] Weights and reported diets from Brown & Amadon (1968) and Haverschmidt (1968). Weights for males on left, females on right.
[d] From observations of the author, C. Janson, R. Kiltie, P. Wright and others.

later we found its remains. In this case, predation was suspected but not proven. In subsequent years at Cocha Cashu, observers have reported numerous attacks on primates by several raptor species (Table 2), and two additional eye-witness kills. Over the same period we have found approximately five fresh carcasses of monkeys that had apparently died of natural causes. Although the sample sizes are pitifully meagre, the hazy picture that emerges is consistent with the notion that predation accounts for something like half the primate mortality at Cocha Cashu. For the sake of my argument, however, it doesn't matter whether the actual figure is as low as a quarter or as high as three-quarters; in any case, predation stands as a major cause of mortality.

The issue can be approached from another point of view by looking at the predators rather than at the prey. Here we gain an important advantage in time, because the predator has to kill every day, or at least every few days. Unfortunately, few of the larger South American raptors have been studied in any detail. There is, however, some data on the prey of *Harpia harpyja*, the largest eagle in the hemisphere (Table 3). Of 85 prey items identified at three nests, 40% were primates (75% of these being capuchins, *Cebus* spp.). Harpy eagles are reliably (though rarely) reported to take adult howlers (*Alouatta* spp.), but I

Table 3. *Identity of 85 prey items brought to three harpy eagle (*Harpia harpyja*) nests in Guyana*[a]

Prey species	Estimated weight (kg)	Number
Didelphis marsupialis (opossum)	3.5	3
Alouatta seniculus (howler)	8.0	1
Cebus sp. (capuchin)	2.5	28
Chiropotes satanus (bearded saki)	1.0	2
Pithecia pithecia (white-faced saki)	1.3	2
Bradypus tridactylus (three-toed sloth)	3.5	6
Cholepus didactylus (two-toed sloth)	5.0	21
Tamandua tetradactyla (tamandua)	4.0	1
Coendou prehensilis (porcupine)	2.5	4
Dasyprocta agouti (agouti)	3.0	8
Bassaricyon beddarti (olingo)	1.0	1
Nasua nasua (coati)	4.5	2
Potus flavus (kinkajou)	2.0	4
Eira barbara (tyra)	1.8	1
Mazama americana (brocket deer)	5.0	1

[a] Data from Fowler & Cope (1964) and Rettig (1978).

have not heard even one report of an attack on spider monkeys (*Ateles* spp.: see McFarland, Chapter IV.3). Spider monkeys, by virtue of their size, strength and agility, may be relatively free of predation, a point that shall gain prominence in arguments to be presented later.

Costs of sociality

I mentioned above that the benefits of group living, in the form of increased protection from predators, tend to increase indefinitely with group size. Yet, clearly, the sizes of wild primate groups vary widely and, for many species, are constrained within quite narrow limits. The arguments developed so far thus do not serve to define an optimum group size, much less to explain between-species variation in the typical size of groups. To understand these aspects of the grouping behavior of primates we must examine the costs of sociality.

Of the possible costs of sociality that are relevant to Cocha Cashu primates, two seem particularly important. These I shall describe as a locomotory cost and a feeding cost. There is, in addition, in the case of *Alouatta seniculus*, a possible reproductive cost paid following group take-overs by infanticidal males (Rudran, 1978). Since the infanticidal behavior of males seems clearly the consequence of a particular social system, rather than a cause, I shall not mention it further.

Individuals that live in groups must pay a cost in increased locomotion. This is because a group must generally travel farther than a solitary individual in order to satisfy the hunger of its members. Waser (1977) convincingly demonstrated this tendency in his comparison of the daily travel distances of small and large troops of white-cheeked mangabeys (*Cercocebus albigena*). Similarly, at Cocha Cashu we found that when brown capuchin (*Cebus apella*) groups were accompanied by squirrel monkeys (*Saimiri sciureus*) they travelled about 50% farther per hour than when unaccompanied (Terborgh, 1983). The presence of squirrel monkeys roughly doubled the biomass of the feeding group, so the effect of adding mouths resulted in less than a proportional, though nevertheless appreciable, increase in travel.

When a group is augmented by additional members, the cost of any required extra locomotion is borne equally by all group members. Such equidistribution of burdens does not, however, typify the second major cost of sociality, which is the denial of access to preferred feeding sites via intragroup competition. Here, the costs paid by dominant or favored individuals, as so elegantly shown by Janson (Chapter IV.2), are minimal, while those paid by subordinate individuals may be unacceptably high, leading to emigration. Where high-quality feeding sites are few or spatially restricted, only a small number of individuals can feed. This is a powerful mechanism for limiting group size. It may well be the most powerful mechanism.

If intragroup competition for feeding sites were to limit the sizes of primate groups, we could expect to find that species which typically live in large groups would mainly use large food resources (capable of accommodating many individuals), while species which live in small groups would be able to use resources of almost any size. Now, let us see whether these expectations are upheld by data gathered at Cocha Cashu (Table 4). Looking first at the size distribution of tree crowns in the forest, it is clear that all eight primate species for which data are available are selecting trees that are larger than average. Nevertheless, there are major differences between the eight species, although the overall pattern is not as simple as the above prediction would have it. In general, the species that live in larger groups (≥10 individuals) use larger trees (mean crown diameter 18 m) than those living in smaller groups (mean crown diameter 11 m), and the species with the largest group size (*Saimiri sciureus*) shows the greatest tendency to feed in

Table 4. *Distribution of feeding time of Cocha Cashu primates in trees of the indicated crown diameter (% feeding time)*

	% Feeding time in trees of crown diameter				
	0–10 m	11–20 m	21–30 m	>30 m	Weighted mean (m)
Ateles paniscus[a]	11	34	33	21	22
Cebus apella[b]	50	33	11	5	13
Cebus albifrons[b]	40	20	37	4	16
Saimiri sciureus[b]	26	23	22	29	21
Aotus trivirgatus[c]	15	52	33		17
Callicebus moloch[c]	78	19	3		8
Saguinus imperator[b]	73	27			9
Saguinus fuscicollis[b]	79	20	1		8
Trees[d]	96	3	<1	≤1	6

[a] From McFarland (Chapter IV.3).
[b] From Terborgh (1983).
[c] From Wright (Chapter IV.1).
[d] Distribution of crown diameters in a sample of trees, ≥10 cm DBH (diameter at breast height). (Unpublished data courtesy of P. Wright.)

trees of the largest crown-diameter class (>30 m). However, the trend is not uniform, and there are some apparent exceptions. *Ateles* feeds mainly in canopy trees in the middle and large diameter classes, but having a fission–fusion social system the groups are of no fixed size. Even so, McFarland (Chapter IV.3) has shown that large aggregations of *Ateles* occur only in large trees, indicating a positive trend intra-specifically. The two capuchins spend more time than might be expected in small trees, but this is due almost entirely to a heavy seasonal use of palms (see Janson: Chapter IV.2). At other seasons they are attracted to trees in the largest diameter class. This is not true of *Callicebus moloch* and the two *Saguinus* species, none of which uses large trees to any appreciable extent. Here we find a contradiction, for our expectation was that species living in small groups would not be restricted as to the sizes of trees they used. The apparent explanation for this anomaly, as suggested in observations on *C. moloch* made by Wright (Chapter IV.1), is that the smaller monkeys are aggressively excluded from big fruit trees by larger monkeys. This conclusion also seems to apply to *Saguinus* spp., as Goldizen and I (unpublished data) have seen tamarins being chased out of feeding trees on numerous occasions, especially by *Cebus* spp. We are now left with *Aotus* and *Saimiri* as apparent exceptions to this rule, in that they use a considerable number of trees of large stature, regardless of their small body size. *Aotus* is able to do so by virtue of its nocturnal habit which enables it to avoid entirely the interference of other monkeys (Wright: Chapter IV.1). Finally, *Saimiri*, although small, gains access to the largest trees through the effect of overwhelming numbers that foil the efforts of larger monkeys to expel them (Terborgh, 1983).

Out of the details of the individual cases emerges a reasonably coherent picture. Large monkeys and/or species travelling in large groups can compete successfully for the richest resource trees. Small monkeys that do not travel in large groups are aggressively excluded from these rich resources, and are thereby obliged to feed in small trees. The two exceptions to a monotonic trend of increasing use of large trees with increasing group size, *Ateles* and *Aotus*, are both explained by special circumstances.

The costs and benefits of sociality: a model

We are now ready to combine the conclusions of the two previous sections regarding the value of groups in deterring predation and the effect of intragroup competition in depressing the feeding success of subordinate individuals. In addition, we need to take note of the empirical observation presented above, that primate species differ in the size of the resources they use (represented by mean crown diameter of feeding trees), and that the interspecific differences are positively correlated with variation in group size. The interaction of these relations is easily represented in a simple graphical model (Fig. 1). In this model, group size is the dependent evolutionary variable through which compromise is achieved between opposing tendencies to increase security and to maximize feeding success. The characteristic size of the resources used by any species is assumed to be the independent variable, this, in turn, being determined by the nature of the species' diet. It should be understood that the use of large resources does not in any sense cause or require the evolution of large groups; it is strictly a permissive condition. The overriding selective force that promotes social cohesion is increased safety from predation. Where feeding conditions permit the aggregation of many individuals without undue strife, large groups will be favored; where feeding conditions mitigate against aggregation, small groups will be the rule,

Fig. 1. Model for the optimization of group size in primates via an evolutionary compromise between benefits derived from enhanced safety from predators and costs in decreased access to preferred feeding sites as a function of increasing group size. The dotted lines labelled S, M and L correspond, respectively, to the group sizes that maximize the differences between benefits and costs of sociality (technically, in fitness units) for species feeding on small, medium and large resources (see text for further details). Modified from Terborgh (1983). A somewhat simpler version of this model has been proposed by van Schaik & van Hooff (1983).

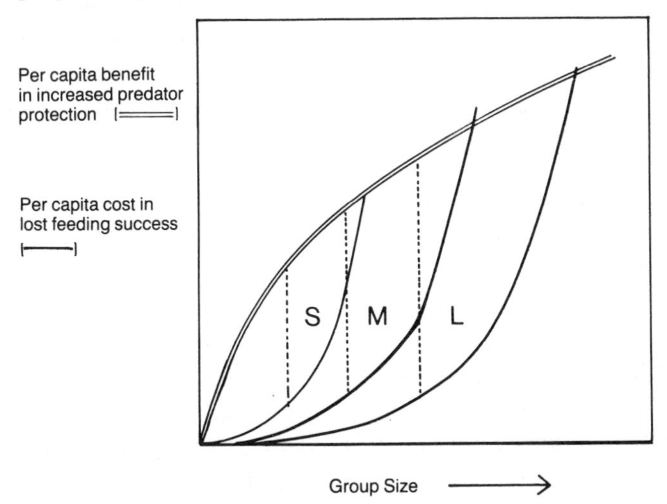

and protection from predators will be achieved through some other means, such as crypsis. In either case, the result is a compromise that maximizes the difference between the benefits and costs of the evolved group's size. Consequently, the compromise is such as to maximize reproductive success without maximizing either security from predators nor individual feeding success.

Tests of the model

The model is amenable to some straightforward tests. First, one can relax the assumption of predation and ask about species that are free of predators, at least as adults. In general these are animals of large body size that are invulnerable to the top predators in their communities. One immediately thinks of rhinoceroses and elephants, animals that live solitarily or in fission–fusion societies. The same can be said of primates. The largest arboreal species in each of the major tropical regions have fission–fusion social systems. These are *Ateles*, *Brachyteles* and *Lagothrix* (South America), *Pan* (Africa) and *Pongo* (Asia). I have found no records of raptors taking any of these species (see McFarland, Chapter IV.3, for a more detailed discussion of *Ateles*), and being arboreal, they are inaccessible to terrestrial predators such as cats. Decreased predation thus appears to lead to small and indeterminate group sizes, as would be expected from the model, as the groups would be moulded primarily by the available feeding opportunities (Wrangham, 1977).

A second test can be found in situations offering reduced intraspecific competition for feeding sites. Here large groups of indefinite size are expected, and one finds them in such places as the open ocean (plankton-feeding fish), mudflats (shorebirds) and grassy plains (ungulates). All these are relatively featureless environments in which food is distributed evenly rather than in discrete patches (e.g. trees). Among primates, terrestrial species typically form larger groups than their arboreal relatives, a correlation that has often been attributed to greater risks of predation away from the cover of trees (Clutton-Brock & Harvey, 1977). While concurring that the risk of predation is likely to be greater in the open, I would also point to release from feeding competition as a contributory factor in the larger group sizes of terrestrial species, as in the feeding of baboons on a lawn of grass corms (Altmann & Altmann, 1970).

Both of the above tests, while admittedly very general, seem to uphold the model. Hopefully, the reader will be encouraged to think of additional tests to satisfy his curiosity in the matter.

Group size and primate social systems

I come now to the final point I wish to make, which is to clarify the relations between group size and primate social systems. To examine this question in detail would require far more than the space allotted, so I shall perforce be limited to pointing out that group size, once fixed by evolution, virtually determines the social system (Fig. 2).

At most group sizes, there is only one social system known to occur. It is mainly in the range of 5–25 individuals that more than one possibility is open. At numbers between 5 and 10 one finds both polyandrous and harem systems, and between 10 and 25 there is overlap between the single and multi-male systems. The line between the latter two is hazily drawn in any case, as both types of organization can sometimes occur within a single species (Oppenheimer, 1977; Curtin, 1980). Why there is so much overlap between the single and multi-male systems is an interesting question that merits special attention (e.g. van Schaik & van Hooff, 1983). Even within a single basic type of social organization there may be many variations of detail, as was, for example, so clearly shown by Janson (Chapter IV.2) in his comparison of two capuchin species. While intragroup competition for food resources does relate to the differences between the

Fig. 2. Primate social systems *vs* group size. Horizontal bars indicate ranges of group sizes for the corresponding social systems. Genera representing each social system are listed above the bars.

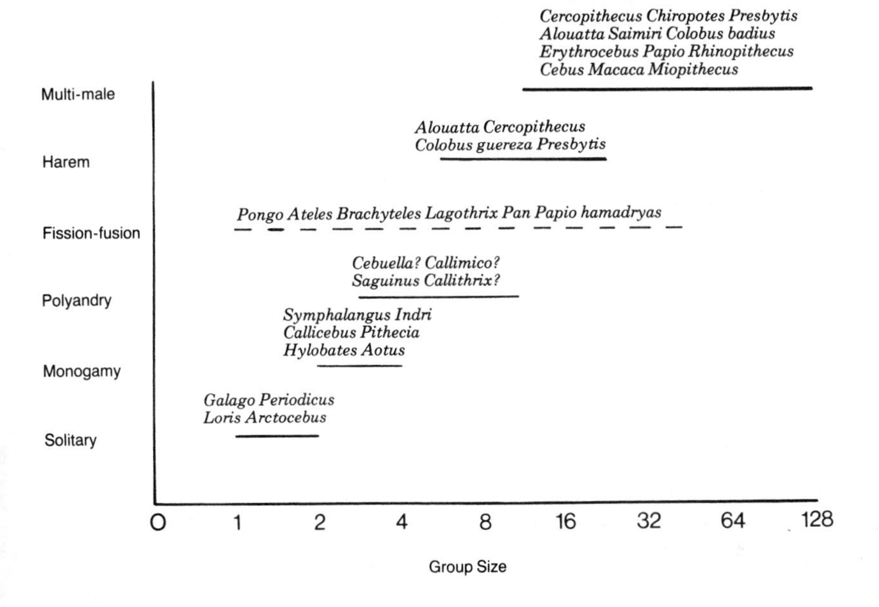

social systems of the two capuchins, the model presented above is not, in general, intended to address the question of the evolution of social systems at so fine a level.

The model makes predictions only about the evolution of group size, and rests on the assumption that natural selection acts primarily to optimize mean group size, and only secondarily on the social system itself. Once group size has been fixed by the interaction of a species with its external environment, additional factors, such as a species' body size and demography, as well as socially mediated forms of selection, such as sexual selection and kin selection, will assume further roles in moulding the details of particular social systems. Thus, understanding the fine details of social systems will surely prove complex, but at the broadest level, the problem of interspecific variation in social systems, when approached from the point of view of optimization of group size, can be seen to be relatively simple.

Acknowledgements

I would first like to thank my colleagues in this symposium for so generously sharing their views and their data. I would also like to acknowledge my deep gratitude to all the Manu monkey watchers, for the ideas expressed here evolved gradually over a number of years while receiving frequent stimulus from conversations with my many co-workers at the Cocha Cashu field station. I wish also to thank the General Directorate of Forestry and Fauna, Peruvian Ministry of Agriculture and Officials of the Manu Park for their continuing interest in our research and timely issuance of authorizations. The National Geographic Society and the National Science Foundation (DEB 76-09831) are thanked for their financial support.

References

Altmann, S. A. & Altmann, J. (1970) *Baboon Ecology: African Field Research.* Chicago: University of Chicago Press

Brown, L. & Amadon, D. (1968) *Eagles, Hawks and Falcons of the World.* New York: McGraw-Hill

Clutton-Brock, T. H. & Harvey, P. H. (1977) Primate ecology and social organisation. *J. Zool. (Lond.)*, **183**, 1–39

Curio, E. (1976) *The Ethology of Predation.* New York: Springer Verlag

Curtin, S. H. (1980) Dusky and banded leaf monkeys. In *Malayan Forest Primates: Ten Years' Study in Tropical Rain Forest*, ed. D. J. Chivers, pp. 107–45. New York: Plenum Press

Fowler, J. M. & Cope, J. B. (1964) Notes on the harpy eagle in British Guiana. *Auk*, **81**, 257–73

Haverschmidt, F. (1968) *Birds of Surinam.* Edinburgh: Oliver & Boyd

Kiltie, R. A. (1981) Application of search theory to the analysis of prey aggregation as an antipredation tactic. *J. Theoret. Biol.*, **87**, 201–6

Landeau, L. & Terborgh, J. (in press) Oddity and the confusion effect in predation. *Anim. Behav.*

Oppenheimer, J. R. (1977) *Presbytis entellus*, the Hanuman langur. In *Primate Conservation*, ed. Prince Rainier & G. H. Bourne, pp. 469–512. New York: Academic Press

Rettig, N. L. (1978) Breeding behavior of the harpy eagle (*Harpia harpyja*). *Auk*, **95**, 629–43

Rudran, R. (1978) The demography and social mobility of a red howler (*Alouatta seniculus*) population in Venezuela. In *Vertebrate Ecology in the Northern Neotropics*, ed. J. F. Eisenberg, pp. 107–26. Washington, D.C.: Smithsonian Institution Press

Schaik, C. P. van & Hoof, J. A. R. A. M. van (1983) On the ultimate causes of primate social systems. *Behaviour*, **85**, 91–117

Terborgh, J. (1983) *Five New World Primates: A Study in Comparative Ecology*. Princeton, N.J.: Princeton University Press

Waser, P. M. (1977) Feeding, ranging and group size in the mangabey *Cercocebus albigena*. In *Primate Ecology: Studies of Feeding and Ranging Behaviour in Lemurs, Monkeys and Apes*, ed. T. H. Clutton-Brock. London: Academic Press

Wrangham, R. W. (1977) Feeding behavior of chimpanzees in Gombe National Park, Tanzania. In *Primate Ecology: Studies of Feeding and Ranging Behaviour in Lemurs, Monkeys and Apes*, ed. T. H. Clutton-Brock, pp. 503–38. London: Academic Press

Part V

Primate–Human Conflict

Editors' introduction

Human population expansion and the resulting increase in land requirements has led to rising pressure on the natural habitats of wildlife. One has only to look at the paucity of wildlife in most developed countries to realise just how important and fundamental this interaction is to any comprehensive conservation programme. So long as there is a genuine need for more land, whether for agriculture, housing or industry, this requirement must be met. The welfare of a country's people will always be considered before that of its wildlife, a fact which must be given serious consideration when launching new conservation efforts.

Closely related to these pressures are the problems encountered at the human–wildlife interface. Wildlife in these areas must be able to exist in an altered environment, while humans must learn the importance of the long-term goals of environmental conservation and whenever possible, adjust their practices to help ensure co-existence with wildlife. Primates, because of their ingenuity and adaptability, are of special concern at this interface, and conflict between humans and primates forms the basis of this section.

In the first paper, Strum (Chapter V.1) presents an excellent review of how baboons adapted to increasing human settlement in their natural range over a period of years. The next two studies examine how primates can alter their behaviour and habits when in prolonged contact with humans and new foods. Forthman Quick (Chapter V.2) approaches this issue by comparing activity budgets and ranging patterns of the baboons observed in Chapter V.1. Lee *et al.* (Chapter V.3) use the same approach in a study of the difficulties caused by vervet monkeys exploiting human foods around a tourist lodge.

Specific management solutions are suggested. The remaining two papers address the problem by examining the habits of the people living in remote areas in close proximity to primates. Eudey (Chapter V.4) looks at how the loss of natural habitats as a result of shifting cultivation adversely affects primate populations. Mitchell and Tilson (Chapter V.5) examine traditional hunting practices and rituals, and how these affect the populations of primates and other wildlife.

Such studies provide the type of information on the primates themselves that must be taken into account when setting up management schemes. Of equal importance is the need for understanding human requirements and the participation of local communities in such endeavours.

V.1

A role for long-term primate field research in source countries

S. C. STRUM

Introduction

Long-term studies of wild primates rely on the hospitality of source countries. Currently, these Third World nations are shifting towards a philosophy and policy of sustainable development (McNeely & Miller, 1984; Strum: Chapter VII.3). Given this orientation, it is understandable that host nations are beginning to question the contribution that long-term primate research makes to sustaining their society.

This paper considers the role of long-term research in source countries by focusing on Kenya as a case study. Historical events help to explain the past lack of utility of long-term field studies. A brief summary of recent activities of one long-term project, the Gilgil Baboon Project, illustrates the potential usefulness of such research.

Kenya history

Kenya has been the focus of primate research since the 1950s. Currently there are several long-term primate studies, including the Gilgil Baboon Project where I have worked since 1972 and have been Director since 1975. Historically, these long-term projects have played no role in Kenya. There are two reasons why this is so.

First, during most of the last three decades of research in Kenya, government agencies did not actively guide primate research. In 1975, Sessional Paper no. 3 (Republic of Kenya, 1975) on the future of wildlife management policy in Kenya clearly stated the objectives of research both for the Wildlife Service and for non-governmental individuals. This position paper also recognized that past research had limited usefulness, primarily because of the absence of clearly stated management questions which research could address. To ameliorate the situation, future priorities were to be established by the National Council on Science and Technology, and wildlife research was to identify and play a role in important issues in management and utilization. All agencies were to help with the coordination of research and the 'inducement' of non-government research into priority areas.

The authors of Sessional Paper no. 3 were quick to point out that 'past failures to plan adequately [and to set priorities — my addition] have not been due to lack of interest, knowledge, or concern by the relevant departments but rather to a lack of sufficient personnel, both to administer day-to-day departmental activities and to consider the future'.

So, for a number of reasons, source country priorities were not filtering through to primate research in Kenya. But scientists, too, played a role. Many had 'ivory tower' attitudes. This was particularly true in primate research since primates were seldom regarded as 'wildlife'. Thus research projects rarely included the more traditional 'wildlife' foci such as management. Additionally, for primarily historical reasons, applied research was sometimes seen as less reputable than 'basic' research. Equally important, most Kenyan *long-term* studies were seldom planned, they just 'happened' as a result of the convergence of good opportunities, the availability of money, the difficulty of research elsewhere (e.g. because of political problems in other source countries). Under these circumstances, what began as short-term research often ended up as a long-term project.

In the past, benign neglect on the part of the government and research priorities set by scientists both contributed to the lack of interaction between a source country's needs and primate field research. But recent years have witnessed 'rapid changes in institutions, increasing population, and development of other activities potentially in conflict with wildlife' (Sessional Paper no. 3) even within the Parks and Reserves. The magnitude of these problems cannot be ignored. The Kenya Government has become more concerned with the nature of research and is more active in setting priorities for research within the country. Scientists are recognizing that in order to continue their study of a specific primate population they will have to confront the problems these animals face in the changing realities of host countries.

Current and future utility

One of the main issues involving primates in source countries is the frequency with which primates become *pests* when near humans. This happens in wildlife refuges, around lodges, along roads, and on the perimeter of Parks where primates easily cross barriers to visit surrounding human settlements. As agriculture expands and forests are utilized, primate habitats suffer. More and more, human and nonhuman primates come into direct contact and direct competition.

The management of primate populations and individual pests poses special difficulties because of the basic nature of primates. The niches of human and nonhuman primates overlap extensively and the two types of primates are more closely matched in competition than is true of the human interaction with other taxa. Primate intelligence, manual dexterity and the opportunism of many nonhuman primate species make containment and control costly, inefficient, and sometimes ultimately ineffective even when the problem is solved in the short term.

This is a general problem in Kenya, one whose magnitude and importance was specifically brought home to me. The Gilgil baboons found themselves in the midst of a new reality in 1976. The private 45 000 acre cattle ranch on which they ranged was sold to an agricultural cooperative in that year. The conflict that developed between the agriculturalists and the wildlife threatened the future of both the baboons and the Baboon Project and forced the research team to reassess Project priorities. Settlement began slowly in 1977, and increased in 1978 and 1979. By August of 1979, the first serious altercations between baboons and farmers over crops occurred. Many months and many baboon deaths later, we initiated a research project that specifically focused on the development of crop-raiding among the Gilgil population (Strum, 1984).

The goals of this research project were augmented as the project proceeded. The primary goal was to find a way for the baboons and the farmers to coexist. To do this we needed to understand crop-raiding, its costs and benefits, from the baboons' perspective and whether we could manipulate the costs or the benefits in innovative ways to bring the problem under control.

It was soon obvious that the animals were only part of the problem and therefore could only be part of the solution. Since people played an active role, we initiated a rural development scheme whose goal was to explore the multiple ways in which the land could be appropriately used.

Finally, as the project grew and we had some major successes, we hoped that the research could contribute to the development of a model for the control of baboon pests elsewhere in Kenya, as well as offer suggestions for rural development in agriculturally marginal areas.

There were several advantages of the Baboon Project's *long-term status* in the conception and execution of the crop-raiding project. First, because we already knew the animals and had information on their behavior before settlement occurred, we were able to study the problem, in depth, from the animals' point of view. We could watch the behavior develop, document the changes and then do experiments. The same project on unstudied animals would be much more difficult, if not impossible. Certainly the quantity and quality of the data would be much reduced if the history of animals was not known. Secondly, we had a palpable physical presence in the area and came to be seen as an arm of the Wildlife Department, by both sides. We handled on-site problems for the Wildlife Department and helped organize government aid and assistance for the farmers. We could function in this way because our longevity gave us both some past experience and a certain status.

The project also played a role in local education, both directly and by facilitating the activities of other agencies. As project activities expanded, more personnel was needed. A cadre of Kenyan assistants was trained to help in various aspects of the project. In the end, the actual research, the exploration and implementation of policy, and our role in education and training were all possible because we were a long-term field research project with demonstrated expertise and a commitment to the animals and the area.

Progress

The crop-raiding project completed its final season in October, 1984, and good progress has been made in both the research and in the rural development scheme. An interim analysis of data provides some important preliminary conclusions (Strum, 1984; Oyaro & Strum, 1984; Musau & Strum, 1984; Forthman-Quick, 1984 and Chapter V.2).

1. Even with the spread of agriculture to traditional primate areas, crop-raiding can be prevented or contained IF the animals still have a reasonable home range refuge in adjoining nonagricultural regions and if certain procedures are followed. Crop-raiding is not inevitable.
2. As the areas with natural foods diminish, crops assume a greater nutritional importance and crop-raiding is increasingly difficult to control.
3. Among the most promising of the new techniques of control is

taste aversion conditioning. Forthman-Quick, working at Gilgil (1984 and Chapter V.2), demonstrated that aversions can be created which influence the willingness of a baboon to eat the 'poisoned' food. Existing chemicals have certain deficiencies and we are currently exploring the possibility of a new emetic which is more effective and easier to use.

The results of the research were of immediate value in helping farmers safeguard their crops.

Our educational activities changed farmers' attitudes towards wildlife; Project assistance to farmers on appropriate agricultural techniques, water and soil conservation, and the like, has created a more productive basis for development of this rural area. Perhaps the most important and successful part, from the farmers' point of view, was the assistance we gave to help broaden their economic base. Kekopey was not really suitable for agriculture; a good harvest might occur every 4–5 years. We initiated a Woolcraft Project which is now in its fourth year and involves individuals from many local families. With a growing market for woolcrafts, the potential yield to any one family from the wool industry far surpasses the returns they might hope to get from agriculture there.

Under the original plan for Kekopey (which has recently changed), half the 45000 acres was to be agricultural and half ranching. Our progress from 1980 through 1983 demonstrated that there could be a workable compromise permitting people and wildlife, including baboons, to coexist.

The Gilgil Baboon Project has tackled a serious problem facing not only the Kekopey farmers but farmers in other parts of the country. Our successes provide a useful demonstration of the utility of long-term primate field studies to source country goals. The results of our research can help in formulating a model for the development of similar rural areas only marginally productive under agriculture and could provide important management recommendations. In retrospect it is clear that much of our success is a direct consequence of the long-term aspects of the Gilgil Baboon Project: understanding baboon behavior, having the right personnel and having a commitment to the animals and the place.

Conclusions

The lessons of the Gilgil case history can be generalized in two ways. Certainly long-term primate studies have resources which can be of considerable help in important wildlife issues in source countries. But these resources are meaningless unless they are directed towards source country needs. At Gilgil, this happened out of necessity, not through planning. In the future, the appropriate role of long-term studies should be defined by host countries and visiting scientists should understand the necessity of these priorities.

The second point is, in a way, to provide an added incentive for scientists to undertake 'applied' research. The necessity of applied projects is obvious. Serious problems face almost all primate species in the wild and the future of scientific research depends on the longevity of wild populations. But, equally important is the need to overcome the historical stigma attached to applied research. Applied projects, like the one at Gilgil, raise important 'basic' science questions, often addressing them in unique ways. The investigation of baboon crop-raiding provided exciting new insights about male reproductive strategies, the evolution of foraging shifts, optimization theory and the interplay between ecological and social factors in baboon behavior.

The benefits of a closer association between source country needs and primate field research extend in both directions. Some countries, like Kenya, are taking an active role in determining priorities for field research. Other source countries will be quick to follow suit. The future contribution of long-term primate research will be greatest if scientists, too, plan projects with Third World realities in mind rather than have these enter into their research *ad hoc*.

Acknowledgements

The research at Gilgil has been sponsored by the Office of the President, the Institute of Primate Research, Nairobi, the National Museums of Kenya and the University of Nairobi. Funding for the work reported in this paper was provided by the L.S.B. Leakey Foundation, the New York Zoological Society, World Wildlife Fund, U.S., and the University of California, San Diego.

References

Forthman Quick, D. (1984) Unpublished doctoral thesis, University of California, Los Angeles

McNeely, J. A. & Miller, K. R. (1984) (eds) *National Parks, Conservation and Development: The Role of Protected Areas in Sustaining Society*. Washington, D.C.: Smithsonian Institution Press

Musau, J. M. & Strum, S. C. (1984) Response of wild baboon troop to the incursion of agriculture at Gilgil, Kenya. *Int. J. Primatol.*, **5**, 364

Oyaro, H. O. & Strum, S. C. (1984) Shifts in foraging strategies as a response to the presence of agriculture in a troop of wild baboons at Gilgil, Kenya. *Int. J. Primatol.*, **5**, 371

Republic of Kenya (1975) *Sessional Paper Number 3: Statement on Future Wildlife Management Policy in Kenya*. Nairobi: Government Printers

Strum, S. C. (1984) The Pumphouse Gang and the great crop raids. *Anim. Kingdom*, **87**, 36–43

V.2

Activity budgets and the consumption of human food in two troops of baboons, *Papio anubis*, at Gilgil, Kenya

D. L. FORTHMAN QUICK

Introduction

There is an ambivalent attitude toward nonhuman primates in many countries where they are endemic. Certainly, nonhuman primates are important natural resources, both intrinsically, and for the purpose of scientific study. They also have aesthetic and monetary value as tourist attractions in their native habitats. However, on the debit side, nonhuman primates are perceived as threats to agriculture and human health in these same areas (Altmann & Altmann, 1970; Groves, 1980; Matsuzawa *et al.*, 1983; Teas *et al.*, 1980). As human populations in Africa and Asia continue to increase, more primates are subjected to predation and habitat destruction by subsistence farmers.

If we hope to alleviate this conflict, it is important to understand the process by which primates adapt to habitat alteration and learn to exploit human foods. Maples (1969) and Maples *et al.* (1976) have described the crop-raiding of baboons, and others have studied the ecology and behavior of established urban populations of rhesus macaques (Neville, 1968; Teas *et al.*, 1980), but little systematic work has been conducted on the phenomenon of human food exploitation until recently (Brennan, Else & Altmann, 1985). The habitat of several baboon troops at the Gilgil baboon site (Harding, 1976; Qvortrup & Blankenship, 1975) has been disrupted recently by the incursion of farmers (Strum & Western, 1982). Beginning in 1978, the baboons nearest to human settlements learned to raid crops on an occasional basis. By 1981, raids had become more systematic and garbage at a nearby boarding school and army base was incorporated into the baboons' diet. This paper reports on differences in the activity budgets of two of the Gilgil troops. The relations between activity budgets, ranging and differential levels of human food consumption are discussed.

Subjects

The troops studied were Pumphouse Gang (PHG) and Wabaya (WBY), which split from PHG at the beginning of the study. From the original troop of 82 animals, 40 adult, subadult and juvenile members were chosen for individual sampling. All baboons which were known to raid local maize crops were chosen, and others were selected according to a stratified random sampling scheme by age/sex class. By this method, approximately equal numbers of raiders and non-raiders were achieved. Baboons less than 2 years old were excluded from the subject pool. During the fission of PHG, subjects split almost evenly between the two troops.

Methods

During the 1981 crop season, which extended from June through October, 240 h of data (6 h/subject) were collected on activity, ranging patterns and food consumption. Animals were randomly assigned a position in each of three time blocks (06.00–10.00; 10.01–14.00; 14.01–18.00 h) and were observed for 1 h during each block until all animals had been observed.

Instantaneous samples of target activity and location (in 1 km^2 quadrats), were scored at 5-min intervals. The activity categories used were: PASSIVE, which included chewing, as well as auto-grooming and other maintenance behaviors; MOVE, defined as any non-social locomotion over a distance of at least one body length; FORAGE/FEED, which included any stationary preparation, manipulation and ingestion of foodstuffs, but excluded chewing or consumption of cheek pouch contents; and SOCIAL, which included both friendly and aggressive interactions with others. One–zero sampling was used during each interval to score the first occurrence of selected categories of foraging and feeding behavior (Altmann, 1974; Lehner, 1979). Of primary interest were analyses of the activity budgets of the two troops, the relative frequency with which wild *vs* crop foods were consumed and the troops' range use in relation to the availability of human foods in local fields and garbage dumps.

Data were summarized over troop and age/sex class, and analyses of variance were performed on the proportion of intervals in which subjects from two troops and five age/sex classes engaged in each of the four activities. For the analyses, proportions were weighted by the number of intervals of data per individual. Data from juveniles of both sexes were lumped; this was done because there was no significant effect of age/sex class on the forage/feed category ($F_{4,38} = 0.66, p > 0.62$).

Results

The activity budgets of the two troops overall are presented in Fig. 1. In comparisons between troops, the overall analyses of variance for passive, move, and forage/feed were significant, with PHG engaged more often in feeding and less often in passive behaviors than WBY ($F_{9,38}$ = 6.97, 3.11 and 3.11, respectively, $p < 0.01$ for all). There were no significant differences overall in the number of intervals spent in social behavior ($F_{9,38} = 1.81$, p > 0.098).

For the passive category, there were significant effects of troop membership ($F_{1,38} = 36.18$, $p < 0.01$) and age/sex class ($F_{4,38} = 6.58$, $p < 0.01$). The only variable which significantly affected movement patterns was age/sex class ($F_{4,38} = 4.36$, $p < 0.005$). Finally, in the forage/feed category, there was a significant effect of troop membership ($F_{1,38} = 19.64, p < 0.01$).

Fig. 1. Comparison of the activity budgets of PHG and WBY troops, collapsed across age/sex classes. PHG was passive 22%, moved 21%, foraged/fed 48% and socialized 9%. WBY was passive 39%, moved 26%, foraged/fed 28% and socialized during 7% of the 5-min intervals of observation. Significant results of analyses of variance are indicated. See text for behavior definitions.

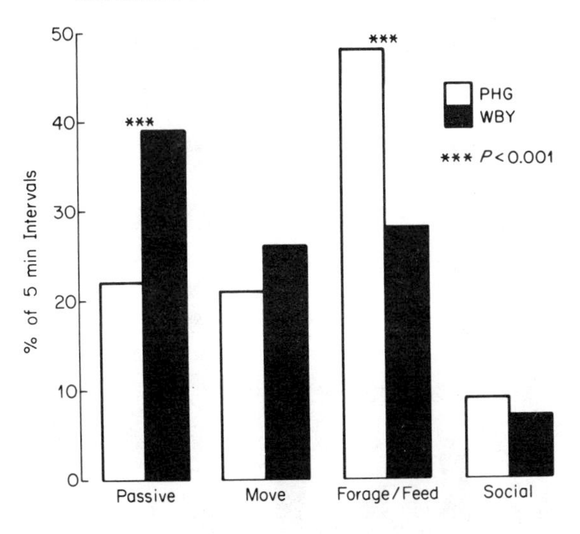

These results are presented by troop and age/sex class in Fig. 2. Significant results of *post hoc* comparisons are indicated in the figure. It can be seen that there were differences between the troops in both passive and forage/feed across every age/sex class except adult male and subadult female. The lack of significant differences is no doubt due to the small number of subjects in those classes in one or the other troop. The only significant between-troop difference in the move category occurred between the subadult males.

In an attempt to account for such striking differences between animals that were formerly members of a single group, a comparison was made of the relative frequency of consumption of wild and human foods in the two troops. There was a significant effect of troop membership on overall consumption of human foods ($F_{1,38} = 24.69$, $p < 0.01$). PHG consumed human foods during only 0.7% of the intervals in which it fed; in contrast, WBY ate human foods during 17% of the intervals in which feeding occurred. There was no significant effect of age/sex class on human food consumption ($F_{4,38} = 1.50$, $p > 0.22$). However, this may have been due in part to the variability within age/sex classes on this measure, and the small number of

Fig. 2. Activity budgets are presented here by troop and age/sex class. Significant results of *post hoc* comparisons are shown. The only age/sex classes which did not show significant differences in both passive and forage/feed categories were adult male and subadult female. These two classes had N's of 1 and 3, respectively, in one of the troops. Variance was also high among the subadult females. Only subadult males differed signficantly in the move category.

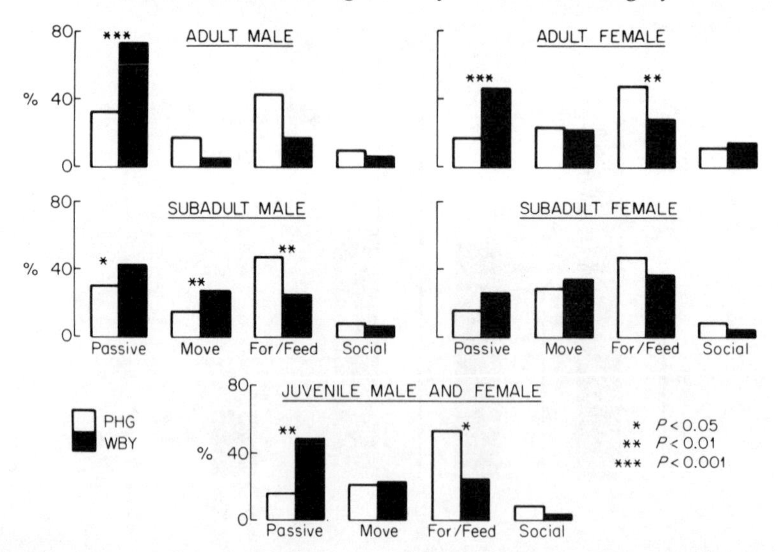

subjects in the above-mentioned two classes. Thus, although human foods constituted a relatively small proportion of the total diet in both troops, PHG was at the time an almost undisturbed troop, while WBY's feeding was significantly altered.

With respect to ranging patterns, PHG was never observed in either of the two quadrats which had the greatest concentration of humans and garbage. However, WBY members spent 26% of their time in those cells, during which they consumed 35% of all the human foods they ate. While only 27% of the raw maize they consumed was eaten there, 63% of other human foods they ate was consumed in the vicinity of the army base and school.

Observations indicated that the food WBY sought most eagerly was large lumps of cooked maize meal. This concentrated food resource was predictable in time and space, and dispersed over a very small area, but it was abundant only for short periods of time during each day. Thus, the baboons' typical pattern was to progress rapidly in the early morning from their sleeping site to the environs of the army base, where they then dispersed into small groups or ranged alone through yards and garbage pits. Throughout these forays, the animals moved quickly, and were more likely to startle or avoid in response to sudden sounds or movements. This was interpreted as behavior used to cope with the increased levels of both human harassment and intraspecific competition for the porridge. Generally, when a baboon discovered a lump of porridge, the food was stuffed into the cheek pouches as rapidly as possible; any excess was carried by tripedal locomotion to the nearest cover behind bushes or rocks on the perimeter of the base.

There was a negative correlation between overall time spent foraging or feeding and the proportion of intervals in which an individual consumed all human foods, but the correlation was not statistically significant ($r = -0.27$, $p > 0.25$). When raw maize consumption was excluded, so that only those foods found in the army base were used in the analysis, the correlation improved slightly ($r = -0.32$, $p > 0.18$). The lack of significance may be due to the fact that frequency of consumption was measured, rather than the actual amount consumed or number of calories obtained. The relation between human food consumption and overall time spent feeding may also have been dampened by the procedure used in analysis, which combined few days in which an animal showed high levels of human food consumption with many days in which no human foods were eaten. Thus, because the effects of porridge consumption are short lived (6–8 h), the measure of overall time spent feeding would not accurately reflect them (Clutton-Brock, 1977; Demment, personal communication).

Discussion

Caloric density and the amount of work required to obtain food are among the variables which influence animals' learned strategies of diet selection (Lang, 1970). The frequency of consumption of maize porridge was relatively low (less than 29% of the intervals in which human foods were eaten), and the duration of ingestion was also short. However, as Hladik (1978) has pointed out, neither is probably closely related to the volume of food ingested. Handling time for the porridge was brief, as it could be rapidly consumed or deposited in the cheek pouches. With respect to digestibility and calories obtained, heat treatment and refinement not only reduce or eliminate toxic elements in raw foods, including maize (Harris, 1970), but heat-treated starches are digested more efficiently by nonhuman primates (Knapka & Morin, 1979). The slower rate of passage associated with decreased food particle size also improves digestibility (Demment, personal communication). As a result, more energy is derived from a given amount of porridge than from the same amount of wild forage or raw maize (Ullrick, personal communication). In summary, maize porridge constitutes a food low in protein, but high in carbohydrates (Inglett, 1975) which requires brief, moderate outputs of energy to acquire in comparison to natural forage or raw maize.

It appears that the microhabitat exploited by WBY troop promotes inactivity by offering a predictable high-energy resource in a limited area. Although the food may be temporarily depleted, as during a morning bout of foraging, the same feeding site will be replenished several hours later, so that extensive travel is no longer required. Activity may also decrease because competition and the threat of human harassment increases feeding rate.

Certain other factors probably contributed to the complexity of the observed feeding and ranging patterns. Before the beginning of this study, the parent troop PHG spent nearly all its time foraging for wild foods, and only occasionally caused serious damage to crops. However, after the baboons discovered the resources available at the school and army base, the tendency of certain animals to seek human foods appeared to have been a major factor in the fission of the parent troop. The splinter group, WBY, was approximately one-third the size of PHG; WBY contained a single adult male unafraid of humans and familiar with the area and resources of the army base. That male became the central figure in a group which otherwise consisted of a cohort of nulliparous or primiparous females (Strum, personal communication). Because of its size, PHG was usually able to displace WBY from rich sources of traditional foods, and thus may have reinforced WBY's tendency to exploit crops and garbage. Since that time, Kekopey has experienced a period of drought which has greatly reduced wild foods and all but eliminated crops. As a result, PHG has since utilized the army base area as well.

Summary

In summary, some of the variables which appear to contribute to alterations in the activity patterns of baboons in disrupted environments are as follows.

1. Nature of the costs and benefits – easily digested, calorie-rich foods may be sought in preference to a high-protein diet which is energetically more costly.
2. Spatial and temporal distribution – spatially concentrated foods which are abundant at predictable times may contribute to a restricted ranging pattern and decreased activity.
3. Seasonality – the pressure to exploit cultivated foods and/or garbage will vary with seasonal availability of foods. Animals often abandon human foods when new growth appears which is high in protein and easy to digest.
4. Age/sex composition – young animals and males appear to be the least conservative age/sex classes. WBY adult females had the lowest percentage of crop consumption. Thus, troops with a large proportion of young animals (as are found in expanding populations) may be more likely to become serious agricultural pests.
5. Individual experience – in the initial stages, dominant troop members familiar with certain areas may tend to utilize those areas despite human harassment.
6. Troop size – smaller troops may benefit more from frequent, systematic exploitation of cultivated foods than larger, more conspicuous troops.

Acknowledgements

I would like to thank the Office of the President and the National Council for Science and Technology of Kenya for permission to conduct this research, and J. Garcia, S. C. Strum, J. G. Else and the Institute of Primate Research for their aid and sponsorship. I would also like to thank E. J. Brennan, M. Demment and C. Menzel for their helpful comments and criticisms. This research was supported by Fulbright-Hays award no. 5048211, Sigma Xi, a Hortense Fishbaugh scholarship and UCLA patent/travel grants awarded to D. F. Quick, and by J. Garcia's NIH no. NS11618.

References

Altmann, J. (1974) Observational study of behavior: sampling methods. *Behavior*, **49**, 227–67

Altmann, S. A. & Altmann, J. (1970) *Baboon Ecology*. Chicago: University of Chicago Press

Brennan, E. J., Else, J. G. & Altmann, J. (1985) Ecology and behavior of a pest primate: vervet monkeys in a tourist lodge habitat. *Afr. J. Ecol.*, **23**, 35–44

Clutton-Brock, T. H. (1977) Some aspects of intraspecific variation in feeding and ranging behaviour in primates. In *Primate Ecology: Studies of Feeding and Ranging Behaviour in Lemurs, Monkeys and Apes*, ed. T. H. Clutton-Brock, pp. 539–56. New York: Academic Press

Groves, C. P. (1980) Speciation in *Macaca*: the view from Sulawesi. In *The Macaques: Studies in Ecology, Behavior and Evolution*, ed. Lindburg, pp. 84–124. New York: Van Nostrand Reinhold

Harding, R. S. O. (1976) Ranging patterns of a troop of baboons (*Papio anubis*) in Kenya. *Folia Primatol.*, **25**, 143–85

Harris, R. S. (1970) Natural vs. purified diets in research with nonhuman primates. In *Feeding and Nutrition of Nonhuman Primates*, ed. R. S. Harris, pp. 251–62. New York: Academic Press

Hladik, C. M. (1978) Adaptive strategies of primates in relation to leaf-eating. In *The Ecology of Arboreal Folivores*, ed. Montgomery, pp. 373–95. Washington, D.C.: Smithsonian Institution Press

Inglett, G. E. (1975) Effects of refining operations on cereals. In *Nutritional Evaluation of Food Processing*, ed. Harris & Karmas, pp. 139–58. New York: Avian

Knapka, J. J. & Morin, M. L. (1979) Open formula natural ingredient diets for nonhuman primates. In *Primates in Nutritional Research*, ed. Hayes, pp. 121–38. New York: Academic Press

Lang, C. M. (1970) Organoleptic and other characteristics of diet which influence acceptance by nonhuman primates. In *Feeding and Nutrition of Nonhuman Primates*, ed. Harris, pp. 263–75. New York: Academic Press

Lehner, P. N. (1979) *Handbook of Ethological Methods*. New York: Garland

Maples, W. R. (1969) Adaptive behavior of baboons. *Am. J. Phys. Anthropol.*, **31**, 107–9

Maples, W. R., Maples, M. K., Greenhood, W. R. & Walek, M. L. (1976) Adaptations of crop-raiding baboons in Kenya. *Am. J. Phys. Anthropol.*, **45**, 309–16

Matsuzawa, T., Hasegawa, Y., Gotoh, S. & Wada, K. (1983) One-trial long-lasting food-aversion learning in wild Japanese monkeys (*Macaca fuscata*). *Behav. Neural Biol.*, **39**, 155–9

Neville, M. K. (1968) Ecology and activity of Himalayan foothill rhesus monkeys (*Macaca mulatta*). *Ecology*, **49**, 110–23

Qvortrup, S. A. & Blankenship, L. H. (1975) Vegetation of Kekopey, a Kenya cattle ranch. *East Afr. Agric. Forestry J.*, **40**, 439–52

Strum, S. C. & Western, J. D. (1982) Variations in fecundity with age and environment in olive baboons (*Papio anubis*). *Am. J. Primatol.*, **3**, 61–76

Teas, J., Richie, T., Taylor, H. & Southwick, C. (1980) Population patterns and behavioral ecology of rhesus monkeys (*Macaca mulatta*) in Nepal. In *The Macaques: Studies in Ecology, Behavior and Evolution*, ed. Lindburg, pp. 247–62. New York: Van Nostrand Reinhold

V.3

Ecology and behaviour of vervet monkeys in a tourist lodge habitat

P. C. LEE, E. J. BRENNAN, J. G. ELSE
AND J. ALTMANN

Introduction

Conflict between non-human primate populations and local human communities is widespread in Kenya and other African countries. Historically, reports of food-raiding have tended to focus on stealing or destroying crops in agricultural areas (Brennan, Else & Altmann, 1985). Increasingly, some primates are creating problems by supplementing their natural diet with food stolen from people or with garbage found around forest reserves, picnic sites, and suburban areas. This study briefly examines the activities of one species of 'pest' primate, the vervet monkey (*Cercopithecus aethiops*), around the lodges, settlements and garbage disposal areas within Amboseli National Park, Kenya. The study was undertaken for the following reasons: (1) to examine and describe the ecology and behaviour of vervets who have become 'pests'; (2) to compare these vervets with unprovisioned groups; and (3) to describe the human–monkey conflict. The overall objective was to understand the existing problem and to help develop management solutions.

Methods

Vervet monkeys in Amboseli are territorial and live in small, stable multi-male groups (Struhsaker, 1967; Cheney, Lee & Seyfarth, 1981). Four groups of vervets lived in the area of Ol Tukai, which included two lodges, self-service cottages for tourists, staff housing and the Park's headquarters. The four Ol Tukai groups were identified as Lodge, South, West and North groups. These Ol Tukai groups had access to water throughout the year from water towers, pipes, garden hoses, and a trough in front of one of the lodges. The number of people resident in the area partially protected these vervets from some of their natural predators.

The Lodge group was the primary study group for EJB during the dry season (July–October) of 1980 and the data presented here are based on over 500 hours of observation. Behavioural and ecological data were collected from focal animal and site sampling (see Brennan *et al.*, 1985, for details). Additional censuses of the Ol Tukai vervets were made by PCL in 1979 and again in late 1982, along with casual observations of behaviour. Comparisons with the behaviour and ecology of unprovisioned groups 20 km to the west (O-lengia) are based on the results of a 2-year study (Lee, 1981). For the purposes of the present report, definitions of activity categories were standardised between the studies. The term 'home range', used here to denote an area used exclusively by one group plus the contested or defended areas of overlap between groups, is synonymous with the term 'territory' used by Lee (1984*a*).

Results

Group size and population dynamics

The provisioned groups in the Ol Tukai area were generally larger than groups elsewhere in the Park. The mean group size was 28 (range 6–53), while that of the unprovisioned groups in O-lengia was 16 (range 9–27) (Table 1). There was an increase in the size of Lodge and

Table 1. *Census data on groups of vervets living in several areas of Amboseli National Park. Ol Tukai groups had access to human food and partial protection from predators. Old Campsite group had access to garbage and was the source of some of the vervets who moved to Ol Tukai after the Trapping scheme. Means of censuses of unprovisioned groups (O-lengia) are provided for comparisons with the provisioned groups*

	Year		
	1979	1980	1982
Ol Tukai			
Lodge	43	48	53
South	–	16	25+
West	–	35–40	25+
North	–	10	6
Old campsite	55	–	68+[a]
O-lengia	16	–	13

[a] Group split 1983

South groups over a 2-year period, while North and West groups decreased slightly in size.

The increases in group size for Lodge and South groups were primarily due to a high birth rate (Brennan *et al.*, 1985), and possibly from the movement of young and adult males into the groups that had constant access to garbage. Decreases in the size of the other groups may have resulted from male emigration. As a result of both protection from predation and the continual dry season access to water, mortality among the provisioned groups appeared to be lower than that of the O-lengia vervets (see Cheney *et al.*, 1981). However, mortality rates of primates that eat garbage should be carefully monitored, because the eating of contaminated foods could result in diseases that might affect nearby people.

Vervets in most areas are seasonal breeders (Rowell & Richards, 1979), but the births among the provisioned groups did not follow the same strictly seasonal pattern observed in the unprovisioned groups. Among the Ol Tukai vervets, at least one and up to three infants were regularly born in Lodge group between June and August. Among the unprovisioned O-lengia groups, births generally occurred only between September and January, and births within a single group were tightly synchronised (Lee, 1984*b*). The Ol Tukai vervets appeared to be less seasonally constrained in their ability to conceive, and this may have led to the observed high birth rates and increases in group size. The 'off-season' breeding may be the result of high levels of nutrition achieved by access to garbage throughout the year.

Habitat use and quality

Naturally occurring foods were derived primarily from the flourishing, well-watered stands of young fever trees (*Acacia xanthophloea*), protected from most browsers within the Ol Tukai area. Foods derived from young and mature *A. xanthophloea* trees provide over 35% of the diet for unprovisioned groups throughout the year (Lee, 1984*a*). Dense stands of young trees made up 4.5% of the 13 ha home range of Lodge group, but only about 3% or less of the home ranges of the unprovisioned groups; both figures are in marked contrast to the 16% cover found in the O-lengia area 15 years ago (Struhsaker, 1967). Home-range size is larger among the unprovisioned groups (mean 25 ha, range 18–50), and the density of vervets in O-lengia and home-range size were both strongly related to the distribution and abundance of *A. xanthophloea* trees (Lee, 1981). Long-term changes in the habitats of vervet monkeys in Amboseli have resulted in the Ol Tukai groups having access to somewhat more 'optimal xanthophloea habitat' (Struhsaker, 1967) than do the unprovisioned groups.

Other natural foods for the Ol Tukai groups were herbs, shrubs and the watered *Cynodon dactylon* lawn of the Lodge, which corresponds to the natural swamp edge habitats. Additional foods were taken from introduced vegetation such as bouganvilia. Of the time spent feeding by the Lodge group, over half was on natural foods (Brennan *et al.*, 1985).

There were eight main feeding sites for the Lodge group where human food was visible and obtainable, and these were sampled by the vervets regularly. However, the frequency with which food was available and the amount of food found at a site were not predictive of the duration of the group's visit to a site (Brennan *et al.*, 1985). It would thus be difficult to discourage vervet visits to food sites by simply altering the amounts available at any time.

In contrast to the unprovisioned groups, the activities of the Lodge vervets did not appear to be seasonally or energetically constrained. The study took place during the peak dry season months, yet less than 20% of total daily time was spent feeding by Lodge vervets, while 43% of time was spent resting and 20% socialising. Among the unprovisioned groups, the time spent feeding during the dry season was 35–45% and social activities occupied approximately 5% of their time. Even during the wet season periods of high food quality and abundance, the unprovisioned vervets were still spending over 30% of their time feeding (Lee, 1983, 1984*a*) (Fig. 1).

Access to garbage and to high quality, abundant natural foods affected both the activity budgets and nature of the social interactions among the Lodge vervets. Rates of aggressive and competitive interactions were high among the Lodge vervets (Brennan *et al.*, 1985). It is interesting to note that the rates of these interactions rose significantly during the wet season for unprovisioned vervets (Lee, 1984*a*), suggesting that when foods of high quality are available, vervets in general are highly motivated and energetically able to act in a more aggressive manner, both towards each other and towards human bystanders.

Discussion

Conflict with humans

The major source of conflict between the vervets and people was the vervets' propensity to attack and seriously bite tourists and Park's staff, and to break into cottages, kitchens and cars to steal food. The destruction of property, danger from bites, and health hazards of regular primate–human contact prompted the Wildlife Conservation and Management Division to permit the trapping and removal of vervets from the Ol Tukai area in late 1983. Unfortunately, the respite from monkey problems was only temporary. As was expected (Brennan *et al.*, 1985), new vervets from the remnants of West group and probably from the nearby Old Campsite group moved into the area within a month, and as of early 1984 these habituated vervets had become a recurrent problem.

Prior to the trapping, we initiated a scheme of frightening the vervets away from direct contact with, and attempts to beg food from, tourists, which reduced the incidence of bites. This was combined with an attempt by Lodge staff to discourage tourists through verbal warnings of the health and safety dangers from approaching and feeding vervets (a main source of the problems). Leaflets of the Park's regulations (funded by researchers and the African Wildlife Foundation, and designed by the Warden) containing warnings not to feed monkeys were given to visitors upon entry to the Park. Finally, we monkey-proofed several of the garbage pits with inexpensive wire

Fig. 1. The percentage of daily time that Lodge group spent in different activities during the dry season (hatched), and that the unprovisioned O-lengia groups spent during the dry season (D) and the wet season (W).

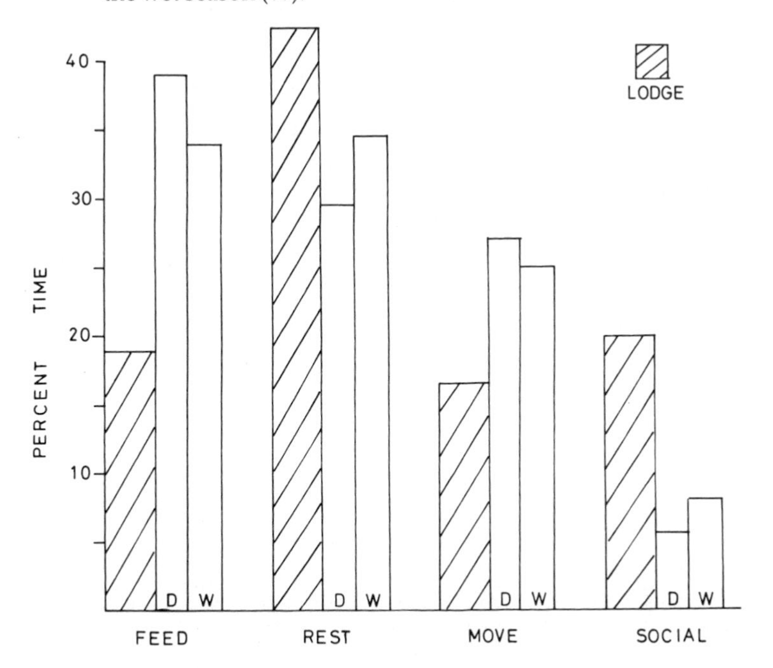

covers and had the garbage burned regularly. These actions had the effect of eliminating much of the garbage in the diet of the vervets.

Vervets around tourist areas are a major attraction for many visitors. However, due to property destruction and the health hazards posed by contact with monkeys, they become a serious management problem once they are accustomed to obtaining human food, and are extremely persistent (and aggressive) in their efforts to obtain this food. Effective management of monkeys around areas of human habitation results from eliminating their access to garbage, ensuring that dwellings and food preparation areas cannot be entered by monkeys, and discouraging the hand-feeding of monkeys through warnings to visitors and through vigilance on the part of Lodge staff in discouraging the entry of monkeys into areas of food and drink consumption.

Merely trapping monkeys in areas where there is a nearby source of primates only appears to perpetuate the problems. With proper management, monkeys can remain in areas of human use, continue to act as a tourist attraction, and the likelihood of their extermination can be reduced.

Conclusions

1. The data presented here and elsewhere suggest that in tourist areas monkeys have high-quality diets and access to human foods.
2. As a result, they live at high densities, are able to reproduce at high rates, spend little time foraging, and are extremely persistent and aggressive in their attempts to maintain access to the source of these benefits.
3. Effective management of monkeys under these conditions requires: (a) that their densities are such that can be sustained by available naturally occurring foods; (b) that access to human food is eliminated through relatively inexpensive alterations in the design of garbage pits and food preparation facilities; and (c) that contact between monkeys and people is discouraged. If these steps are taken, we expect that the problems of monkeys living in areas of human habitation can be reduced without jeopardising the future of the monkey populations.

Acknowledgements

We thank the Warden of Amboseli, B. R. O. Oguya, for his help and many ideas on dealing with the monkey problem, and the Lodge managers for their assistance. Support of EJB was from Animal Resources Branch of the National Institutes of Health through the U.S. Regional Primate Centers Program and PCL from an L. S. B. Leakey Studentship (Leakey Trust) at New Hall, Cambridge. We thank the Wildlife Conservation and Management Division, Ministry of Tourism and Wildlife, and the Office of the President for research permission. K. Lindsay, M. Hauser, C. Moss, J. Poole and A. Samuels all contributed towards funding for garbage pit covers. We thank the Ol Kajiado County Council for permission to cover the pits.

References

Brennan, E. J., Else, J. G. & Altmann, J. (1985) Ecology and behaviour of a pest primate: vervet monkeys in a tourist lodge habitat. *Afr. J. Ecol.*, **23**, 35–44

Cheney, D. L., Lee, P. C. & Seyfarth, R. M. (1981) Behavioural correlates of non-random mortality among free-ranging female vervet monkeys. *Behav. Ecol. Sociobiol.*, **9**, 153–61

Lee, P. C. (1981) Ecological and social influences on the development of vervet monkeys. Ph.D. thesis, University of Cambridge

Lee, P. C. (1983) Ecological influences on relationships and social structure. In *Primate Social Relationships: An Integrated Approach*, ed. R. A. Hinde, pp. 225–30. Oxford: Blackwell

Lee, P. C. (1984*a*) Ecological constraints on the development of vervet monkeys. *Behaviour*, **91**, 254–62

Lee, P. C. (1984*b*) Early infant development and maternal care in free-ranging vervet monkeys. *Primates*, **25**, 36–47

Rowell, T. H. & Richards, S. M. (1979) Reproductive strategies of some African monkeys. *J. Mammal.*, **60**, 58–63

Struhsaker, T. T. (1967) Ecology of vervet monkeys (*Cercopithecus aethiops*) in the Masai-Amboseli Game Reserve, Kenya. *Ecology*, **48**, 891–904

V.4

Hill tribe peoples and primate conservation in Thailand: a preliminary assessment of the problem of reconciling shifting cultivation with conservation objectives

A. A. EUDEY

Introduction

According to the most recent figures made available by the Government, there are about 500000 hill tribe people in Thailand (Thailand, Division of Development & Public Welfare, 1982), although this number, like earlier ones, may be an underestimate. Most of these people are found in the highlands of the north and west, peripheral to the main region of irrigated or wet rice production in central Thailand. They tend to practice slash-and-burn or swidden agriculture, making use of the hoe but not the plough, and some of the highlands that they exploit, in spite of the recent rapid increase of the entire upland population (Kunstadter, Chapman & Sabhasri, 1978; McKinnon & Bhruksasri, 1983*b*), may still contain relatively intact stretches of forest habitat with primates and other wildlife.

Through the use of ERTS (satellite) images it is known that the forest cover of Thailand, a country which encompasses a total area of almost 520000 km^2, declined from more than 53% in 1961 to less than 28% in 1981, and during the 1970s the rate of forest loss may have been nearly 10% a year (Secter, 1982). Slash-and-burn agriculture, which is practiced by rural Thai people as well as hill tribe peoples, and illegal logging, especially of tropical hardwoods, have been identified as the major causes of this destruction, but timber concessions, dam construction, highways, resettlement programs, mineral exploration and recreation also are contributing factors (Jintanugool, Eudey & Brockelman, 1984).

Hill tribe peoples in Huai Kha Khaeng Wildlife Sanctuary

In 1973 I began a field study of the five species of macaque monkeys (*Macaca* spp.) that are sympatric in Huai Kha Khaeng Wildlife Sanctuary in the western mountain subregion of the continental highlands of Thailand (Eudey, 1980, 1981). The sanctuary encompasses about 165 000 h of a mosaic of forest types dominated by moist deciduous forest between 15° and 16°N latitude on the eastern flank of the Dawna Range, near the border between Thailand and Burma (Fig. 1). The Dawna Range is part of the series of mountain ranges radiating outward to the south and west from the Yunnan Plateau in southern China (Fig. 2). Huai Kha Khaeng Wildlife

Fig. 1. Huai Kha Khaeng Wildlife Sanctuary, as indicated by the Khao Nang Rum Research Station, in the Dawna Range in the western mountain subregion of the continental highlands in Thailand.

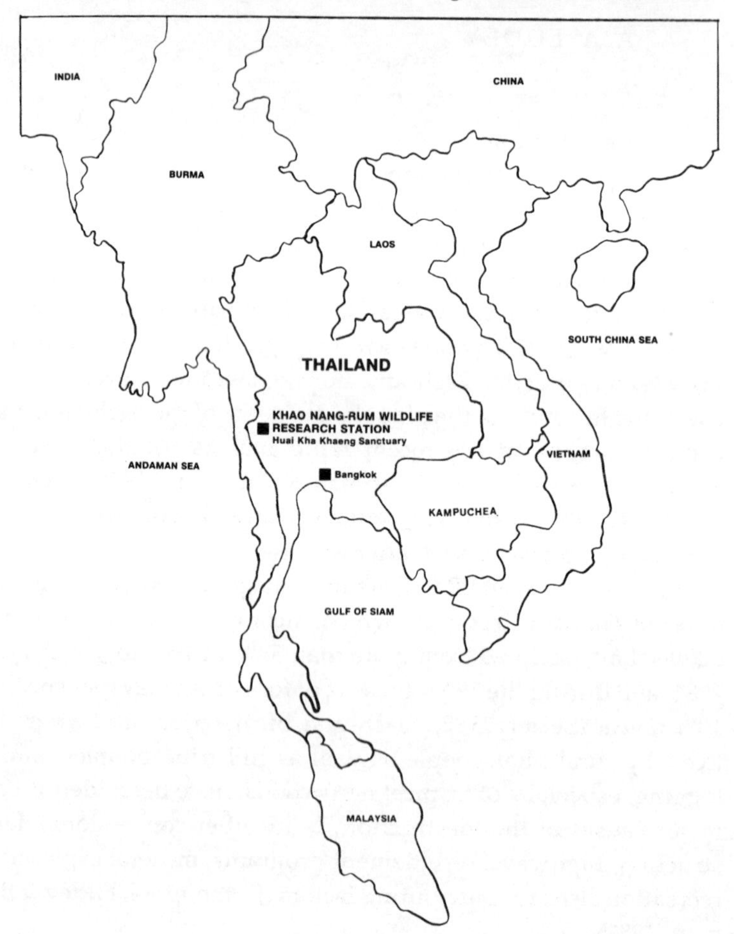

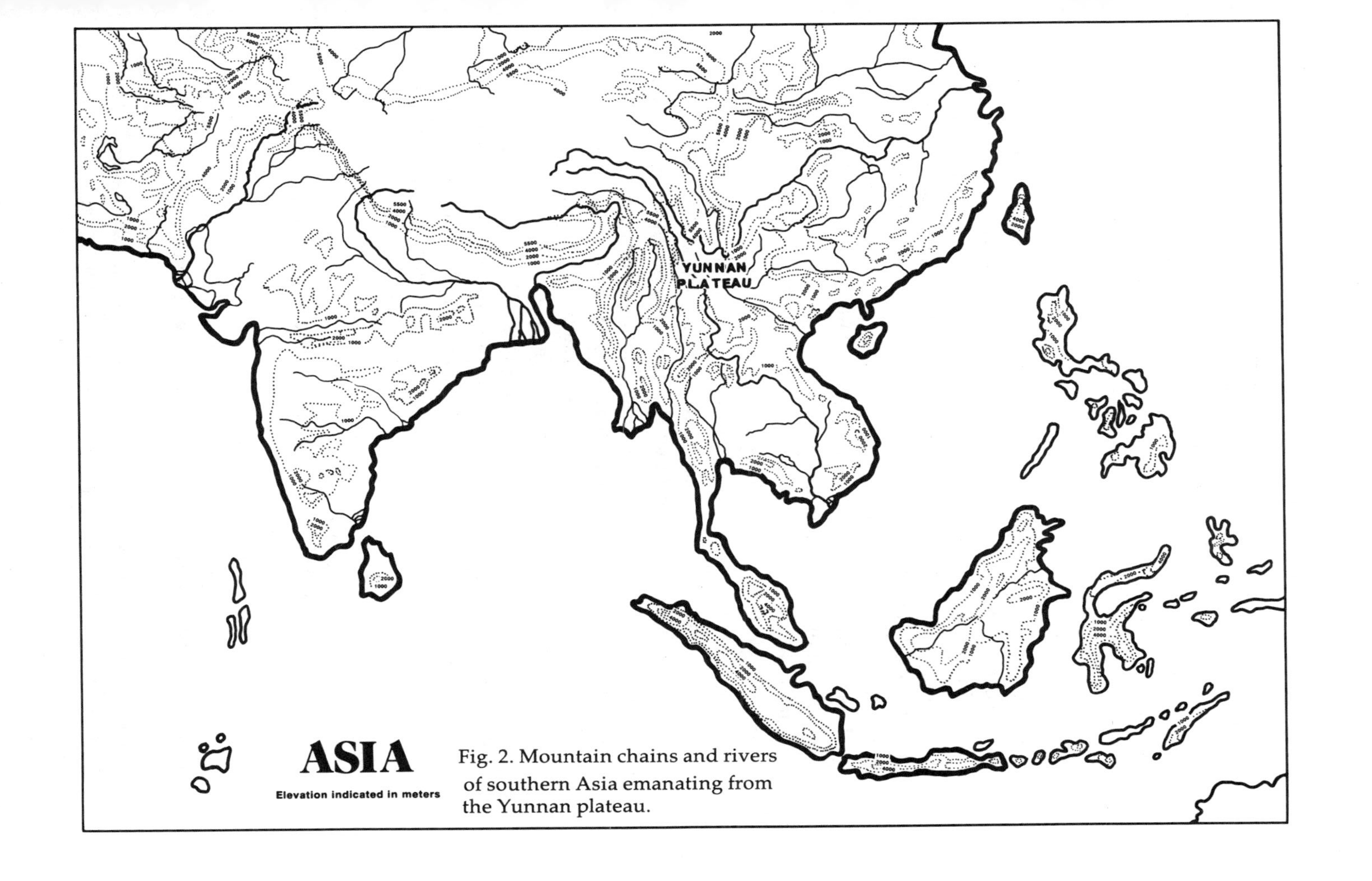

Fig. 2. Mountain chains and rivers of southern Asia emanating from the Yunnan plateau.

Sanctuary is under the jurisdiction of the Wildlife Conservation Division of the Royal Forest Department, and all of my research has been carried out in cooperation with the Division.

Huai Kha Khaeng Wildlife Sanctuary was declared in 1972, only one year before I first entered the sanctuary. At the time of its declaration, Karen hill tribe people living within the boundaries of the sanctuary withdrew voluntarily or were incorporated into resettlement (development) projects of the Government. The Karen are the most numerous hill tribe people in Thailand, numbering around 200000 (Thailand, DD & PW, 1982). They began to move eastward from Burma into Thailand about 250 years ago, and the population remains concentrated in the westernmost part of Thailand, with many settlements occurring at elevations below 500 m (see Young, 1982; McKinnon & Bhruksasri, 1983*b*: map of highland ethnic village settlements). However, all land in Thailand technically belongs to the state, and areas such as protected forests or sanctuaries and hill land legally cannot be occupied or owned by anyone (Ratanakorn, 1978).

Some of the former residents of *ban* (= village) Similak near *huai* (= river) Ai Yo within Huai Khaeng Wildlife Sanctuary eventually settled near the southern boundary of the sanctuary and founded *ban* Jawat ('Temple Village'). The residents deliberately try to maintain a traditional life style, which includes the gathering of wild bananas and the hunting of small game such as wild pig, but the hunting of large game such as elephant, which characterized their former existence, has been abandoned. Elsewhere in the immediate region the Karen live in resettlement villages of various degrees of development. The political unrest of the Karen people to the west in Burma has caused the Royal Army to regard the local Karen as potential insurgents, and, as a consequence, a military garrison is maintained in the area. Some Karen from the resettlement villages may be involved in commercial or 'market' hunting of protected wildlife and in some small-scale opium traffic, and there is some suspicion of complicity in these activities on the part of Army personnel.

At the time that Huai Kha Khaeng Wildlife Sanctuary was declared in 1972, another hill tribe people, the Hmong (who also have been called the Miao or Meó), had penetrated into the mountains inside its western boundary, although the Royal Forest Department and other agencies of the Government appear to have been unaware of this at the time.

The most detailed study on the Hmong people of Thailand was undertaken by the anthropologist W. R. Geddes (1967, 1976) in *changwat* (= province) Chiang Mai at approximately 18°N latitude during the later 1950s and the 1960s. Observations were also made on Hmong ecology in western *changwat* Tak, in a habitat more nearly approximating that occupied by the Hmong in Huai Kha Khaeng Wildlife Sanctuary, by F. G. B. Keen (1978) during November 1963 to March 1964. The Hmong appear to have entered Thailand only within the last 100 years (Mandorff, 1967; Young, 1982; Kunstadter, 1983), although they are the secondmost numerous hill tribe people within the country, numbering over 37000 (Thailand, DD & PW, 1982). Their presence in Huai Kha Khaeng and adjacent Thung Yai Wildlife Sanctuaries, in *changwat* Uthaithani and Tak, respectively, may represent their southernmost expansion in western Thailand, to less than 300 km northwest of Bangkok. In 1965, according to Government figures (see Geddes, 1976), there was no record of Hmong in Uthaithani, although a population of 5600 was estimated for Tak. Likewise, in 1973 the Tribal Research Centre in Chiang Mai (see Kunstadter, 1983) estimated the presence of 7072 Hmong in Tak but none in Uthaithani, and a map showing the distribution of major highland ethnic group village settlements prepared by the Tribal Research Centre in November 1978 (and published in McKinnon & Bhruksasri, 1983*b*) does not record any Hmong villages in Uthaithani. In August 1983, the most recent date for which I have information, the Department of Public Welfare had no record of the number of Hmong living in Uthaithani, although it is recorded that during that year the Royal Army had moved 643 Hmong, in 110 families, to Uthaithani, but later removed them to some other place, perhaps a resettlement village.

The Hmong hill tribe people and the cultivation of opium poppy

The Hmong practice true shifting, not cyclical, agriculture, which contributes to the destruction of forest habitat, specifically the montane forests in which they establish their settlements. Opium is the primary source of income or cash for the Hmong, and the intensive cultivation of poppy, which grows at elevations between 900 and 1500 m, determines their way of life. They continue to exploit an area until actual decline in crop productivity makes further efforts unrewarding, especially in the cultivation of opium. Opium poppy fields may be used for 4–10 or perhaps even 20 years. Corn is grown on the same fields as poppy, although this reduces the fertility of the soil, but rice is grown on other fields that are abandoned after 1 or 2 years, primarily due to the invasion of unwanted and uncontrollable grasses

(Geddes, 1967, 1976; Keen, 1978, 1983). This system has been described as one of 'long cultivation' and 'very long fallow', in which the forest cover almost never recovers within a human lifetime (Kunstadter & Chapman, 1978). As a consequence, the desire for productive land (for poppy) has caused continuous movement among the Hmong people (Geddes, 1976).

The migration of the Hmong may be a continuation of a process begun within their homeland of China. About 2 680 000 were estimated to be present in China in 1957 (Geddes, 1976). Their southward penetration into Indochina appears to have occurred not much less than 200 and probably not more than 400 years ago. Although it is ultimately commitment to poppy that makes the Hmong true shifting cultivators, they may have grown it for only a few centuries. Opium-growing was legalized in China in the mid-nineteenth century, and cultivation of it expanded significantly only after 1919, as warring military leaders resorted to poppy to support private armies. The cultivation of opium on mountain slopes may be an accentuation of a pre-existing agricultural pattern, for the Hmong have the reputation of being the ones 'who cut down the majority of the old forests of Asia' (F. M. Savina, 1930, *Histoire de Miao*, Société des Missions Etrangères de Paris, second edn, Hong Kong, as translated from the French by Geddes, 1976).

Opium-smoking was made legal in Thailand in 1852 by King Rama IV, but only for Chinese. By 1907 the Government had created a monopoly on the sale of opium, and in 1950 the Government realized a revenue in excess of 5.5 million US dollars from its sale (Geddes, 1976). However, between 1955 and 1959, as a consequence of pressure exerted by the United Nations, largely at the urging of the United States, the Government banned the sale, smoking, and growing of opium and began to encourage the Hmong (and other hill tribe peoples) to seek alternative crops to opium (Mandorff, 1967; Geddes, 1976; Sabhasri, 1978).

Observations on the Hmong people in Huai Kha Khaeng Wildlife Sanctuary

In 1982 I was first given permission by the Hmong to enter the mountains under their occupation in the western region of Huai Kha Khaeng Wildlife Sanctuary. Effective contact between personnel of the Royal Forest Department and the Hmong had been established only a short time earlier when failure of their rice crops necessitated that local Hmong travel through the sanctuary to purchase rice and other supplies at the nearest market, which is approximately 30 km from sanctuary headquarters. I undertook two brief surveys in the Hmong area in 1982 and 1983, respectively, in order to search for stumptail macaques (*Macaca arctoides*), which appear to be threatened throughout their disjunct distribution in south Asia, and Assamese macaques (*M. assamensis*), which are poorly known throughout their habitat range in the same region (Eudey, 1984).

Although stumptail macaques were reported by the Hmong to be relatively abundant locally, the species was found to be rare. The Hmong are not attempting to exterminate the stumptail macaques, which they claim not to hunt for food, but loss of their natural habitat to agriculture is forcing the monkeys to rely increasingly on crops such as corn for food and results in them being persecuted as agricultural pests. Assamese macaques, leaf monkeys (probably *Presbytis phayrei*), white-handed or lar gibbons (*Hylobates lar*), and other wildlife, including great hornbills (*Buceros bicornis*), if encountered in forest adjacent to fields, or within the fields themselves, may be shot for food to supplement the chickens and pigs raised for consumption by the Hmong. Forest 'animals and birds' also are described as diet supplements for the Hmong in western Tak (Keen, 1978). However, in spite of the deleterious effects of shifting cultivation and associated hunting activities, some primary forest with evidence of the above wildlife populations does remain in the Hmong region in Uthaithani and immediately adjacent areas of Tak.

The area occupied by the Hmong in Huai Kha Khaeng Wildlife Sanctuary may be marginal in both elevation and soil type for growing opium poppy. Within the sanctuary the fields cleared by the Hmong appear to be between 800 and 1000 m. The short grain rice of the Hmong is reported to grow best at from 610 to 1067 m, but poppy is reported to be progressively better from 914 to 1524 m, as increasing altitude discourages the growth of unwanted grasses (Geddes, 1976). Reddish-brown earths of low acidity, related to the weathering products of limestone, are favored for growing poppy. Limestone outcroppings characterize the local habitat, but the slightly sticky and slippery soil that results in the area with precipitation is not considered very good for either poppy or rice, and fields require width (and direct exposure to sunlight) to become warm (Geddes, 1976). However, in similar habitat in western Tak the Hmong are reported to consider that good primary forest, such as that found in the sanctuary, indicates fertile soil and impedes the establishment of uncontrollable grasses for several years (Keen, 1978, 1983). The Hmong within Huai Kha Khaeng

Wildlife Sanctuary also may have exhausted the fertility of their older opium fields and perhaps are entering a crash phase in what appears to be an alternation between 'plenty and poverty' or 'feast and famine' that is characteristic of their life cycle (Geddes, 1976). Viewed from another perspective, the Hmong traditional system of land use is steadily destroying the forest which is vital to the continuation of their lifeway (Keen, 1978).

Conservation and the Hmong people

During the latter 1950s and the 1960s, the Government of Thailand appears to have criticized the shifting cultivation of the Hmong (and similar hill tribe peoples) on three counts: (1) that it destroys land for further agricultural use; (2) that it depletes timber resources; and (3) that its interference with watersheds endangers the agriculture of the river plains critical to the economy of Thailand (Geddes, 1976). According to Geddes (1976), the major grounds for criticizing the Hmong method of cultivation were to be found in the effects that it has on the welfare of the people themselves, that is, it leads to the fields of the Hmong being invaded by imperata grass (*Imperata cyclindrica*), which is difficult to control given their level of technology.

The fact that the shifting cultivation of the Hmong is contributing to the destruction of the western mountain ecosystem through the clearing of lower montane forest and the hunting of wildlife seems to have received little or no consideration even into the 1970s. Geddes (1976), for example, does not once refer to the hunting of wild animals, for food or as agricultural pests or for trade, by the residents of the Hmong village that he studied in northern Thailand, which suggests that the local fauna may have been extirpated before he began his study. Statements even appear in the literature to the effect that in the highlands 'wild animals are almost gone' and the human population pressure 'has reached the point where it is no longer possible to have extensive game preserves' (Kunstadter, 1978). The lack of concern with the conservation of the ecosystem itself is not surprising since Thailand's first national park was declared only in 1962 and the first wildlife sanctuary was declared 3 years later in 1965. General recognition of the intrinsic importance of forest habitat and wildlife is of recent origin (Kunstadter, 1983; Jintanugool *et al.*, 1984).

Given the fact that the Hmong's traditional ecological adaptation is contributing to the loss of Thailand's natural heritage, what options are available to the Hmong people and to the Government of Thailand?

Until recently, the Government through the Border Patrol Police and the Department of Public Welfare and a special program for the hill tribe peoples under the patronage of the King, has pursued a gradual policy of substitution, encouraging the Hmong to develop alternative means of income to opium, especially the growing of other cash crops (Geddes, 1967, 1976; Mandorff, 1967; Sabhasri, 1978). The points of contention between the Government and hill tribe peoples such as the Hmong are not just opium but the endless demand for new land that its cultivation creates. In 1976 Geddes considered that with 'time and patience' the situation might resolve itself, that is, the interests of both parties in a program of crop replacement would converge as the land appropriate for poppy ultimately becomes exhausted.

The requisite time for gradual ecological change on the part of the Hmong people is not available from the standpoint of conservation. From the standpoint of poppy itself, time also may be running out for the Hmong and other hill tribe peoples engaged in its cultivation. The mayor of New York City, Edward Koch, who is involved in combating drug abuse in the United States, stated on National Public Radio early in 1984 that the Government of Thailand has the means to eradicate the opium crop in one season. What might have precipitated such action is a political consideration, the fear of insurgency in the area, a sentiment which waxed and subsequently waned following the defeat of the United States in Indochina in the mid-1970s. As it is, in 1982 the Government proposed the development of a new, highly structured program for the hill tribe peoples, and the first stated objective of this program is to bring them under political control.

For the Hmong people within Huai Kha Khaeng Wildlife Sanctuary, at best two kinds of possible action might be entertained. On the one hand, the Hmong might be able to remain in a remote and protected area such as the sanctuary with the introduction of appropriate modern agricultural technology or a commitment to more labor-intensive practices, such as the development of irrigated paddy terraces, either of which would facilitate more permanent land tenure. A conservation ethic contrary to the prevailing attitude of unlimited resource exploitation held by the migratory Hmong would have to be fostered deliberately among them. An appropriate cash-crop substitute for poppy in a remote area such as the sanctuary presents a problem because of transport; lowland merchants now travel to the Hmong to purchase opium. One partial solution to this problem might be the legalization and modernization of opium production for use in the pharmaceutical industry in Thailand (see Sabhasri, 1978): most opium that enters the international market may even be grown in neighboring Burma and its derivates smuggled into Thailand for distribution (McKinnon & Bhruksasri, 1983*a*). This would necessitate new methods of marketing under government control, however. The Royal Army has proposed building a ring road in western Thailand to increase communication with and control over the hill tribe peoples as long as they remain in remote areas. The construction of such a road would have the effect of opening up all the remaining wilderness in western Thailand to rapid human exploitation. One proposal already has been made to open up large tracts of National Reserve Forest in the region for logging and settlement (see Lekagul, 1983).

On the other hand, it might be possible to relocate the Hmong outside the boundaries of Huai Kha Khaeng Wildlife Sanctuary. As mentioned above, the area which they currently exploit within the sanctuary may be nearing exhaustion (or may be marginal) relative to their present agricultural methods. The Hmong are reported to have no attachment to any specific lands, and groups on the periphery, such as those in the sanctuary, are thought to be the least stable (Geddes, 1976). But simple relocation is not a sufficient solution since resettlement itself has been implicated as contributing to forest loss in Thailand. A successful relocation program would have to be based once again on modernization of agriculture and permanency of land tenure.

The success of any program of change will be enhanced to the extent that the Hmong people are permitted to participate in every step of its design and implementation. Elsewhere in Thailand the Hmong themselves have responded successfully to the problem of land pressure by developing new cash crops, such as Irish potato, or by constructing terraces for irrigated production of subsistence rice (Keen, 1983). The very fact that they have been able to maintain their cultural identity as a minority people for several thousand years, first in China and subsequently in Southeast Asia, may mean that they have the necessary plasticity to enter the modern world with that identity intact.

Acknowledgements

Since 1977 research and conservation activities in Huai Kha Khaeng Wildlife Sanctuary have been made possible by three awards from the New York Zoological Society, with some additional support from the Fauna and Flora Preservation Society in 1977. I wish to thank the Wildlife Conservation Director of the Royal Forest Department and Mr Phairot Suvanakorn, its former Director and now Deputy Director General of the Royal Forest Department, for sponsoring and facilitating my research in Thailand. Field work among the Hmong people would not have been possible without the assistance of my Thai counterpart Mr Tosaporn Leopad, formerly chief of the Khao Nang Rum Wildlife Research Station, Huai Kha Khaeng Wildlife Sanctuary.

References

Eudey, A. A. (1980) Pleistocene glacial phenomena and the evolution of Asian macaques. In *The Macaques: Studies in Ecology, Behavior and Evolution*, ed. D. G. Lindburg, pp. 52–83. New York: Van Nostrand Reinhold

Eudey, A. A. (1981) Morphological and ecological characters in sympatric populations of *Macaca* in the Dawna Range. In *Primate Evolutionary Biology*, ed. A. B. Chiarelli & R. S. Corruccini, pp. 44–50. Berlin: Springer-Verlag

Eudey, A. A. (1984) Distribution and habitat preference of Assamese macaques and stump-tail macaques in the Dawna range, Thailand. Paper presented at the 53rd Annual Meeting of the American Association of Physical Anthropologists, Philadelphia, Pennsylvania, April 1984

Geddes, W. R. (1967) The Tribal Research Centre, Thailand: an account of plans and activities. In *Southeast Asian Tribes, Minorities, and Nations*, vol. 2, ed. P. Kunstadter, pp. 553–81. Princeton, N.J.: Princeton University Press

Geddes, W. R. (1976) *Migrants of the Mountains*. London: Clarendon & Oxford University Press

Jintanugool, J., Eudey, A. A. & Brockelman, W. Y. (1984) Species conservation priorities in Thailand. In *Species Conservation Priorities in the Tropical Forests of Southeast Asia*, ed. R. A. Mittermeier & W. R. Konstant. Occasional Papers of the IUCN Species Survival Commission (SSC), no. 1, pp. 41–51

Keen, F. G. B. (1978) Ecological relationships in a Hmong (Meo) economy. In *Farmers in the Forest*, ed. P. Kunstadter, E. C. Chapman & S. Sabhasri, pp. 210–21. Honolulu: An East–West Center Book. University Press of Hawaii

Keen, F. G. B. (1983) Land use. In *Highlanders of Thailand*, ed. J. McKinnon & W. Bhruksasri, pp. 293–306. Kuala Lumpur: Oxford University Press

Kunstadter, P. (1978) Alternatives for the development of upland areas. In *Farmers in the Forest*, ed. P. Kunstadter, E. C. Chapman & S. Sabhasri, pp. 289–308. Honolulu: An East–West Center Book. University Press of Hawaii

Kunstadter, P. (1983) Highland populations in northern Thailand. In *Highlanders of Thailand*, ed. J. McKinnon & W. Bhruksasri, pp. 15–45. Kuala Lumpur: Oxford University Press

Kunstadter, P. & Chapman, E. C. (1978) Problems of shifting cultivation and economic development in northern Thailand. In *Farmers in the Forest*, ed. P. Kunstadter, E. C. Chapman & S. Sabhasri, pp. 3–23. Honolulu: An East–West Center Book. University Press of Hawaii

Kunstadter, P., Chapman, E. C. & Sabhasri, S. (1978) (ed.) *Farmers in the Forest*. Honolulu: An East–West Center Book. University Press of Hawaii

Lekagul, B. (1983) Message from the editor. *Conservation News*, **5**, 35–6

Mandorff, H. (1967) The hill tribe program of the Public Welfare Department, Ministry of Interior, Thailand: research and socio-economic development. In *Southeast Asian Tribes, Minorities, and Nations*, ed. P. Kunstadter, vol. 2., pp. 525–52. Princeton, N.J.: Princeton University Press

McKinnon, J. & Bhruksasri, W. (1983*a*) Preface. In *Highlanders of Thailand*, ed. J. McKinnon & W. Bhruksasri, pp. ix–xii. Kuala Lumpur: Oxford University Press

McKinnon, J. & Bhruksasri, W. (1983*b*) (ed.) *Highlanders of Thailand*. Kuala Lumpur: Oxford University Press

Ratanakorn, S. (1978) Legal aspects of land occupation and development. In *Farmers in the Forest*, ed. P. Kunstadter, E. C. Chapman & S. Sabhasri, pp. 45–53. Honolulu: An East–West Center Book. University Press of Hawaii

Sabhasri, S. (1978) Opium culture in northern Thailand: social and ecological dilemmas. In *Farmers in the Forest*, ed. P. Kunstadter, E. C. Chapman & S. Sabhasri, pp. 206–9. Honolulu: An East–West Center Book. University Press of Hawaii

Secter, R. (1982) Southeast Asia fast losing lush forests. *Los Angeles Times*, 22 March

Thailand, Division of Development and Public Welfare (1982) *Problems of the Hill Tribes and Opium Planting*. Bangkok: Ministry of the Interior (in Thai)

Young, G. (1982) *The Hill Tribes of Northern Thailand*. New York: AMS Press (Reprinted from the edition of 1962, Bangkok)

V.5

Restoring the balance: traditional hunting and primate conservation in the Mentawai Islands, Indonesia

A. H. MITCHELL AND R. L. TILSON

Introduction

The origins of the Mentawai Islands and the human and non-human primates inhabiting them are cloaked in mystery. The indigenous Mentawaians were originally part of the Neolithic migrations of proto-Malay peoples from southern China into Indonesia, arriving in their islands between 2000 and 4000 years ago (Loeb, 1929; Wirz, 1929–30; Schefold, personal communication). The islands remained largely undisturbed by outside influence until the turn of the century. Isolation enabled the people to retain many cultural practices once common throughout archaic Indonesia (Suzuki, 1958). Today the Mentawaians continue to hunt primates, among other animals, with bow and poisoned arrows, yet typically show their reverence for them by incorporating the primates into their art, music, dances and folklore (Schefold, 1980). The primates that the original immigrants found in pristine Mentawai forests had evolved in isolation from mainland Sumatra since the Pleistocene. The Mentawai rituals and taboos that surrounded all activities, including the hunt, kept people and forest resources in an equilibrium that is now being disrupted.

Our observations refer primarily to events on Siberut, the largest (4480 km^2) and most northern of the four Mentawai Islands lying from 85 to 135 km off the west coast of Sumatra, Indonesia, and just south of the Equator (Fig. 1). Available geomorphological and bathymetric evidence (Tjia, 1940; Fairbridge, 1966; Jongsma, 1970) and comparisons of the Mentawai mammifauna with that of adjacent Sumatra (Thomas, 1894; Miller, 1903; Lyon, 1916; Chasen & Kloss, 1927; Banks, 1931; Brandon-Jones, 1978) suggests that the Mentawai Islands were isolated from Sumatra before the last Pleistocene separation of 'Sundaland'. Other distinguishing characteristics of the islands include a high endemism of mammalian fauna, although impoverished by comparison with mainland forests, and an absence of large, natural predators (Chasen & Kloss, 1927; World Wildlife Fund, 1980). In Mentawai, primates are the most abundant large mammals and are thus most frequently hunted (Tilson, 1977). Against this background of biogeographical specialization and cultural isolation, field work began in the islands. Here we attempt to integrate our observations with those of others (e.g. Hanbury-Tenison, 1974; Whitten, Whitten & House, 1979; Whitten & Sardar, 1981) to provide a newer perspective

Fig. 1. Map of the Mentawai Islands, Indonesia, showing location of conservation area on Siberut.

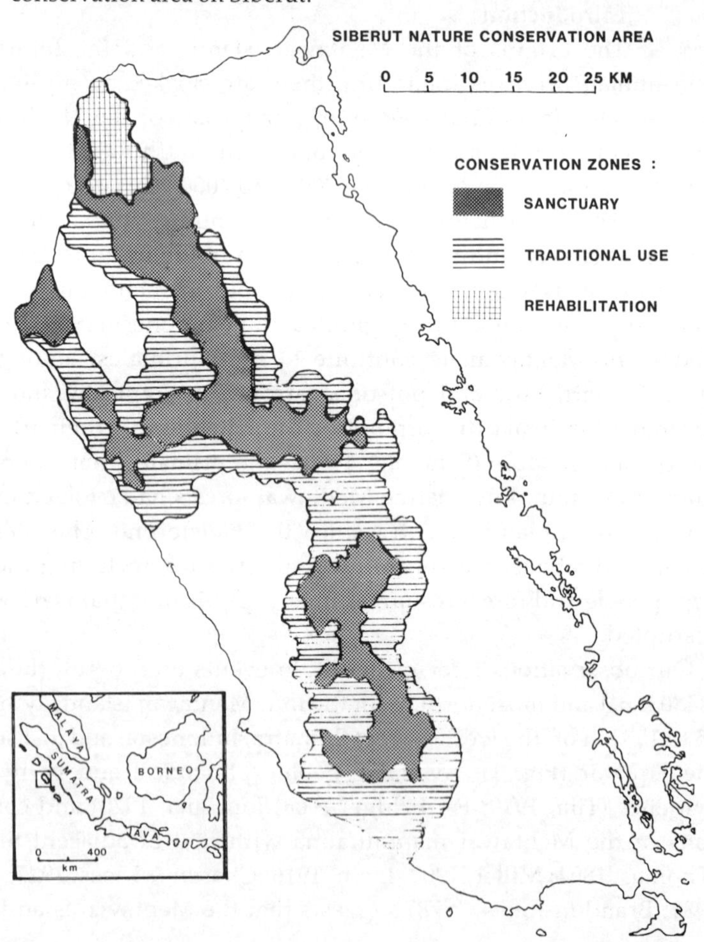

and clearer image of factors that now threaten the survival of the Mentawai Island primates.

The Mentawai primates

Four species of primate inhabit the Mentawai Islands. The only ape is a monomorphic black gibbon (*Hylobates klossi*), known locally as *bilou* (Tenaza, 1974; Tilson, 1980; Whitten, 1980). The two species of langurs are the sexually monomorphic Mentawai langur (*Presbytis potenziani*), known as *joja* (Tilson & Tenaza, 1976; Watanabe, 1981) and the asexually dichromatic (dark grey or golden) but sexually dimorphic pigtailed langur (*Nasalis concolor*), sometimes placed in a genus of its own (*Simias*), and known locally as *simakobu* (Tilson, 1977; Watanabe, 1981). The fourth species is the Mentawai, or Pagai Island, pigtail macaque (*Macaca nemestrina pagensis*), sometimes referred to as an endemic species (*M. pagensis*) and known as *bokkoi* (Tenaza, 1975; Watanabe, 1979; Whitten & Whitten, 1982).

The people and traditional religion of Siberut

Mentawai society was traditionally organized into exogamous, patrilineal, patrilocal clans, called *uma*, each consisting of about 5–10 families. Clan social life was centered around the communal long-house, also called *uma*. Each family had field houses, *sapo*, near cleared fields in addition to *lalep*, where free-foraging pigs were tended. Some local descent groups within a river basin consider themselves related to certain *uma* in other valleys, and together they constitute a sib, *sirubeiteteu*, but originally there were neither villages nor chiefs (Schefold, 1973). Their traditional economy was based on the sago palm (*Metroxylon* spp.) as the major source of carbohydrate (see Whitten & Whitten, 1981) along with the swidden cultivation of taro and bananas. Fishing, raising pigs and chickens, and hunting game provided protein.

The Mentawai religion, in common with other indigenous religions of Indonesia, is based on the soul concept (Loeb, 1929). Its basis is a belief in internal harmony in the environment, with a single overriding force, addresssed as *kina ulau*, or 'the beyond'. The religion concentrates on manifestations of this force: the souls (*simagere*) and the protective spirits of the air, water and earth (*sabulungan* or *tai*), including those that guard the animals of the forest (*taikaleleu*). The prevailing concepts of spirits and soul are far from clear and have in this century been influenced largely by Christianity, making it difficult to obtain precise information about traditional beliefs (Nooy-Palm, 1968).

For Mentawaians who adhere to the traditional religion, each thing lives and has an individual *simagere* separate from its host and free to wander at will. Souls of people, animals, plants, objects and even weather can communicate and mutually affect one another (Schefold, 1973). All acts of the soul have repercussions for the body, or host, and vice versa. All things possessing a soul emit a sort of radiation called *badjou*, the intensity of which varies with circumstances and the host. Souls and spirits are believed to be in a state of harmonious contact, but every human activity is a threat to this harmony. When interfering with the environment, man's actions can cause dangerous disturbances in the interplay of forces. In order to reduce human disturbance and restore balance in the environment, people living according to custom accompany most of their activities with various religious ceremonies, known as *punen* (or *puliaijat*). Various kinds of *punen* may involve an individual family, clan or whole village. The *punen* and *lia*, ceremonial rites on a smaller scale, may be seen as the integrating factors of the traditional culture: 'The "punen" dominates the entire life of the Mentawaian from cradle to grave' (Börger, 1932).

The rites of *punen* provide the means by which a person can render exposure to unfamiliar *badjou* harmless (Schefold, 1973). Man is thus able to influence everything in his environment through the souls and spirits, and to do this special mediators (*gaut*), usually certain plants, must be used to convey man's wishes (Schefold, 1973).

The *kerei* (medicine man or seer) has the ability to both prophesy and perform exorcisms (Loeb, 1929). The *kerei* is said to perceive both spirits and souls and can be taught by them, such as the cause of a particular illness. Through the use of *gaut*, a *kerei* may accustom two souls to each other in anticipation of initial confrontation with unfamiliar *badjou*. Mediators are also used to lure together helpful forces and banish evil, and are used to conjure the powers responsible for an illness brought about by contact with too-powerful *badjou* (Schefold, 1973).

Taboos and the control of hunting

In many cultures where animals are believed to possess souls, the killing and eating of animals may be fraught with spiritual hazards (e.g. Rasmussen, 1929). Ceremonial rites required prior to the hunt indicate relationships between humans and animals that are of a reciprocal interactive nature (Marsh, 1954; Hallowell, 1960; Schefold, 1973). Thus, the hunter often must spend more time utilizing his knowledge of animal behavior rather than improving his hunting equipment:

> Appropriate behavior toward animals is prominently based upon familiarity with animal behavior and includes ways of living peacefully with animals, of maintaining a discourse with them, as well as the appropriate behaviors, the highly coordinated movements of the hunter proceeding toward a kill, and appropriate social behavior where other hunters are involved (Laughlin, 1968, p. 305).

The traditional Mentawai culture is an example of one governed by rituals and taboos (or *kekei*) that have managed sustainable use of resources for thousands of years. Their rituals and taboos govern social relations, land use, forest gathering and hunting to produce a balance between their society and the environment, a 'ritually regulated ecosystem' (Rappaport, 1968). Hunting is a major social activity for the men of Siberut and is closely linked with traditional religion. Perhaps more than any other socio-religious community building in Southeast Asia and Oceania, the traditional Mentawai *uma* has the character of a 'ceremonial hunting lodge' (Nooy-Palm, 1968). Nooy-Palm has noted that because the hunt, the main pursuit of men, is surrounded with much ritual and taboo, it is considered more important to have men living in the *uma* than women. Men are associated with the earth, forest, forest animals and forest spirits, the *taikaleleu*, while women are linked with water, tubers and the water spirits, the *taikaoinan*.

The concept of restoring harmony to an environment disturbed by man's activities is fundamental to the Mentawaian apprehension of taboos (Schefold, 1972). Taboos, observed during periods of *punen*, govern the hunting of game. During the hunt, those not hunting also must observe various taboos at home. Schefold (1972) notes:

> . . . often a relation exists between such psychic characteristics and the form of the objects: e.g., one should not eat anything sour before going on a hunt with weapons since sour and sharp are considered to be equivalent and a man must not at the time he lays claim to the sharpness of the weapons lay claim to other things having similar characteristics. The weapons would be offended and wound the hunter.

The hunt was a regular feature of every *punen* which, however, stood virtually apart from the other ceremonies (Nooy-Palm, 1968). For example, Loeb (1929) noted how a successful hunt was necessary for a successful *punen* that was required for the opening of a new *uma*:

> Now the men had to go hunting in order "to strengthen" ("tugege") the hard wood used in construction. If they obtained monkeys, their work was correct. When they returned, coconuts were sliced with monkey meat, and the

> priest sacrificed. "This is for you, spirits of the monkeys which were shot. If we go out hunting again, summon your friends and let them be fat ones."

Similarly a monkey was hunted after floor and wall boards were in place and, in coastal areas, a sea-turtle was hunted before the roof of the *uma* could be completed (Loeb, 1929). Prior to the hunt, the *rimata*, a clan elder, prayed and brought into the *uma* a chicken with which he touched the heads of all those present as well as the weapons and dogs. In the evening the *kerei* danced to please the guardian spirits of the animals so that the hunt would be successful (Nooy-Palm, 1968).

Skulls and mandibles of monkeys and other game are hung in the *uma* and are decorated with leaves of plants believed to possess magical properties. Over the years some *uma* have amassed hundreds of skulls (see below). In this way the souls of the dead animals are persuaded to stay in the *uma* rather than wander into the forest. The souls that reside in these skulls are then ritually implored to summon the souls of their living relatives into the *uma*. 'If the hunters then come across one of these animals, then – again, "subconsciously" – the animal will desire nothing better than to be shot, so that it can be reunited with its soul' (Schefold, 1973). Such a procedure may have been believed to be necessary for the rebirth of the slaughtered animals so that they are preserved as game (Eliade, 1964; Nooy-Palm, 1968).

Impact of human predation on Mentawai primates

Primates are the most abundant large mammals in Mentawai and are thus the hunting mainstay (Tilson, 1977; WWF, 1980). The survival of the primates in Mentawai has to a great extent been due to the taboos governing hunting. For example, taboos generally forbade the hunting of *bilou* (Kloss's gibbon) and the golden-phase *simakobu* (pigtailed langur), two animals believed to have extremely strong *badjou*. When *bilou* is hunted in some areas it may be eaten only by young men and only outside the *uma*, it may not be hunted while it is calling, and, in most regions, may never be taken into an *uma* (WWF, 1980). Rarely are *bilou* skulls found in the *uma*, gibbon hair tufts are not placed on quiver lids, and in areas where the *bilou* is now hunted, its flesh is still considered distasteful, an opinion that is probably rooted in the abandoned taboo (Tilson, 1974).

In the Sarareiket and Sakuddei regions of Siberut, traps (*leleupleup*) made of bound saplings or palm trunks are set for *bokkoi* (pigtail macaque); occasionally terrestrial foraging *simakobu* are also caught in them (Tilson, 1974). Usually baited with sago or taro, traps have restricted use because they are regarded as effective only in connection with an intricate ritual known only to a few and require the observance of strong taboos; several years may pass with no traps being used in an area (Schefold, personal communication).

Other than humans, the only potential predators are large pythons and raptors (Chasen & Kloss, 1927), none of which are major predators upon Mentawai or mainland populations of primates. Mentawai Islanders thus constitute the major predator on primates. We base this conclusion on three lines of evidence: (1) *uma* skull and quiver hair-tuft counts; (2) personal observations of hunting forays and hunter interviews; and (3) observed patterns of primate anti-predator behavior in response to approaching humans.

Uma skull counts for 18 different clans living in four river basins in southeastern and central Siberut provided estimates of the composition of primates among mammalian prey. In these *uma*, primates constituted 77% (598) of 777 large mammalian skulls taken by hunters (Tilson, 1977). Subsequent counts of skulls within the same river basins by others (Bakar, 1979; WWF, 1980; Weitzel & Mitchell, unpublished) revealed similar relative proportions. At two other sites in southwestern Siberut, the Sakuddei clan accounted for 432 primate skulls. Of the total skulls (1030) from all 20 sites, 536 (52.0%) were *simakobu*, 285 (27.7%) were *bokkoi*, 206 (20.0%) were *joja* (Mentawai langur), and only 3 (0.3%) were *bilou*. It is also customary for a hunter to attach a tuft of hair to the lid of his quiver for each monkey killed. Hair tufts from 21 quiver lids accounted for 366 primates: 208 (56.8%) were *simakobu*, 84 (23.0%) were *joja* and 74 (20.2%) were *bokkoi* (Tilson, 1977). These tufts, together with the skulls, may represent as many as 1396 primates! However, neither the period of time nor the range over which hunting occurred can be determined accurately.

Additional evidence of human predation was derived from personal hunting observations and interviews. During 28 months (1972–74) of field study in central Siberut, 17 primates disappeared from Tenaza's and Tilson's study population and another 18 primates were killed outside but near the study area (Tenaza & Tilson, 1985). Whitten (1980) and Watanabe (1981) reported losses under similar circumstances from their Siberut field sites and attributed them to human predation. Tenaza (1974) accompanied five Mentawai hunters during a 6-day hunt during which four primates were killed (Tenaza & Tilson, 1985).

The long history of human predation on Siberut primates has resulted in: (1) specialized *bilou* alarm calls that specifically signal the presence of humans (Tenaza & Tilson, 1977); (2) distraction displays produced by male *joja* in response to humans (Tilson & Tenaza, 1976); and (3) competition between *bilou* and *joja* for liana-free sleeping trees, presumably to avoid nocturnal predation attempts by humans (Tenaza & Tilson, 1985). A disturbing trend seen throughout the island was the high proportion of *simakobu* killed for meat, which is considered by many to be the best tasting of the four primates. Also, it appears to be the easiest of the four primates to hunt successfully (Tilson, 1979; WWF, 1980). It is our opinion that *simakobu* (pigtailed langur) must continue to be considered one of the world's most endangered primates.

Management and conservation

Resettlement of clans into centralized villages, prohibition of traditional dress, ceremony and religion, first by missionaries and then by the Indonesian government, together with the spread of commercial logging, have all contributed to the disintegration of long practiced social relations that kept the Mentawai people in balance with their resources. The degree of disintegration varies considerably throughout the islands. The effects of acculturation on hunting are mixed. In many parts of the islands, traditional ceremonies associated with hunting are now performed by very few people, and primate skulls are frequently discarded or, if placed within a house, often lose their earlier significance. The timing and location of hunting are now largely a matter of whim or opportunity of the individual. Relaxation of taboos, for example, has resulted in more frequent gibbon hunting, and the use of traps for other primates may increase. A once acceptable balance between hunter and hunted has thus shifted, with local or island-wide primate extinctions likely without the implementation of new restrictions. Prohibiting primate hunting, as decreed by the Indonesian authorities, is not the solution. Controls that are too strict, too quickly implemented, and which do not take into account the reality of local values and attitudes, are likely to lead to resentment and a 'backlash' of excessive hunting; just the opposite of the original intention. Proper management of Mentawai primate populations thus will require maintenance of good relations with the traditional hunting society.

Commercial logging has dramatically altered the natural ecosystems and culture of the southern Mentawai islands. In contrast, much of Siberut remains relatively undisturbed, though logging is present and an increasing threat (Table 1). The Indonesian Government has renegotiated some timber concessions and has given high priority to the establishment of nearly one-third of this 4500 km^2 island as a conservation area zoned for both strict protection and traditional use by local people (Fig. 1). Logging concessions have been restricted to areas outside of this reserve. To assist the government's Department of Nature Conservation (PPA), a 5-year (1983–1987) management plan for the reserve was prepared by WWF/IUCN as a follow-up to an earlier, preliminary master plan (WWF, 1980; Mitchell, 1982). The primary aims of reserve management were:

1. To maintain the ecological and genetic diversity of Siberut's lowland tropical rainforest
2. To benefit local people by respecting customary rights to forest resources within the reserve and traditional use which does not threaten the conservation value of the reserve
3. To integrate the reserve into a regional land-use plan for the Mentawai Islands that ensures both long-term conservation and local socioeconomic development
4. To utilize the qualities of the reserve to their best advantage through training and research that promote the first three objectives.

Effective conservation will not be achieved on Siberut solely by setting aside a 'protected area' and preparing a plan for its management. Although these are important initial steps, they unfortunately remain the only steps taken so far. To avoid yet another 'exercise in futility', a follow-up to the management plan with serious implementation and revision, if necessary, to suit changes in local conditions, must now be made. Renewed international interest and attention to Siberut, which had mounted in the 1970s and early 1980s (Hanbury-Tenison, 1974; Tilson, 1979; Whitten *et al.*, 1979; WWF, 1980; Whitten

Table 1. *Categories and areas of disturbance on Siberut Island (1982). Category descriptions are referable to WWF (1980) and Mitchell (1982)*

Category	Area (ha)	% of total
I. Undisturbed	282 900	63
II. Moderately disturbed	71 700	16
III. Very disturbed	26 700	6
IV. Disturbed by logging	57 700	13
V. Totally disturbed (e.g. village areas)	9 000	2
Totals	448 000	100

& Sardar, 1981; Mitchell & Weitzel, 1982) is vital to the successful implementation of the management plan. But of even greater importance is a recognition of the original stewards of the Siberut forests. Taboos that assumed protection of the individual served to prevent over-exploitation of resources and kept people, forests and game in a sort of balance for thousands of years. Yet, a continual, and ultimately self-defeating, problem with many conservation plans is that 'they have been done for the people rather than *with* the people (Hanks, 1984). Any plans to achieve conservation goals on Siberut must be suited to local attitudes and conditions, which are continually changing and which are in many ways unique. Securing a future for the Mentawai endemic primate species will require the creation of ecologically sustainable systems of land and resource use that extend beyond the boundaries of the nature reserve (Dasmann, 1982; Mitchell, 1982). The fate of the endemic Mentawai primates is inexorably linked to local patterns of land use and hunting practices. Without the full involvement and cooperation of local people in the management of their homeland, destruction will become the unifying theme for Mentawai ecosystems and traditional culture.

Acknowledgements

Field work was supported by funds from the World Wildlife Fund, the American Committee for International Wildlife Protection, the Alleghany Fund for the Study of Wildlife, the Carnegie Museum of Natural History and the Fauna Preservation Society. We were sponsored in Indonesia by the Indonesian Institute for Sciences (LIPI), the Bogor Biological Museum and the Directorate of Nature Conservation and Wildlife Management (PPA). We wish to thank Ir. Soetomo Soepangkat and Ir. Sumin of PPA, Sumatra, for their valuable assistance in the preparation of the Siberut management plan. We also thank Dr Harold Conklin and Dr Richard Potts for their critical reviews of this paper.

References

Bakar, A. (1979) Morphological study of mandibles of primates living on Siberut Island, Indonesia. In *A Comparative Sociological Study on Coloboid Monkeys in Tropical Asia*, Report of overseas scientific survey in 1976–1978, pp. 95–141. Kyoto University, Inuyama: Primate Research Institute

Banks, E. (1931) The distribution of mammals and birds in the South China Sea and West Sumatran Islands. *Bull. Nat. Mus. St. Singapore*, **30**, 92–6

Börger, F. (1932) Wie man sicht freuet in der Ernte. *Das Missionsblatt*, (*Barmen*), 71–3

Brandon-Jones, D. (1978) The evolution of recent Asia colobinae. *Recent Adv. Primatol.*, **3**, 327–30

Chasen, F. N. & Kloss, C. B. (1927) Spolia Mentawiensa, mammals. *Proc. Zool. Soc., Lond.*, **53**, 797–840

Dasmann, R. F. (1982) The relationship between protected areas and indigenous peoples. Paper presented at World National Parks Congress, Bali, Indonesia, 17 October 1982

Eliade, M. (1964) *Shamanism: Archaic Techniques of Ecstasy*. London

Fairbridge, R. W. (1966) Mean sea level changes, long-term eustatic and other. *Encyclopedia of Oceanography*, pp. 479–85

Hallowell, A. I. (1960) Ojibwa ontology, behavior and world view. In *Culture in History: Essays in Honor of Paul Radin*, ed. S. Diamond. Columbia, N.Y.: Columbia University Press

Hanbury-Tenison, R. (1974) Last chance for the handsome Mentawaians. *Geogr. Mag.*, **46**, 276–81

Hanks, J. (1984) Conservation and rural development: towards an integrated approach. In *IUCN* Committee on Ecology, Paper No. 7, *Traditional Life-styles, Conservation and Rural Development*, pp. 60–7

Jongsma, D. (1970) Eustatic sea level changes in the Arafura Sea. *Nature*, **228**, 150–1

Laughlin, W. S. (1968) Hunting: an integrating biobehavior system and its evolutionary importance. In *Man the Hunter*, ed. R. B. Lee & I. DeVore, pp. 304–20. New York: Aldine

Loeb, E. M. (1929) Mentawaei religious cult. *Univ. Calif. Publ. Am. Arch. Ethn.*, **25**, 185–247

Lyon, M. R., Jr. (1916) Mammals collected by Dr. W. L. Abbott on the chain of islands lying off the western coast of Sumatra, with descriptions of twenty-eight new species and subspecies. *Proc. U.S. Nat. Mus.*, **52**, 437–62

Marsh, G. H. (1954) A comparative study of Eskimo-Aleut religion. *Anthropol. Papers Univ. Alaska*, **3**(1), 21–36

Miller, G. S., Jr. (1903) Seventy new Malayan mammals. *Smithsonian Misc. Coll.*, **45**, 1–73

Mitchell, A. H. (1982) *Siberut Nature Conservation Area: Management Plan 1983–1988*. Bogor, Indonesia: World Wildlife Fund/IUCN Report

Mitchell, A. H. & Weitzel, V. (1982) Monkeys and men in the land of mud. *Hemisphere*, **27**, 308–14

Nooy-Palm, C. H. M. (1968) The culture of the Pagai-Islands and Sipora, Mentawei. *Trop. Man*, **1**, 153–241

Rappaport, R. A. (1968) *Pigs for the Ancestors: Ritual in the Ecology of a New Guinea People*. New Haven: Yale University Press

Rasmussen, K. (1929) Intellectual culture of the Hudson Bay Eskimo, I: Intellectual culture of the Iglulik Eskimos. In *Report of the Fifth Thule Expedition, 1921–24*, vol. 7(1). Copenhagen: Gyldendalske Boghandel

Schefold, R. (1972) Religious conceptions on Siberut, Mentawai. *Sumatran Res. Bull. (University of Hull)*, 12–24

Schefold, R. (1973) Schlitztrommeln und Trommelsprache in Mentawai. *Z. Ethnol.*, **98**, 36–73

Schefold, R. (1980) *Spielzeug für die Seelen: Kunst and Kultur der Mentawai-Inseln (Indonesien)*. Zurich: Museum Rietberg

Suzuki, P. (1958) Critical survey of studies on the anthropology of Nias, Mentawai and Enggano. *Koninklijk Inst. Tall-Landen Volkenkunde, Bibl. Ser.* 3

Tenaza, R. (1974) I. Monogamy, territory and song among Kloss' gibbons in Siberut Island, Indonesia. II. Kloss' gibbon sleeping trees relative to human predation: implications for the socio-ecology of forest-dwelling primates. Ph.D. thesis, University of California, Davis

Tenaza, R. R. (1975) Territory and monogamy among Kloss' gibbons (*Hylobates klossii*) in Siberut Island, Indonesia. *Folia Primatol.*, **24**, 60–80

Tenaza, R. R. & Tilson, R. L. (1977) Evolution of long-distance alarm calls in Kloss's gibbon. *Nature*, **268**, 233–5

Tenaza, R. R. & Tilson, R. L. (1985) Human predation and Kloss's gibbon (*Hylobates klossii*) sleeping trees in Siberut Island, Indonesia. *Am. J. Primatol.* (in press)

Thomas, O. (1894) On some mammals collected by Dr. E. Modigliani in Siberut, Mentawai Islands. *Annali Mus. Cir. Stor. Nat. Giocomo Doria*, **14**, 660–72

Tilson, R. L. (1974) *Man and Monkey: Concepts in Conservation for the Mentawai Islands' Primates.* Bogor, Indonesia: World Wildlife Fund/IUCN Report

Tilson, R. L. (1977) Social organization of Simakobu monkeys (*Nasalis concolor*) in Siberut Island, Indonesia. *J. Mammal.*, **58**, 202–12

Tilson, R. L. (1979) Der unbekanute Affe. *Tier*, **5**, 20–3

Tilson, R. L. (1980) Monogamous mating systems of gibbons and langurs in the Mentawai Islands, Indonesia. Ph.D thesis, University of California, Davis

Tilson, R. L. & Tenaza, R. R. (1976) Monogamy and duetting in an Old World monkey. *Nature*, **263**, 320–1

Tjia, H. D. (1940) Quaternary shore lines of the Sunda Land, Southeast Asia. *Geologie Mijn.*, **49**, 135–44

Watanabe, K. (1979) Some observations of the Mentawai pigtail macaques, *Macaca nemestrina pagensis* on Siberut Island, West Sumatra, Indonesia. In *A Comparative Sociological Study on Coloboid Monkeys in Tropical Asia*, Report of overseas scientific survey in 1976–1978, pp. 86–92. Kyoto University, Inuyama: Primate Research Institute

Watanabe, K. (1981) Variations in group composition and population density of the two sympatric Mentawaian leaf-monkeys. *Primates*, **22**, 145–60

Whitten, A. J. (1980) The Kloss Gibbon in Siberut Rain Forest. Ph.D thesis, Cambridge University, Cambridge

Whitten, A. J. & Sardar, Z. (1981) Masterplan for a tropical paradise. *New Scientist*, July, 230–5

Whitten, A. J. & Whitten, J. E. J. (1981) The sago palm and its exploitation on Siberut Island, Indonesia. *Principles*, **25**, 91–100

Whitten, A. J. & Whitten, J. E. J. (1982) Preliminary observations of the Mentawai macaque on Siberut Island, Indonesia. *Int. J. Primatol.*, **3**, 445–59

Whitten, A. J., Whitten, J. & House, A. (1979) Solution for Siberut? *Oryx*, **15**, 166–9

Wirz, P. (1929–30) Het eiland Sabirut en zijn bewoners. *Nion*, **14**, 131–9, 187–92, 209–15, 241–8, 337–40, 377–97

World Wildlife Fund (1980) *Saving Siberut: A Conservation Masterplan*. Bogor: WWF Indonesia Programme

Part VI

Conservation: Trends and Practice

Editors' introduction

Over the years, conservation practices have changed with new knowledge and many new approaches have been attempted. The importance of a diverse range of activities to conservation efforts is still just beginning to be understood. If a true impact is to be made, conservation must be approached from all sides and the programmes and solutions must constantly be evaluated and reassessed with the knowledge gained. In many respects, this has been the theme of the entire volume. In this section, we have integrated some of these approaches, but in doing so, have touched briefly upon relatively few areas.

Kortlandt (Chapter VI.1) addresses the importance of being able to identify accurately the natural habitats of the primates under study. The ability of primates to exploit changing environments, with resulting alterations in behaviour, was reported in the previous section. As Kortlandt points out here, similar changes might be expected in habitats that are natural but may have undergone cyclical or abrupt alteration. The importance of taking the nature of the habitat into account when interpreting results is stressed. Green (Chapter VI.2) reviews the state of Landsat (remote sensing satellite) technology and its use as a tool for assessing primate habitats on a large (even global) scale. Such techniques enable direct comparisons of changes over time, which may be difficult to ascertain accurately through ground observations. They also allow the assessment of pressures on habitats at distances from the areas under study, the impact of which on the areas of interest may not be considered or understood from a restricted, earthbound perspective.

The problem of protected areas developing into isolated pockets and the effects of this on populations of primates are discussed by Fedigan (Chapter VI.3). This isolation of protected species is of increasing concern for wildlife reserves in general, since it is often easier to establish a reserve than to understand the long-term impact of isolation on the genetic diversity of different species. Fedigan uses population counts in combination with studies of behaviour to assess the future of protected primates in Costa Rica.

The need for zoos to spend more time studying captive breeding is addressed by Lindburg *et al.* (Chapter VI.4), with examples of how such research can assist in the perpetuation of endangered species both in captivity and the wild. Stevenson (Chapter VI.5) covers the importance of collating and evaluating information on the reproductive biology of primate species, and how this can be used to improve management practices and increase breeding efficiency at captive facilities. Experimental manipulations of conditions in order to improve the reproductive potential of different prosimian species are presented by Izard and Simons (Chapter VI.6). It is important that captive propagation of all primate species which are required for experimental studies be of a level to meet such needs. The improvement of primate management and breeding practices in captivity to the extent where all biomedical requirements could be supplied without drawing from natural populations would be a very significant step in primate conservation.

The final paper by Mittermeier (Chapter VI.7) presents a brief overview of the conservation problems facing primate species on a global scale. The major threats to populations are listed and the priority areas and species are defined. He also suggests possible general programmes for primate conservation. The need to understand the status and distribution of primate species is highlighted by this paper.

VI.1

Studying the treescape

A. KORTLANDT

In the past we have been told that 'the apes brachiate' as an adaptation to the continuous canopy of the rain forest. Seen from a vantage point the canopy may indeed appear to be continuous, owing to the perspective. However, when one climbs a tall tree by means of a long ladder tied to the trunk, or by using a catapult in combination with mountaineering equipment and a rope ladder, the overall picture of the vegetation is quite different. Most rain forests show a stratified architecture. Between the different layers the forest is often very open. The so-called C-layer, at about one-third of the height of the tallest trees, is as a rule more or less continuous. Above it, however, in the B-layer, the tree crowns tend to keep distances of at least 0.5–1 m from one another, apparently as a protection against the movements of herbivorous insects, rodents, etc. ('Crown-shyness': Ng, 1977). Finally, the tall emergents, the A-layer, stand widely apart from one another and tower high above the B-layer. The same picture emerges also from stereoscopic photographs taken from an ultra-light aircraft flying low over the forest (Vooren & Offermans, 1984). Thus there is no continuous canopy above the C-layer. Brachiation from one large tree crown to the next is impossible as a rule.

In fact, the essence of gibbon brachiation is a special technique allowing acceleration during repeated ricochet in order to throw the body through the air across wide gaps, rather than a way of moving around by arm-swinging at constant speed. The essence of orang-utan suspensory locomotion is the special ability to make thin trees, long branches or hanging lianas sway or swing with increasing amplitude, in order to be able to reach out and grasp the next tree, rather than to climb like a sloth. Chimpanzees (especially the adults) use their long arms primarily to semi-embrace thick tree trunks and in this way vertically to climb tall trees that are too thick and too smooth to be climbed by monkeys. Finally, those forest-dwelling monkeys that live at the higher levels are adapted for jumping quite far. All this shows that studying the treescape can reveal many things of which anatomists have rarely been aware.

Differences between rain forests in different geographical and ecological conditions may be analyzed by comparing the various profiles drawn by foresters (e.g. Eggeling, 1947; Richards, 1952; Germain & Evrard, 1956; Hallé, Oldeman & Tomlinson, 1978; Hladik, 1978). For instance, the Southeast Asian Dipterocarp climax forests are characterized by an abundance of phallic tree shapes (Fig. 1), whereas the African primary and mature-secondary forests show a predominance of umbrella- and funnel-shaped tall trees (Fig. 2). This might possibly explain the difference in adaptation between, on the one hand, the Asian apes and, on the other hand, the two species of chimpanzee. Furthermore, seen from a plane at low altitude, the evergreen equatorial lowland rain forests in Africa show a dense stand of the tall trees, like sheep huddled together (Fig. 3), whereas the semi-deciduous lowland rain forests on the same continent have a much more open stand of the tall trees, like sheep standing somewhat apart, so that from an oblique angle one can see the tree trunks as if

Fig. 1. Tree shapes in a Southeast Asian primary (Dipterocarp) rain forest (Mt Dulit, Borneo).

Fig. 2. Tree shapes in an African mature colonizing (*Maesopsis*) rain forest (Budongo Forest Reserve, Uganda).

2

Fig. 3. The dense stand of the tall trees in an African evergreen lowland rain forest (Réserve de Faune de Douala-Edéa, Cameroon, annual rainfall ~4000 mm).

these were the legs of the sheep (Fig. 4). These and other features can easily be noticed by an alert, ecology-minded observer, but very little research has been done on them. The tropical forest canopy is indeed the last unexplored part of our world. Yet the implications for primate locomotion are obvious.

The shape of trees often reveals their ecological adaptation. Rain forest trees compete for height. Consequently they form long, thin trunks with a small crown on top until they have penetrated through the C-layer and can fan out (Fig. 5). The trees in the few remaining relicts of the once widespread Sudanian dry forest show enormous ball-shaped crowns adapted to the open stand of this type of arboreal vegetation (Fig. 6). Sahelian trees strive chiefly for width and, when growing old, may develop crowns several times as wide as their height (Fig. 7). Thus the differences in size and anatomy of the hands and feet between, for example, chimpanzees and baboons partly reflect the difference between predominantly vertical and predominantly horizontal pathways in the trees. They also reflect the difference between the thin and smooth bark in the forest and the thick rough bark that protects savanna and woodland trees against sun heat and bushfires.

The difference between primary, secondary and secondarized forests is not only a difference in botanical composition, but also in physiognomy and architecture of the treescape. The primary rain forest has a regular profile, very straight trunks, usually a quite open

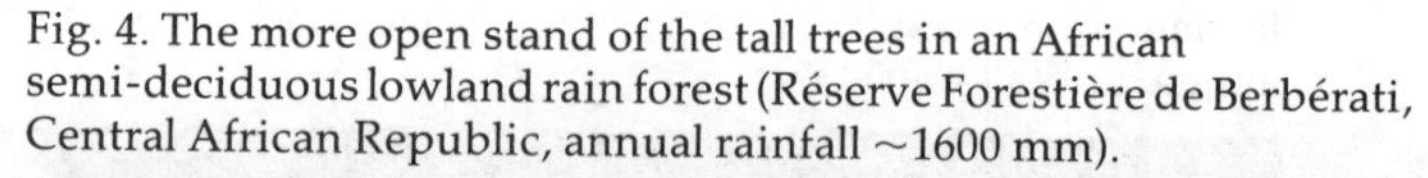

Fig. 4. The more open stand of the tall trees in an African semi-deciduous lowland rain forest (Réserve Forestière de Berbérati, Central African Republic, annual rainfall ~1600 mm).

Fig. 5. The result of competition and adaptation for height in a rain forest relict tree (Benu-Mbau road, Zaïre, annual rainfall 1700 mm).

Fig. 6. The ball-shaped crowns of trees in relicts of Sudanian dry forest (Forêt Classée de Kabo, Ivory Coast, annual rainfall 1250 mm).

Fig. 7. The result of competition and adaptation for width in a Sahelian relict tree (Ndutu, Tanzania, annual rainfall 600 mm).

Fig. 8. Architecture of a primary lowland rain forest (foot of the Nimba Mts, Liberia).

Fig. 9. Architecture of a secondary lowland rain forest, with three primary relict trees in the background (Liberian peneplain, Buchanan–Nimba railroad).

Fig. 10. Architecture of a secondarized rain forest resulting from selective logging (Budongo Forest Reserve, Uganda).

undergrowth and relatively few lianas, which may, however, be thicker than a human leg (Fig. 8). The secondary forest that results from slash-and-burn agriculture, on the other hand, shows a rather flat canopy (occasionally interrupted by tall relict trees of the former primary forest), more irregular trunks, a very dense undergrowth and an abundance of young lianas (Fig. 9). Finally, the secondarized forest which is the result of selective logging develops a quite irregular pattern of vegetation, with an equally irregular canopy, a dense undergrowth, and many young (and a few old) lianas (Fig. 10). This third type of forest seems to be the most densely populated primate habitat. All these differences in vegetation can be easily observed, but they have rarely been accounted for in research on primate locomotion and ecology.

Another aspect which is often overlooked by field workers who study primates in pseudo-natural habitats is the absence of the effect caused by the action of large herbivores as occurs in a really natural ecosystem. Elephant, hippo, rhino, buffalo, giraffe, zebra, wildebeest, antelopes and pigs damage or kill trees and bush, each in their own way, by browsing, breaking, debarking, uprooting, digging and trampling. Thus they open up, thin out, rejuvenate and may eventually 'parklandize' the existing forest vegetation (Fig. 11). Under certain conditions the interaction between the 'bulldozer her-

Fig. 11. Thinning of a rain forest due to the action of elephant, hippo and buffalo (Rabongo Forest, Murchison Falls National Park, Uganda).

bivores' and the vegetation leads to a cyclical process: forest → grassland → bushland → forest, which may result in a mosaic landscape (Fig. 12).

In a really natural forest habitat the botanical composition may be affected owing to selective feeding by the large herbivores (e.g. Jackson, 1956; Mitchell, 1960; Guillaumet, 1967; Wing & Buss, 1970; von Gadow, 1973; Laws, Parker & Johnstone, 1975; Olivier, 1978; Merz, 1981; Short, 1981; Janzen & Martin, 1982). For instance, fig trees (an important source of primate food) tend to be highly resistant to debarking and killing by elephants. In woodlands and savannas, on the other hand, it is the presence of a dense population of grazers that decreases the frequency and the heat of man-made grassfires, and thus may facilitate the regeneration of bushland and forest. Such factors are absent when the large game has been exterminated.

For primate field workers it is not easy, of course, to estimate which types of landscape represent a more or less natural equilibrium between vegetation and herbivores. Ungulate experts disagree even on the question of whether the natural condition is a state of equilibrium or an oscillating system. Yet the primate field worker who wishes to study his favourite species under really natural conditions (rather than under pseudo-natural conditions) needs some criteria by which to judge whether or not his study site may be considered at least as a

Fig. 12. The mosaic landscape resulting from 'bulldozer herbivores' in a mountain forest (Aberdares National Park, Kenya).

valid approximative model of a natural habitat. The following guidelines may assist him:

1. White's (1983) memoir to the new UNESCO/AETFAT/USNO (1983) *Vegetation Map of Africa* provides descriptions of the natural vegetation in the different floristic belts distinguished on the map.
2. Areas of secondary forest cannot be considered as natural habitats. However, so-called 'old mature' secondary forests in the evergreen forest belt can be considered more or less as a model for the natural forests in the semi-deciduous forest belt (e.g. in Southwest Ivory Coast, South Nigeria, Cameroon and Gabon).
3. In the semi-deciduous rain forest belt, forests in which a moderate amount of tree felling for timber has taken place provide a model for the physiognomy of such forests before the extermination or decimation of the elephants, even though the botanical composition may have changed (e.g. the Budongo Forest).
4. Areas which are subject to annual or frequent bushfires and areas which have a moderate or dense population of cattle are covered by a vegetation resulting from human action. However, in hilly and mountainous areas in such regions, one may occasionally find small, steep valleys in which the original vegetation has been protected against woodcutting, bushfires and/or overgrazing by the topography and a rocky soil. Such places can be recognized by the presence of a wide diversity of tree species in all age classes, including many tall trees with straight trunks and large crowns, a dense undergrowth, a rich flora and the absence of charred and warped tree trunks (Fig. 13). Furthermore, in some regions, the original vegetation is carefully protected by the native population at certain sacred sites, initiation places, graveyards, etc. (e.g. Beni and Bossou).
5. Most national parks and game reserves in eastern and southern Africa suffer (or suffered until recently) from overpopulation by large game, especially elephants. The causes are (or were) the expulsion of game from elsewhere, the suppression of human hunting (poaching) and, in prehistoric times, the extinction of the large sabretooth cats. Indications of overpopulation by large game can be seen by the presence of many damaged, dying and dead trees, the absence or scarcity of

young trees, excessive bush encroachment, and floristic poverty. Conversely, the presence of a lush and diversified undergrowth in a forest or dense woodland may indicate a sort of equilibrium between the vegetation and the large game (e.g. in the northern part of Manyara National Park).

6. In regions having a very long dry season the game (or the cattle) will spend much of their time in the gallery forests along brooks and rivers. An intact gallery forest with a continuous canopy therefore indicates, in such regions, an underpopulation of game (or cattle), and consequently cannot be considered to represent a really natural habitat.

Keeping these six points in mind may help to forestall erroneous extrapolations from pseudo-natural to natural habitats with regard to the ecology and behaviour of primates.

Note

The author's lecture given at the Congress consisted essentially of a presentation of 75 slides showing various aspects of primate habitats, and interlinked with one another by a spoken commentary. The present text should, therefore, be considered as a summary with incomplete presentation of the visual data. However, a photocopied version of the original, spoken text containing black-and-white prints of all slides may be obtained by sending to the author's department either a postal cheque (postal money order) of 5 Dutch guilders, or an approximately equivalent bank note (e.g. US $2).

Fig. 13. A park landscape shaped by grazing cattle (circled), without overgrazing, due to the protection caused by the hilly topography and a rocky soil (Télimélé–Gaoual road, Guinea).

Photographs

References

Eggeling, W. J. (1947) Observations on the ecology of the Budongo rain forest, Uganda. *J. Ecol.*, **34**, 20–87

Gadow, K. von (1973) Observations on the utilization of indigenous trees by the Knysna elephants. *Forestry S. Afr.*, **14**, 13–17

Germain, R. & Evrard, C. (1956) Etude écologique et phytosociologique de la forêt à *Brachystegia laurentii. Publ. INEAC, Sér. Scient.*, **67**, Bruxelles

Guillaumet, J.-L. (1967) *Recherches sur la Végétation et la Flore de la Région du Bas-Cavally (Côte-d'Ivoire).* Paris: ORSTOM

Hallé, F., Oldeman, R. A. A. & Tomlinson, P. B. (1978) *Tropical Trees and Forests.* Berlin: Springer

Hladik, A. (1978) Phenology of leaf production in rain forest of Gabon: distribution and composition of food for folivores. In *The Ecology of Arboreal Folivores*, ed. G. G. Montgomery, pp. 51–71. Washington, D.C.: Smithsonian Institution Press

Jackson, J. K. (1956) The vegetation of the Imatong Mountains, Sudan. *J. Ecol.*, **44**, 341–74

Janzen, D. H. & Martin, P. S. (1982) Neotropical anachronisms: the fruits the gomphotheres ate. *Science*, **215**, 19–27

Laws, R. M., Parker, I. S. C. & Johnstone, R. C. B. (1975) *Elephants and their Habitats.* Oxford: Clarendon

Merz, G. (1981) Recherches sur la biologie de nutrition et les habitats préférés de l'éléphant de forêt, *Loxodonta africana cyclotis* Matschie, 1900. *Mammalia*, **45**, 299–312

Mitchell, B. L. (1960) Ecological aspects of game control measures in African wilderness and forested areas. *Kirkia*, **1**, 120–8

Ng, F. S. P. (1977) Shyness in trees. *Nature Malaysiana*, **2**(2), 34–7

Olivier, R. C. D. (1978) On the ecology of the Asian elephant (*Elephas maximus* Linn.): with particular reference to Malaya and Sri Lanka. Ph.D thesis, University of Cambridge

Richards, P. W. (1952) *The Tropical Rain Forest.* Cambridge: Cambridge University Press

Short, J. (1981) Diet and feeding behaviour of the forest elephant. *Mammalia*, **45**, 177–85

UNESCO/AETFAT/USNO (1983) *Vegetation Map of Africa – Carte de Végétation de l'Afrique.* Paris: UNESCO

Vooren, A. P. & Offermans, D. M. J. (1985) An ultra-light aircraft for low-cost, large-scale stereoscopic serial photographs. *Biotropica*, **17**, 84–8

White, F. (1983) *The Vegetation of Africa.* Paris: UNESCO

Wing, L. D. & Buss, I. O. (1970) Elephants and forests. *Wildl. Monogr.*, 19

General literature

Andrews, P. & Aiello, L. (1984) An evolutionary model for feeding and positional behaviour. In *Food Acquisition and Processing in Primates*, ed. D. J. Chivers, B. A. Wood & A. Bilsborough, pp. 429–66. New York: Plenum

Chivers, D. (1980) (ed.) *Malayan Forest Primates.* New York: Plenum

Fleagle, J. G. (1984) Primate locomotion and diet. In *Food Acquisition and Processing in Primates*, ed. D. J. Chivers, B. A. Wood & A. Bilsborough, pp. 105–17. New York: Plenum

Hallé, F., Oldeman, R. A. A. & Tomlinson, P. B. (1978) *Tropical Trees and Forests*. Berlin: Springer

Kortlandt, A. (1983) Marginal habitats of chimpanzees. *J. Hum. Evol.*, **12**, 231–78

Kortlandt, A. (1984) Vegetation research and the 'bulldozer' herbivores of tropical Africa. In *Tropical Rain-Forest*, ed. A. C. Chadwick & C. L. Sutton. Leeds: Leeds Phil. Lit. Soc.

Mitchell, A. W. (1982) *Reaching the Rain Forest Roof*. Leeds: Leeds Phil. Lit. Soc.

Perry, D. R. (1984) The canopy of the tropical rain forest. *Sci. Am.*, **251**(5), 114–22

Richards, P. W. (1952) *The Tropical Rain Forest*. Cambridge: Cambridge University Press

Ripley, S. (1979) Environmental grain, niche diversification, and positional behavior in neogene primates: an evolutionary hypothesis. In: *Environment, Behavior, and Morphology: Dynamic Interactions in Primates*, ed. M. E. Morbeck, H. Preuschoft & N. Gomberg, pp. 37–74. New York: Fischer

Schnell, R. (1970/71/76/77) *Introduction à la Phytogéographie des Pays Tropicaux. (1: Les Flores – Les Structures*, 1970; *2: Les Milieux – Les Groupements Végétaux*, 1971; *3: La Flore et la Végétation de l'Afrique Tropicale*, 1re partie, 1976; *4: La Flore et la Végétation de l'Afrique Tropicale*, 2e partie, 1977.) Paris: Gauthier-Villars

Stern, J. T. & Oxnard, C. E. (1973) Primate locomotion. *Primatologia*, **4**(11), 1–93

VI.2

Assessing primate habitat using Landsat technology

K. M. GREEN

Introduction

In general, the information available on the status of tropical forests in conjunction with changes in land use of converted forest areas is far from satisfactory. Nonetheless, the problem of loss of tropical forests has received considerable national and international attention during the past several years.

The Interagency Primate Steering Committee of the National Institutes of Health recognized the potential usefulness of digitized Landsat data for analyzing primate habitat and provided funding to establish IMAGES (Image Analysis and Graphic Facility for Ecological Studies) at the National Zoological Park, Smithsonian Institution, in October 1980 and the continued effort at the University of Maryland.

The advantages of Landsat over other remote sensing products include:

- An orbiting space platform that provides a synoptic view of a very large portion of the Earth in one scene. A Landsat image covering 34 000 km^2 can provide essential data for global, national, and regional inventorying and monitoring of resources. The scale of the imagery also is often suitable for subregional investigations.
- Landsat potentially acquires data over the same spot on Earth every 18 (or 16) days, which presents a unique opportunity for resource monitoring.
- Data are available in a digital format, which can be processed by computers to facilitate the rapid extraction of useful information. Conventional aerial photographs are quite difficult to digitize.

Satellite remote sensing

Information from environmental satellites concerning vegetation analysis used for making inventories of large sections of tropical forest is extremely useful when desiring up-to-date and extensive information on the declining forest ecosystems in developing tropical countries. Ultimately, once ground cover classification is established and verified through intensive field work, long-term monitoring of forest conversion indicating trends can be accomplished by satellite.

Landsat's multispectral scanner system (MSS) and thematic mapper (TM) have shown potential for land use analysis in the tropics. The sensors contain sets of electron optical sensors that record light from a separate part of the electromagnetic spectrum and in turn either store the data on magnetic tapes or transmit the data via electrical impulse to ground receiving stations. These data are then converted to computer compatible tapes (CCTs).

Since Landsat MSS has a resolution of 0.45 h (1.1 acres), the selection of land cover categories must be compatible with ground features. For example, Landsat cannot distinguish individual tree canopies, distinct storeys in tropical forests, or measure tree heights. In view of these limitations, certain cover types may not be identified by Landsat if they are consistently 1–3 ha in size. Since vegetative cover influences the radiation measured by the MSS, large uniform vegetation types such as crops, planted grasses, forest plantations, and orchards are logical land cover categories. In the tropics, one has the problem of selecting appropriate vegetation cover types from an extremely heterogeneous community.

Interpretation of Landsat data

Preliminary procedures

To assist potential users, a report, 'Landsat in the tropics: Guideline for habitat evaluation and monitoring', was written. This report was sent to all individuals who inquired about the project and was also distributed at the primate meetings. Upon receipt of a user inquiry, the guideline report was distributed so that the scientist could provide the necessary details. Generally, the request for habitat analysis was followed by a geographic search, assessing via modem the INORAC data base management system at EROS DATA CENTER.

Altogether, the P.I. responded to over 50 requests for information concerning images. Unfortunately, several study areas had absolutely no Landsat coverage. In these instances, a special request to NASA mission's control was made to acquire that scene. Prior to October 1982, a letter detailing the reason for requesting coverage was sufficient, but after that date new policies also required paying a surcharge. Although such requests were made for several regions, no new Landsat data were actually acquired due to excessive cloud cover, problems in copying master data tapes, or apparent lost images at the GSFC computer processing center. For example, discussions were held with Drs McGrew and Tutin in 1980 with regard to their Gabon field work. For a 3-year period we were not able to acquire any data to supplement their excellent survey work. Another frustrating situation was encountered with Landsat data acquired from Brazil, for the golden lion tamarin's habitat, Poco das Antas. A CCT was ordered, received, but never analyzed because none of the three systems used could read the Brazilian formatted tapes.

Nonetheless, digital analysis was performed on six regions as shown in Table 1. For each study site, the following results were provided: location of study site, field investigations, description of habitat, primate species, and processing results.

The habitat analyses conducted under the project used three image processing systems: IMPAC, the I^2, and ERDAS. Funds from the Interagency Primate Steering Committee (IPSC) of the National Institutes of Health were provided to purchase IMPAC (Image Analysis Package for Microcomputers), available through Egbert Scientific Software and established the Image Analysis and Graphic Facility for Ecological Studies (IMAGES). This facility operated within the Department of Zoological Research, National Zoological Park, Smithsonian Institution, between 1980 and 1982. IMAGES used the Smithsonian Institution's Honeywell 6680 as a mainframe. IMPAC provides full digital computer analytic capabilities and is capable of creating and displaying full-colour multi-spectral classification maps of Landsat data. The full statistical analysis capabilities include histogram generation and image ratioing. Interactive digital systems such as IMPAC are a cost-effective image transformation process.

In the summer of 1982, the IMAGES facility was relocated to the Remote Sensing Systems Laboratory (RSSL) at the University of Maryland. This move was undertaken to take advantage of utilizing a more sophisticated image processing system. The Model 70 is manufactured by International Imaging Systems (I^2S). This system is linked to a host PPP 11/23 computer system. Expanded capabilities include a tape drive, enabling direct reading of the Landsat CTT to the image processing system, filter terminal display at 512 × 512 pixels,

Table 1. *Summary information for the case studies*

Location	Landsat Scene Information			
(site, country)	I.D. no.	Date	Path/Row	Field investigator
Tumbes, Peru	22159-14503	Dec. 20, 1980	011,063	Saveda
Silent Valley, India	31175-04275	May 23, 1981	155,052	Kumar
Djerem, Camaroon	21015-08132	Nov. 2, 1977	198,056	Gartlan
Mamfe, Camaroon	2352-0856300	Jan. 9, 1976	201,056	Gartlan
Kibale, Uganda	10196-0744	Feb. 4, 1973	185,060	Struhsaker/Van Orsdal
Gombe, Tanzania	10358-07412	Jul. 16, 1973	185,063	Bauer/Pussy
Madhupur, Bangladesh	2748-03294	Feb. 8, 1977	148,043	Green

capability to display three bands of an image, subsampling capabilities, and production of 8 × 10 mm and 33 mm color products via a matrix causes system. Finally the Earth Resources Data Analysis System (ERDAS) was also used to provide analysis. This microcomputer based image processing system has the added feature of a very useful Geographic Information System (GIS).

Remote sensing application to primate habitat analysis

In total, seven different scenes were analyzed for this project. Table 1 provides the summary information for each case study, listing location, Landsat scene identification, and name of field investigator. A brief review of the results for the Silent Valley analysis will be presented.

The rare lion-tailed macaque lives only in the Western Ghat Mountains of South India, in the tropical evergreen rainforests known as *shola*, and is one of India's most endangered species. Among the macaques, only the lion-tailed monkey is an obligate rainforest dweller.

Habitat (from Green & Minkowski, 1977)

Although the Indian term *shola* properly applies to remnant patches of forest found in sheltered sites along watercourses among rolling grassy hills or downs, it is now generally used to connote any of the wet broad-leaved evergreen forest formations of the Western Ghats. These occur in regions with a minimum annual rainfall as low as 1750 mm, but usually above 2500 mm. Their most luxurious growth occurs at the middle elevations of 500–1500 m altitude. Below 300–600 m the rain forest usually grades imperceptibly into the lower-lying moist deciduous type, but may infrequently occur as narrow belts of riverine forest to as low as 100 m. Above 1500–1800 m, they give way to more stunted montane forests.

All of the areas where lion-tailed monkeys have been reported contain the lofty dense evergreen forests characterized by a large number of tree species reaching 30–50 m or more in height and forming a dense canopy. The vast majority of trees have large simple leaves. Giant climbers, epiphytic ferns, mosses, and orchids are numerous at all levels from the canopy to the forest floor. The woody understoreys include saplings, smaller species of trees, shrubs, and often a tangle of cane or bamboo-like reeds. Except in the ravines, the undergrowth is relatively free of herbaceous plants. In the southern *shola* forests, the plant communities favored by *Macaca silenus* are abundant in *Cullenia exarillata*, a very large, slightly buttressed tree with fluted bole, occurring at elevations from 500 m to 1400 m in areas of over 3000 mm rainfall. The typical dominant canopy tree associations are *Cullenia–Palaquium*, *Cullenia–Calophyllum*, *Poeciloneuron–Cullenia*, and *Palaquium–Mangifera*.

The Ghat forests are not uniform in structure. The highly variable terrain includes steep hillsides topped with rocky precipices reached by sharply ascending slopes. The major river valleys are usually very deep, and the monsoon torrents in the feeder streams have cut abrupt gorges. Between the major valleys are a few gently sloping interior hills with nearly flat stream banks. This landscape, together with edaphic conditions, factors of climate and exposure, especially the two monsoons, and different biotic influences yield localized, distinct vegetation formations within each hill range.

Numerous field studies have attempted to study and/or survey the primate population throughout the Western Ghats. The Landsat analysis done here was in conjunction with recent field work conducted by A. Kumar in the Silent Valley Region.

Fig. 1. Close up of the Silent Valley, enhanced bands 7 and 4 using every other pixel and scan line (500 × 500 pixels).

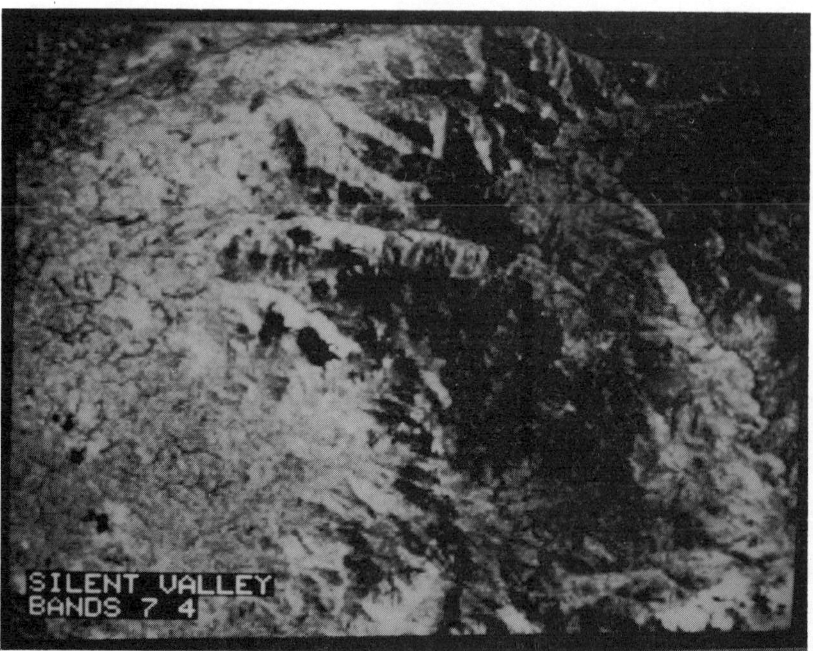

Habitat analysis

The close up enhanced image (Fig. 1) of the area shows the hilly terrain (with associated slope shadows), drainage pattern and flatter land to the east and west. Unsupervised classification results with a cluster algorithm yielded 27 distinct computer classes that were aggregated into the more ecologically meaningful categories of: agriculture fallow; agriculture; scrub; deciduous forest; riparian forest; evergreen forest; and two non-vegetation categories of high reflectance and cloud cover/shadow (Table 2). These results show spatially that the higher regions are covered with continuous as well as patchy corridors of good evergreen forest. A concentration of agricultural activity can be clearly seen at the western edge of the region as well as moving down one of the valleys in the northeast. Other hard-copy products produced were a gray scale classified image and the forest pixels (deciduous, riparian, and evergreen) plotted in black from a line printer.

Conclusions

It is necessary to point out that this short report represents one of several case study results available in Green (1984).

Products from these analyses have been made by several hardware and software systems. As noted earlier, output capabilities are not necessarily equivalent. For the general tropical field biologist this means that the color image products will have only a limited usefulness. Maximum impact would be possible if these images were geometrically corrected map-like products. Unfortunately, the image processing systems used did not have this service. Future products should be produced with this capability.

Table 2. *Merged cluster results for Silent Valley (301 × 401 pixels subsection, 2 bands)*

Class	Description	Color	No. pixels	Area (ha)	%
1	Agriculture fallow	Gray	4593	2021	3.9
2	Agriculture	Purple	27965	12305	23.9
3	Scrub	Yellow	12109	5328	10.3
4	Deciduous	Tan	29092	12800	24.6
5	Riparian	Lt. Green	18052	7943	11.1
6	Evergreen	Red	24487	10074	20.9
7	High reflectance	White	5732	2522	4.9
8	Cloud/shadow	Black	3651		

Another discouraging feature of this project was the limited number of study sites that could actually be supported. This was due to several factors: no Landsat coverage; poor quality of coverage; technical problems in acquiring a CCT; and foreign station CCT data that was formatted differently than U.S. data, and consequently unreadable by any of the image analysis systems used.

Another negative impact upon this project was the significant price increases for the Landsat products. The original budget contained funds to purchase the CCTs at US $200 each. Into the second year of the effort, prices increased to US $300 and by the third year they had reached US $650. These added costs had to be absorbed in the budget.

On the positive side, the project discussed provided a mechanism to disseminate technical remote sensing information to the scientific community, both in the literature and at professional meetings. At least for the case studies discussed in the complete report, this information is used as a tool for assessing primate habitat. Hopefully, this information will also serve as a catalyst for asking how should we continue to use this technology for improving our efforts at inventorying and monitoring tropical primate habitat.

Acknowledgements

The research reported here was supported through contract N01-RS-0-2125 from the National Institutes of Health (NIH). Specifically, the funds were provided through the Interagency Primate Steering Committee. Both Drs Robert Whitney and Joe Held of NIH were instrumental in supporting efforts to undertake this work. NIH administrative cooperation for this project was handled for the last several years by Dr Tom Wolfle. I appreciate his tolerance and desire to help in what, sometimes, were extremely confusing administrative procedures.

References

Green, K. M. (1984) *Monitoring Primate Habitat in the Tropics using Landsat Technology*. (Prepared for the Interagency Primate Steering Committee, National Institutes of Health.) 180 pp. Washington, D.C.: NIH

Green, S. & Minkowski, K. (1977) The lion-tailed monkey and its south Indian rain forest habitat. In *Primate Conservation*, ed. Prince Rainier & G. H. Bourne, pp. 290–338. New York: Academic Press

VI.3

Demographic trends in the *Alouatta palliata* and *Cebus capucinus* populations of Santa Rosa National Park, Costa Rica

L. M. FEDIGAN

Introduction

With 30% of its land in national parks, the small Central American country of Costa Rica is a leader in conservation in Latin America. Three species of neotropical primates occur in Santa Rosa National Park in the northwestern province of Guanacaste: *Ateles geoffroyi*, the red spider monkey; *Alouatta palliata*, the mantled howling monkey; and *Cebus capucinus*, the white-faced cebus monkey. Guanacaste Province has undergone extensive deforestation that has negatively affected all three primate species, in particular spider monkeys, which are not reliably reported outside of the national parks, and cebus monkeys, which have disappeared from many previously known ranges. Thus, along with a number of other public and privately owned and protected areas, Santa Rosa represents a major refugium for the primates in Guanacaste.

Demographic trends in the Santa Rosa howling and cebus monkey populations are compared in this paper with the population and group composition data from two well-studied sites in Central America: La Pacifica in Guanacaste (Glander, 1980; Clarke & Glander, 1984) and Barro Colorado Island in Panama (Carpenter, 1964; Froehlich & Thorington, 1981; Milton, 1982; Oppenheimer, 1982).

Methods

An exhaustive census of monkeys in all parts of the 100 km^2 park was conducted by a team of three to four researchers during the wet season of 1983 and the dry season of 1984. A detailed description of census techniques and results will be provided in Fedigan, Fedigan & Chapman (in preparation).

Twenty-two groups of the total 25 *Alouatta* groups counted in the park, and 22 of the total 28 *Cebus* groups counted, are used in this analysis. These are the groups for which the most reliable counts were obtained, and all of the group counts used in the major analysis were made in April–May 1984. As a first step, group compositions were compared between the wet season of 1983 and the dry season of 1984 to test for birth seasonality. Then population parameters for *Alouatta* and *Cebus* at Santa Rosa in 1984 were compared with findings from other sites. Finally, demographic characteristics of *Alouatta* and *Cebus* at Santa Rosa were compared with each other for species differences.

Results and discussion

The infant to adult female ratios in 10 of the better-known groups of *Alouatta* were compared between the wet season of 1983 and the dry season of 1984. A similar analysis was performed for 10 groups of *Cebus*. In neither case was there a significant difference between the two seasons in the ratio of infants to adult females (*Alouatta*: Wilcoxon $T = 25$, $P > 0.05$; *Cebus*: Wilcoxon $T = 17$, $P > 0.05$). Thus, given the presently available evidence, there does not appear to be birth seasonality in howling or cebus monkeys at Santa Rosa, a finding confirmed for these species by Clarke & Glander (1984) and Oppenheimer (1982) at nearby Central American sites.

Table 1 summarizes some population parameters for *A. palliata* at Santa Rosa, compared to the same species at other sites. Although we have not yet monitored the primate populations in the park over a sufficient time period to establish definite patterns of growth, there are three lines of evidence to suggest that the Santa Rosa howler population is stable, and that it may be stationary or only slowly increasing in size. First, Freese (1976) censused the monkeys in the park in 1972, and although he found fewer howlers in fewer places in the park than we did, the age/sex composition in 1972 was very similar to values from 1984. A stable age distribution is characteristic of a stable population. Secondly, there is a wealth of data on mantled howler populations from other research sites, including census information covering a 50-year period for Barro Colorado Island (BCI) in Panama, which has beeen summarized by both Milton (1982) and Froehlich & Thorington (1981). From these data one can extract the demographic characteristics of *A. palliata* populations which are known to be declining or increasing in size.

For example, data are available which were obtained prior to, soon after, and some years following, a 1951 population crash at BCI. From

Table 1. *Population parameters for* Alouatta palliata *at Santa Rosa and other sites*

Study site	Mean group size	Mean group composition (%)				Mean sociometric ratios		
		AdM	AdF	Juv	Inf	M:F	F:Inf	F:Imm (Juv & Inf)
Santa Rosa 1984 (This study)	13.6	22	44	20	14	1:2.03	1:0.31	1:0.77
Santa Rosa 1972 (Freese, 1976)	8.1	20	44	25	11	1:2.23	1:0.29	1:0.79
BCI, declining 1951 (Collias & Southwick, unpublished)	7.8	15	58	13	14	1:3.75	1:0.24	1:0.46
Los Tuxtlas, Mexico 1982 (Estrada, 1982)	9.1	33	45	9	13	1:1.37	1:0.28	1:0.48
BSI, expanding 1959 (Carpenter, 1964)	18.5	18	49	17	16	1:2.75	1:0.33	1:0.77
Modeled stable population (Heltne *et al.*, 1976)	15.0	–	–	–	–	–	1:0.25	1:0.75

AdM = adult males; AdF = adult females; Juv = juveniles; Inf = infants; Imm = immatures.

this information, both Carpenter (1964) and Heltne, Turner & Scott (1976) proposed that there are four indicators of a declining or recently declined *A. palliata* population: a small mean group size, of less than eight; an elevated number of adult females per male; a low proportion of females with infants; and a low proportion of immatures to adult females. Further, Estrada (1982) has argued that the low proportion of immatures in the howler population at Los Tuxtlas, Mexico, and the skewed sex ratio (although in the opposite direction from BCI) both suggest a population in decline and under stress.

The 1984 Santa Rosa values for mean group size, sex ratio, and female to infant ratio, fall between those of declining mantled howler populations and BCI values for time periods when that population was expanding. In the last few years at BCI, the birth rate has been unusually high, but the total population size appears to have remained stationary, and Milton (1982) has argued that the control of population size there is achieved through the counter-balancing effects of high juvenile mortality against high birth rates. I will return to the seemingly low numbers of juvenile howlers at many study sites later in this paper.

The third line of evidence for a stable howler population in Santa Rosa is the relatively good fit between empirical values obtained for this population in 1984 and the theoretical values derived from a model proposed by Heltne *et al.* (1976) for a stable population of *A. palliata*. The assumption used in their '11 year model' seems closest to what is known today about reproductive and life-history parameters in female howlers. From this model, they argued that a stable mantled howler population should have mean group sizes of around 15, a female to infant ratio of at least 1:0.25, and a female to immature ratio of 1:0.75; all values which are close to those found in our study population.

Table 2 summarizes data for *Cebus* in the park, and compares them to the only other published demographic data on white-faced cebus monkeys, those of Oppenheimer's (1982) 4-year study of two groups on BCI. These two sets of demographic data are quite similar, especially the female to immature ratios. During Oppenheimer's study in Panama, his local population was increasing at a mean rate of 23% a year, that is, growing rapidly. Thus, although only future censuses will finally determine the growth pattern of the *Cebus* population in Santa Rosa, the comparative data and the current structure do suggest an expanding population.

Table 2. *Comparison of demographic data for* Cebus capucinus *at Santa Rosa and Barro Colorado Island*

Study site	Mean group size	Mean group composition (%)				Mean sociometric ratios		
		AdM	AdF	Juv	Inf	M:F	F:Inf	F:Imm (Juv & Inf)
Santa Rosa 1984	14.1	16	37	35	12	1:2.24	1:0.33	1:0.27
BCI 1966–70 (Oppenheimer, 1982)	14.4	17	32	36	15	1:1.71	1:0.43	1:1.38

Table 3 compares some howler demographic characteristics to those of cebus at Santa Rosa. One major difference lies in the proportions of juveniles, particularly young or 'small' juveniles in the two populations. Although both species are reported to have similar maturation rates, and both currently experience similar birthrates at Santa Rosa, there is a greater proportion of juveniles in the cebus population than in the howler population. Lifetable statistics cannot yet be calculated for these animals. However, a simplified estimation of survivorship compares the observed to the expected number of juveniles in each group, assuming that these monkeys are juveniles for 3 years, and that the mean number of adult females in each group and birthrate have remained constant over the last 3 years. Milton (1982) originally proposed this measure, and she found that at BCI 39% of the expected immature howlers were 'missing'. Similarly, the observed number of juvenile howlers in each group at Santa Rosa is much lower than expected (2.8 compared to 5.6, or almost 50% 'missing'). However, there are nearly as many cebus juveniles observed as expected. Low numbers of juvenile howlers also have been reported from several other studies (e.g. Rudran, 1979; Froehlich & Thorington, 1981; Estrada, 1982).

Thus, the question presents itself: where have all the juvenile howlers gone? One obvious answer is that the juveniles have dispersed and are living as solitaries, rather than as group members. Various studies have reported that female as well as male howlers transfer between groups (e.g. Rudran, 1979; Glander, 1980), and one study of red howlers found that many immatures of both sexes may leave their natal group without immediately joining a new group (Crockett, 1984). However, we did not encounter temporary heterosexual associations made up of emigrants, like those described for red howlers, nor did we find sufficient numbers of solitaries to begin to correct for their low representation in groups.

If patterns of emigration do not fully account for the low proportion of juveniles, a second possibility is that of high mortality in the late-infant to early-juvenile stages. Rudran (1979) argued that this is indeed the pattern that occurs in red howlers, and that the high mortality is the result of frequent episodes of infanticide. The killing of infants by adult males has not been seen at Santa Rosa or BCI, but it has been reported by Clarke & Glander (1984) to occur in mantled howlers. In her consideration of juvenile mortality at BCI, Milton (1982) argued that the effects of parasites and unpredictable shortages of high-quality foods are far more important than infanticide in

Table 3. *Comparison of demographic data for howlers and cebus monkeys at Santa Rosa, 1984*

Species	Mean group size	Mean group composition (%)					Mean sociometric ratios		
		AdM	AdF	Large juv	Small juv	Inf	M:F	F:Inf	F:Imm (Juv & Inf)
Howlers	13.6	22	44	10	10	14	1:2.03	1:0.31	1:0.77
Cebus	14.1	16	37	14	21	12	1:2.24	1:0.33	1:1.25

Mean no. juv. howler per group: observed = 2.8, expected = 5.6
Mean no. juv. cebus per group: observed = 4.9, expected = 5.1
Expected howlers = 3 yr × 6F × 0.31I per F per yr
Expected cebus = 3 yr × 5.2F × 0.33I per F per yr

reducing numbers of immatures. In addition, Froehlich & Thorington (1981) argued that juvenile howlers suffer disproportionately high mortality during nutritionally poor years. All of these factors, as well as disease and predation, could limit the number of individuals that survive through the juvenile stage to reproductive maturity.

In conclusion, our preliminary findings suggest that the creation of Santa Rosa National Park 14 years ago, and the subsequent relaxation of some of the human-induced pressures on its wildlife, has resulted in local abundance of primate species. The howler population appears stable, and either stationary or only slowly increasing in size, whereas the cebus monkey population is likely an expanding one. While the park itself will continue to provide a secure habitat, with little hunting or encroachment pressure, the continuing deforestation and cattle-grazing in the areas surrounding the park will result in the increasing isolation of the primate populations, with new attendant demographic pressures. In the coming years, this investigation of Santa Rosa demography will continue, in addition to the monitoring of life-histories of selected individuals and groups in the park.

Acknowledgements

This research was supported by an NSERC (Canada) Operating Grant no. A7723. Colin Chapman, Larry Fedigan and Jeff Bullard helped to collect the census data. Daniel Janzen provided helpful criticisms of an earlier draft.

References

Carpenter, C. R. (1964) *Naturalistic behavior of nonhuman primates.* University Park, Pa: Pennsylvania State University Press

Clarke, M. R. & Glander, K. E. (1984) Female reproductive success in a group of free-ranging howling monkeys (*Alouatta palliata*) in Costa Rica. In *Female Primates: Studies by Women Primatologists*, ed. M. F. Small, pp. 111–26. New York: Alan R. Liss

Crockett, C. M. (1984) Emigration by female red howler monkeys and the case for female competition. In *Female Primates: Studies by Women Primatologists*, ed. M. F. Small, pp. 159–73. New York: Alan R. Liss

Estrada, A. (1982) Survey and census of howler monkeys (*Alouatta palliata*) in the rain forest of "Los Tuxtlas", Veracruz, Mexico. *Am. J. Primatol.*, **2**, 363–72

Freese, C. (1976) Censusing *Alouatta palliata. Ateles geoffroyi* and *Cebus capucinus* in the Costa Rican dry forest. In *Neotropical Primates: Field Studies and Conservation*, ed. R. W. Thorington, Jr & P. G. Heltne, pp. 4–9. Washington, D.C.: National Academy of Sciences

Froehlich, J. W. & Thorington, R. W., Jr (1981) The demography of howler monkeys *Alouatta palliata* on Barro Colorado Island, Panama. *Int. J. Primatol.*, **2**, 207–36

Glander, K. E. (1980) Reproduction and population growth in free-ranging mantled howling monkeys. *Am. J. Phys. Anthropol.*, **53**, 25–36

Heltne, P. G., Turner, D. C. & Scott, N. J. Jr (1976) Comparison of census data on *Alouatta palliata* from Costa Rica and Panama. In *Neotropical Primates: Field Studies and Conservation*, ed. R. W. Thorington, Jr & P. G. Heltne, pp. 10–19. Washington, D.C.: National Academy of Sciences

Milton, K. (1982) Dietary quality and demographic regulation in a howler monkey population. In *The Ecology of the Tropical Forest. Seasonal Rhythms and Long-term Changes*, ed. E. G. Leigh, A. Stanley Rand & D. M. Windsor, pp. 273–89. Washington, D.C.: Smithsonian Institution Press

Oppenheimer, J. R. (1982) *Cebus capucinus*: home range, population dynamics and interspecific relationships. In *The Ecology of the Tropical Forest: Seasonal Rhythms and Long-term Changes*, ed. E. G. Leigh, A. Stanley Rand & D. M. Windsor, pp. 253–72. Washington, D.C.: Smithsonian Institution

Rudran, R. (1979) The demography and social mobility of a red howler (*Alouatta seniculus*) population in Venezuela. In *Vertebrate Ecology in the Northern Neotropics*, ed. J. F. Eisenberg, pp. 109–26. Washington, D.C.: Smithsonian Institution Press

VI.4

The contribution of zoos to primate conservation

D. G. LINDBURG, J. BERKSON
AND L. NIGHTENHELSER

Introduction

The history of zoological gardens is one of bringing exotic animals into captivity to serve human needs for entertainment and education. In recent years many zoological institutions have re-evaluated their mission, with conservation through captive breeding emerging as the basis on which their continuing existence is legitimized. In reality this shift in mission has been unavoidable as commerce in exotics has become severely restricted. Faced with reduced augmentation of captive stock through continued harvesting of wild populations, the future of zoos depends increasingly upon the development of self-sustaining captive populations from brood stock on hand.

Between 1972 and 1984 there have been four international conferences on the breeding of endangered species. These conferences have provided a forum for discussion of programs for species whose survival is in doubt and for reporting on efforts undertaken in specific cases. However, there have been very few summaries of zoo breeding efforts since conservation began to be taken more seriously and, therefore, very limited opportunity to provide baselines against which to measure the success or failure of captive propagation efforts.

In an earlier report (Lindburg, Berkson & Nightenhelser, 1984) we summarized birth and mortality figures for 126 primate species which reproduced in zoos during 1971–80. Data for this summary were drawn from the records of mammals bred in captivity which are published annually in the *International Zoo Yearbook* (*IZY*). Using the same data base, we provide here a more detailed analysis of those species at present classified by the International Union for the Conservation of Nature (IUCN) as *endangered, vulnerable, rare* and *indeterminate*.

Methods

As previously noted (Lindburg *et al.*, 1984), zoo reports to the *IZY* are voluntary. They consist of numbers born and surviving to 30 days, listed by species or subspecies and by sex where ascertainable. There are many problems with these data, frequency and accuracy of reports being two of the most obvious ones. However, despite their limitations, the *IZY* records are the only ones available from zoos on a global basis.

IZY records for 1971–80 were computerized at the University of California–San Diego. Species hybrids and primates born at identifiable research facilities were excluded. In matters of taxonomy the *IZY* system was followed. Other methodological details are reported in Lindburg *et al.* (1984).

Results

The computerized records from the *IZY* indicate that a grand total of 24366 primates were born in zoos during 1971–80 (Table 1). Births have occurred to 126 different species, and the number of offspring born to any one of these ranges from 1 (six cases) to 2284 in the case of *Papio hamadryas*. Since the *IZY* records consist only of numbers born and surviving to 30 days, we are unable to specify the number of species in zoos having no reproduction or the size of the breeding population for those that have reproduced.

The IUCN lists 76 different primates in the categories of *endangered*, *vulnerable*, *rare*, and *indeterminate*. Zoos have reported births for 41 of these during the decade surveyed, apparently occurring at a lower rate

Table 1. *Birth and survival of zoo-living primates in relation to IUCN status*

IUCN status	No. of different primates[a]	No. born	Percent of total born	No. dying in ≤30 days (%)
Endangered	14	2252	9.3	744 (33.0)
Vulnerable	19	2493	10.2	559 (22.4)
Rare	3	158	0.7	30 (19.0)
Indeterminate	5	130	0.5	57 (43.8)
Unclassified	127	19333	79.3	4474 (23.1)
Total	168	24366	100.0	5864 (24.1)

[a] Includes all subspecies listed by *IZY* during the 1971–80 period of reporting

than for other zoo-breeding primates ($\bar{x}$ of 12.3 *vs* 15.2 infants per species each year).

Of 31 primates (including 12 races) classified as *endangered* by the IUCN, 14 reproduced in zoos in numbers ranging from a single, non-surviving Sanford's lemur to 776 cotton-top tamarins (Table 2). The cotton-top and golden lion tamarins and the orang-utan together account for 71.5% of all births to primates in this class. For 24 species or subspecies classified as *vulnerable*, 19 reproduced in zoos in numbers ranging from 1 to 968 births (Table 3). One species, the Barbary macaque, accounted for 38.8% of all infants born to vulnerable primates, and over two-thirds of the species in this category produced less than 20 infants over the 10-year period analyzed. Three of 10 primates on the IUCN's *rare* list and 5 of 11 classified as *indeterminate* reproduced in zoos (Table 4). As in the previous categories, relatively few species accounted for the majority of births.

Survival of progeny must be considered in evaluating the contribution of captive breeding programs to conservation. In general, prosimians and New World monkeys appear to have fared less well in this regard than Old World monkeys or apes. Using only those species which produced 20 or more infants ($N = 83$), infant survivorship of 70% or higher occurred in less than two-thirds of all prosimians and New World monkeys, whereas the comparable figures for Old World

Table 2. *Infants produced by primates on the IUCN endangered species list*

Name	No.
1. Cotton-top tamarin (*Saguinus oedipus oedipus*)	776
2. Golden lion tamarin (*Leontopithecus rosalia*)	419
3. Orang-utan (*Pongo pygmaeus*)	416
4. Lion-tailed macaque (*Macaca silenus*)	270
5. Black lemur (*Lemur macaco macaco*)	111
6. Douc langur (*Pygathrix nemaeus*)	74
7. Drill (*Papio leucophaeus*)	48
8. Red-backed squirrel monkey (*Saimiri oerstedi*)	47
9. Red-fronted lemur (*Lemur macaco rufus*)	39
10. Javan gibbon (*Hylobates moloch*)	18
11. Pileated gibbon (*Hylobates pileatus*)	16
12. White-eared marmoset (*Callithrix aurita*)	14
13. Philippine tarsier (*Tarsius syrichta*)	3
14. Sanford's lemur (*Lemur macaco sanfordi*)	1
Total	2252

Table 3. *Infants produced by primates classified as vulnerable by IUCN*

Name	No.
1. Barbary macaque (*Macaca sylvana*)	968
2. Chimpanzee (*Pan troglodytes*)	537
3. Geoffroy's spider monkey (*Ateles geoffroyi*)	358
4. Gorilla (*Gorilla gorilla*)[a]	183
5. Mongoose lemur (*Lemur mongoz*)	105
6. Woolly monkey (*Lagothrix lagothricha*)	103
7. Black spider monkey (*Ateles paniscus*)	86
8. Proboscis monkey (*Nasalis larvatus*)	53
9. Long-haired spider monkey (*Ateles belzebuth*)	24
10. Pygmy chimpanzee (*Pan paniscus*)	18
11. Grey gentle lemur (*Hapalemur griseus*)	12
12. Uakari (*Cacajao calvus*)[b]	12
13. Tassel-eared marmoset (*Callithrix humeralifer*)	11
14. Nilgiri langur (*Presbytis johnii*)	7
15. Fat-tailed dwarf lemur (*Cheirogaleus medius*)	6
16. White-footed tamarin (*Saguinus leucopus*)	6
17. Coquerel's mouse lemur (*Microcebus coquereli*)	2
18. White-nosed saki (*Chiropotes albinasus*)	1
19. Kloss's gibbon (*Hylobates klossi*)	1
Total	2493

[a] Lowland subspecies only.
[b] Only the red uakari (*C. c. rubicundus*) has reproduced in zoos.

Table 4. *Infants produced by primates classified as rare and indeterminate by IUCN*

Name	No.
Rare	
1. Goeldi's marmoset (*Callimico goeldi*)	143
2. Snub-nosed langur (*Rhinopithecus roxellanae*)	8
3. Golden langur (*Presbytis geei*)	7
Total	158
Indeterminate	
1. Emperor tamarin (*Saguinus imperator*)	65
2. Brown-headed spider monkey (*Ateles fusciceps*)	33
3. Black gibbon (*Hylobates concolor*)	24
4. Guatemalan howler monkey (*Alouatta villosa*)	6
5. Bare-faced tamarin (*Saguinus bicolor*)	2
Total	130

monkeys and apes were 92 and 83%, respectively (Table 5). In addition, mortality for primates on the IUCN lists (Table 1) was significantly greater than for other zoo-born infants (Chi-square = 26.24; d.f. = 1, $p < 0.005$). Of those primates classified as *endangered* (Table 3), mortality in species producing 20 or more infants was highest for the red-fronted lemur (54%), but was also high for the cotton-top (41%) and golden lion (44%) tamarins. For primates considered vulnerable, the woolly monkey at 50% experienced the highest rate of infant loss.

Discussion

In a somewhat different analysis of captive breeding in zoos, Pinder & Barkham (1978) concluded:

> Zoos . . . should seek their philosophical *raison d'être* in their ancient and creditable origin, the Chinese emperor's education and culture park, rather than in goals which are set too high for achievement. (p. 235)

If the past, even the recent past, is an adequate predictor of future performance, this pessimistic conclusion may be warranted. The record clearly indicates that the preponderance of offspring born as far as primates are concerned has been to relatively small numbers of the more common species. In addition, the rarer primates have reproduced at a lower rate than those whose future is relatively secure, and have suffered a significantly higher rate of infant loss. This performance suggests that breeding of exotics in zoos has been allowed to pursue its own course with only occasional assistance from human managers. Yet there is significant change in the form of increased research, of improved husbandry and technological advances, of increased financial resources, and improved living environments in zoos today, all of which should help to bring those populations presently in captivity closer to being self-sustaining. Zoos are now

Table 5. *Survivorship by taxon for species producing 20 or more offspring during 1971–80*

Percentage of infants surviving to 30 days	Number of species reproducing			
	Prosimians	New World monkeys	Old World monkeys	Apes
<70	4	10	3	1
70–79	5	13	18	3
⩾80	1	6	17	2

applying approaches developed in human and animal science to the propagation and management of exotics. As the individual subject and its potential contribution to the future captive population become the focus of attention for scientists and clinicians working in zoos, genetic, physiological and behavioral information becomes available for use in reversing the seemingly inevitable trend toward extinction. Perhaps the most exemplary case of a successful zoo program is that of the golden lion tamarin which, after a concerted captive breeding effort, is now in the process of being reintroduced to the wild (Kleiman, 1984). Our findings indicate that this pattern will be repeated only if zoos assign a higher priority to carefully designed breeding programs for the rarer primates now consigned to their collections.

Acknowledgements

We thank Stuart Nightenhelser for writing the computer programs used in data analysis for this report.

References

Kleiman, D. (1984) The National Zoo's role in an international primate conservation program. *IUCN/SSC Primate Specialist Group Newsletter*, 4, 45–6

Lindburg, D. G., Berkson, J. M. & Nightenhelser, L. K. (1984) Primate breeding in zoos: a ten-year summary. In *One Medicine*, ed. O. A. Ryder & M. L. Byrd, pp. 162–70. Berlin: Springer-Verlag

Pinder, N. J. & Barkham, J. P. (1978) An assessment of the contribution of captive breeding to the conservation of rare mammals. *Biol. Conserv.*, **13**, 187–245

VI.5

Captive breeding of callitrichids: a comparison of reproduction and propagation in different species

M. F. STEVENSON

Introduction

Marmosets and tamarins comprise one of the families of New World Primates, the Callitrichidae. Goeldi's monkey (*Callimico*) is at present considered by Hershkovitz (1977) to be more accurately placed in a family of its own, the Callimiconidae, but as its captive management is similar to that of the true callitrichids, it is included in this review. The marmosets and tamarins are each divided into two families (Hershkovitz, 1977): *Callithrix* and *Cebuella*, the marmosets; and *Saguinus* and *Leontopithecus*, the tamarins. The main point of taxonomic difference is found in dentition: only the marmosets have incisiform lower canines which form a cup shape with the four incisors. Field work has shown that this enables marmosets to gouge gums from the barks of trees (Coimbra-Filho & Mittermeier, 1977), which may result in differences in feeding behaviour and socioecology between the two groups.

Callitrichids have been recorded in captivity since the sixteenth century; they were part of the fauna on Prospero's Island in *The Tempest* and have been popular as pets through the ages. Landseer featured the common marmoset in one of his paintings. Few of these animals could have lived for long in captivity. The early records of Edinburgh Zoo serve as an example. Out of 37 marmosets that were kept in the collection between 1914 and 1965, 21 died within 6 months of arrival and none lived for more than six years. It was only relatively recently that the correct conditions for maintaining callitrichids in captivity began to be evolved (Stevenson, 1984).

It has been noted that some species of callitrichid are easier to breed and adapt more readily to the captive environment than others (Hampton, Hampton & Levy, 1972; Wolfe *et al.*, 1975). For example,

Callithrix jacchus adapts well and breeds regularly whereas *C. argentata* is difficult to maintain (Omedes, 1981). Similarly *Saguinus fuscicollis* appears to be more easily kept and bred than *S. oedipus* (Wolfe *et al.*, 1975; Evans, 1983). These difficulties could be related to differences in environmental and/or dietary requirements and can really only be completely understood by comparing the ecology and behaviour of species in the wild.

Callitrichids are becoming increasingly used as animal models in biomedical research. A species is usually chosen for this purpose because of its availability at the time and because it exhibits a particular medical condition, without reference or thought to its ease of captive breeding or the conservation status of the species (Wolfe & Deinhardt, 1978). Some species of callitrichid are highly endangered (Thornback & Jenkins, 1982), and successful captive breeding programmes may be the only alternative to total extinction. It is thus essential that we try to understand the needs of different species of callitrichid in captivity and evaluate their reproductive potential.

This paper presents data on reproductive potentials and interbirth intervals of some callitrichid species, using data obtained through questionnaires and also from published work. Using the data published each year in the *International Zoo Yearbook*, the relative success of captive propagation of different species can be examined.

Materials and methods

A questionnaire was sent to institutions known to keep and breed callitrichids, asking for dates of birth of successive sets of offspring from family groups where the same adult pair had produced more than one set of infants. The time from pairing to first parturition and the fate of the offspring was also requested. When possible, interbirth intervals were calculated in days. Published material from callitrichid workers was also referred to. High interbirth interval values which occur at low frequencies are probably not of any great significance but they do have the effect of raising the mean value. For this reason median values were taken as being of more significance when comparing values for different species. Modal values were used when there were sufficient data.

The annual inventories in the appendices of the *International Zoo Yearbook*, (vols 10–22) have been used for all species of callitrichid plus *Callimico*. For each year the numbers of animals bred and the number that survived for 30 days is given. For those species designated as 'rare', additional information is provided, and for each year the total number of animals in captivity and the proportion of those that were captive-born is listed. There are problems in analysing these data but it is the most complete information available (for a more detailed discussion see Stevenson, 1983). For all species listed for a sufficient number of years, data on surviving births were divided into three 4-year periods. These time periods were compared with each other using Mann-Whitney *U* tests (two-tailed; Siegel, 1956) to ascertain whether surviving births increased or decreased significantly ($P < 0.05$) between time periods.

For those species listed in the 'rare' section additional information was available. For the period 1977–1980 the average population was calculated and used to estimate the viability of the captive population of that species. Unfortunately the total captive population of a species may be very different from the effective breeding population (Frankel and Soulé, 1981), and must be estimated as less than the total population.

Questionnaire returns

Information was collected from 11 institutions, which provided data on 10 races of callitrichid. Data were not collected for the common marmoset, *Callithrix jacchus*, as sufficient information on its gestation and rate of breeding has been published. Unfortunately there was not sufficient information available on intervals between pairing of animals and first parturition to provide analysable data. Table 1 summarises the results of the present study and, where possible, the results from other published studies. The number of interbirth intervals used for a species obviously affects the significance of the results; however, in the case of *Leontopithecus rosalia rosalia*, where as few as eight were available, the lowest value indicates that this species may have the shortest gestation, being the lowest interbirth interval in the Table. Thus, even a low number of values does provide important information. The results indicate that the silvery and black-tailed marmosets (*Callithrix argentata*), although having gestations of less than 154 and 161 days, respectively, breed less regularly in captivity than the common marmoset and may require specific conditions of husbandry. *Cebuella*, under good conditions, probably breeds at 5-month intervals. Members of the genus *Saguinus* are known to be irregular breeders (Evans, 1983). Unfortunately the true gestations of members of this genus are not known although data in Table 1 indicate that *S. fuscicollis* has a gestation of less than 155 days. *S. oedipus* has recently been thought to have a slightly longer

Table 1. *Reproductive data on some species of Callitrichid*

Species	Interbirth interval (days)		Gestation		Reference
	Range	Median	Range	Mean	
Cebuella pygmaea	158–286	160	138–140	–	Hershkovitz (1977) Vertovec (1981)
Callithrix jacchus	145–190	155	141–146	144	Poole & Evans (1982)
C. argentata argentata	154–404	199	?	?	Present study ($n = 11$)
C. a. melanura	161–320	245	?	?	Present study ($n = 17$)
Saguinus fuscicollis	155–396	225	?	?	Present study ($n = 57$)
	162–367	222($\bar{X}$)			Gengozian *et al.* (1978)
S. oedipus oedipus	192–751	224	?	?	Present study ($n = 60$)
	197–255	*	?	?	French (1983)
	182–374	238	?	?	Evans (1983)
	–	–	–	166	Brand (1981); Cross & Martin (1981)
	172–430	–	–	–	Carroll (1983)
S. o. geoffroyi	302–365	240	–	–	Present study ($n = 8$)
S. labiatus	200–556	299	–	–	Present study ($n = 22$)
S. imperator	163–355	225	–	–	Present study ($n = 9$)
S. midas niger	169–285	201	?	?	Hershkovitz (1977)
Leontopithecus rosalia rosalia	135–442	165			Present study ($n = 8$)
			125–132	128	Kleiman (1977)
Callimico goeldi	153–341	169	–	–	Present study ($n = 28$)
	–	–	150–165	154	Heltne *et al.* (1981)

$\bar{X}$ = mean value
* = mean value given as 208.5 ± 16.3
n = number of interbirth intervals
– = not measured in this study
? = unknown generally

Table 2. *Data from* International Zoo Yearbook. *For those species in the 'rare' section additional information is given on the average total population*

Species		Average total population (1977–1980)	Young per year[a]	Significant increase[b]
Callithrix jacchus[c]	Common marmoset		66	+
Saguinus fuscicollis nigricollis[c]	Mantled tamarin		29	+
Callithrix pencillata	Black-eared marmoset	8		+
C. geoffroyi	Geoffroy's marmoset		15	+
C. argentata	Silvery/black-tailed marmoset		28	+
Saguinus labiatus[c]	Red-bellied tamarin		15	+
S. mystax[c]	Moustached tamarin		7	n.s.
S. midas midas	Red-handed tamarin		<1	n.s.
S. oedipus geoffroyi	Geoffroy's tamarin		6	n.s.
S. o. oedipus[c]	Cotton-top tamarin	436	80	+
S. imperator	Emperor tamarin	60	7	+
S. bicolor	Pied tamarin	2	0	
S. leucopus	White-footed	6	<1	
Cebuella pygmaea	Pygmy marmoset	41	11	+
Leontopithecus rosalia rosalia	Golden lion tamarin	165	40	+
L. r. chrysomelas	Golden-headed tamarin	9	?	
L. r. chrysopygus	Golden-rumped tamarin	17	?	
Callithrix aurita	Buffy-tufted marmoset	5	<1	
C. humeralifer	Tassel-eared marmoset	2	0	
Callimico goeldi	Goeldi's monkey	104	24	+

[a] average number of young which survived each year 1977–1980

[b] + = significant increase ($P < 0.05$) 1977–1980 over 1973–1976; n.s. = not significant

[c] species kept in laboratories, so *IZY* numbers are underestimates

gestation of 166 days (Brand, 1981; Cross & Martin, 1981) and these data might also indicate this to be the case. The same may be true for *S. labiatus*. These two species are used extensively in medical research laboratories at present.

In Fig. 1, histograms of interbirth intervals (IBI) for the six species with the most available data are presented. The IBIs are shown in 10-day blocks. In three of the races, the modal value is within a 10-day range of the median value. These modal values are obviously of great importance when examining the biology of the species, as they show the most frequent IBI.

Yearbook analysis

When analysing the results from the *IZY*, considerable taxonomic problems are encountered which preclude full analysis for some races of callitrichid. *Saguinus fuscicollis* is sometimes split into subspecies and sometimes not. *S. nigricollis* and *S. fuscicollis* are frequently confused, hence are expressed together in Table 2. Most of the data will be from *S. fuscicollis*, *Callithrix argentata argentata* and *C. a. melanura* have had to be lumped together as *C. argentata*. It is also probable that in some years *C. aurita* is actually misidentified as *C. jacchus*.

Most callitrichids can be expected to rear two offspring each year and *Callimico*, one. For those species not listed in the 'rare' section of the *Yearbook* the effective captive breeding population must be deduced from the figure for surviving young per year. A viable number of breeding females could reasonably be set at 40 (Frankel & Soulé, 1981), and therefore even assuming that only one young per year is raised, a minimum figure for a reasonable population would be 40 year pear. Only three of the species listed comply with this criterion. Of those listed in the 'rare' section only three have an average population of over 100 animals. Some of these species, however, have captive populations that are increasing through captive breeding. The pygmy marmoset, although with low total numbers, has a significantly increasing birth rate and the population is growing. The population of *Callimico* is now becoming established (Fig. 2), and the wild-caught proportion decreasing. However, no captive population can be considered self-supporting until there is a viable breeding second generation, and it is still too early to deduce this in most callitrichids. Fig. 3 illustrates the remarkable increase in the golden lion tamarin since 1976. Although part of this is due to the inclusion of additional, previously unrecorded animals, much of the recent growth is due to

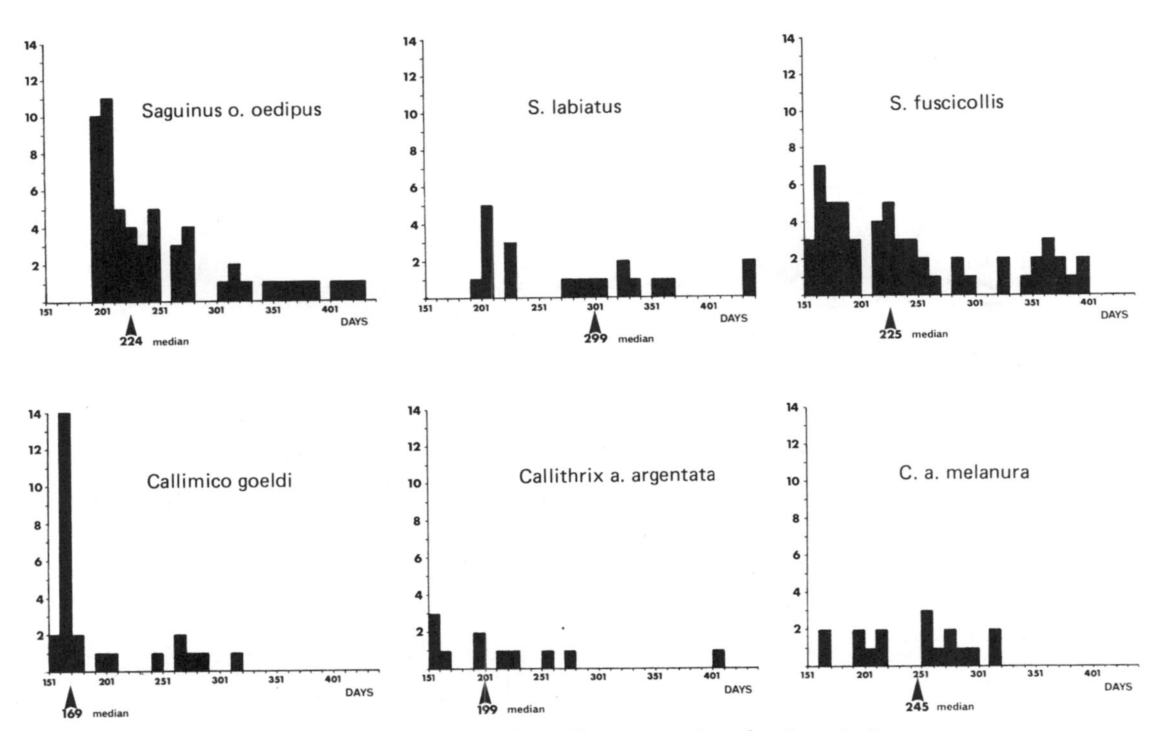

Fig. 1. The frequency of interbirth intervals for different species of callitrichid.

efficient organisation of the captive population through the studbook (Kleiman & Evans, 1981).

Eight of the species listed in Table 2 produce less than 10 young per year on average, which probably reflects small captive populations, although in the case of the emperor tamarin one would expect better results for the population size.

Discussion

Although it has been suggested by several workers (Hershkovitz, 1977) that data on interbirth intervals of different callitrichids provide invaluable information on their biology, little comparative

Fig. 2. The status of captive populations of Goeldi's monkey.

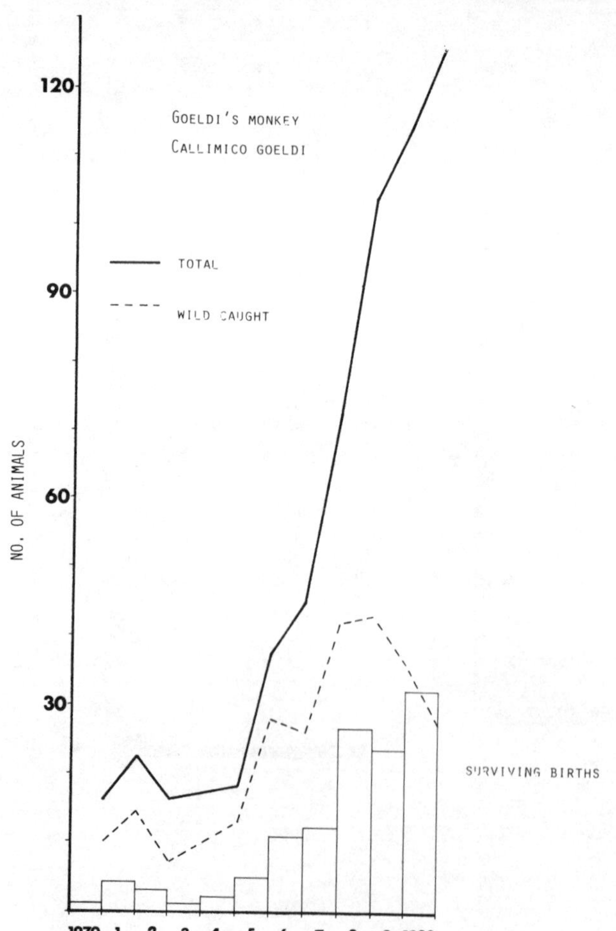

work has been carried out. IBIs not only provide information on possible gestation periods and on fecundity, but can also give an indication on whether lactation affects ovulation and conception of the next litter. Most of the data used in this study were from IBIs which resulted in a reared litter, so it was not possible to compare the lactating and non-lactating females. Lunn & McNeilly (1982) demonstrated that there is no lactation anoestrus in *Callithrix jacchus*, but this may not be true for other callitrichids. Some workers suggested that there is a lactation anoestrus effect in *S. oedipus oedipus* (Carroll, 1983; Evans, 1983) but French (1983), using larger sample sizes, recently demonstrated that there is no significant effect. The data received in this present work, especially those from established breeding pairs at Banham Zoo, suggest that once breeding pairs are established they breed at regular intervals. However, even allowing for a

Fig. 3. The status of captive populations of golden lion tamarins.

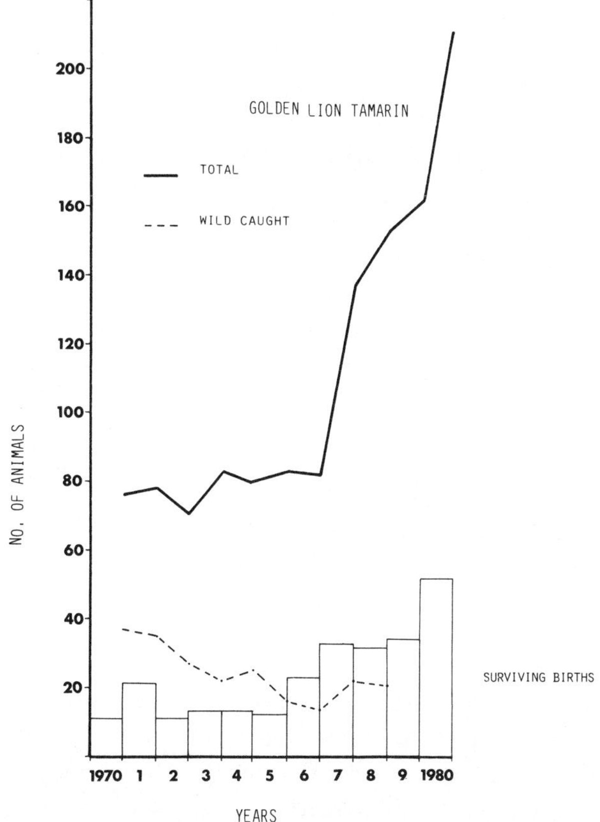

gestation of 166 days, it is rare to find an IBI in *Saguinus oedipus oedipus* of less than 190 days. This suggests a longer period between birth and subsequent ovulation than in *Cebuella* and *Callithrix jacchus*. It is difficult to compare the different members of the genus *Callithrix* due to lack of information. However, results from *Callithrix argentata* suggest that although gestation cannot be more than a few days longer than for the common marmoset, ovulation following birth is less predictable, and indeed this species, especially the *C. a. melanura* race, presents problems in captive breeding (Omedes, 1981). Rylands found that *Callithrix humeralifer intermedius* could breed at 5-monthly intervals in the wild (Stevenson & Rylands, in press) but there is little captive information on this species. Kingston & Muniz (1983) cited two interbirth intervals for *C. h. humeralifer* as 30 and 36 weeks.

A seasonal distribution of births, which obviously influences IBIs, thus far has been demonstrated in only three races, in *S. o. oedipus* (Brand, 1981), in *S. o. geoffroyi* (Dawson & Dukelow, 1976) and in *Leontopithecus rosalia rosalia* (Kleiman, 1977).

The median values for all members of the genus *Saguinus* are greater than 200 days and most fall in the range of 220 to 240 days, whereas the median values for *Leontopithecus* and *Callimico* fall within 160–170 days. This suggests that the reproductive potential of *Saguinus* species may be the lowest of the group as a whole, even allowing for the fact that some species may have the longest gestations.

The goal of any captive breeding programme is to reach a state where the captive population is self-sustaining. This is especially true in the case of a species which is or may become endangered in the wild and also for those species (Table 2) commonly used in medical research where the turnover of animals tends to be high. One species, *S. mystax*, used in biomedical research may be one of the most difficult to breed in captivity. Box & Morris (1980) stated that it presents problems in captive breeding and Kim & Wolfe (1980) listed causes of death in imported animals in their laboratory.

Table 2 lists 20 callitrichids maintained and bred in captivity. Of these only 5 can be said to occur in any numbers: *Callithrix jacchus, S. fuscicollis/nigricollis, S. o. oedipus, L. r. rosalia* and *Callimico goeldi*. For the *S. fuscicollis* group the taxonomic problems make it difficult to ascertain exactly what the captive population consists of.

It is unfortunate that without proper management and cooperation between institutions a species can die out very quickly in captivity. Many *Callithrix argentata melanura* were imported into the UK through Bolivia in the 1970s; although initial breeding was good it was not sustained and few institutions in the UK now hold the species (Stevenson, unpublished). It is important to try to stop this occurring again, especially as it is now known that there are over 30 *L. r. chrysomelas* in captivity through illegal exports (Mallinson, 1984). It is important that those maintaining a species in captivity are aware of its reproductive potential and management problems, and refer to other institutions maintaining the same species. Fig. 2 shows that the captive population of *Callimico goeldi* is currently healthy, but only through cooperation such as that in process for *L. r. rosalia* can this be maintained.

It is to be hoped that those maintaining captive Callitrichids understand the importance of maintaining and referring to studbooks, of correct identification of the species maintained, of considering the captive population of the species as a whole, and of close liaison with others working with the same and similar species.

Acknowledgements

I should like to thank the staff of the following zoos and other institutions for contributing information: Cleveland Zoological Society, Helsinki Zoo, Marwell Zoological Park, Penscynor Wildlife Park, Jersey Wildlife Preservation Trust, Banham Zoo, Bristol Zoo, Zoological Society of London, Edinburgh Zoo, Zoology Department, U.C.W. Aberystwyth. Pat Neyman allowed me to use data on cotton-top tamarins that she has previously collected. I should also like to thank Chris Morris for drawing the figures.

References

Box, H. O. & Morris, J. M. (1980) Behavioural observations on captive pairs of wild-caught tamarins (*Saguinus mystax*). *Primates*, **21**, 53–65

Brand, H. M. (1981) Husbandry and breeding of a newly-established colony of cotton-topped tamarins (*Saguinus oedipus oedipus*). *Lab. Anim.*, **15**, 7–17

Carroll, J. B. (1983) Breeding the cotton-topped tamarin *Saguinus oedipus oedipus* at Jersey Wildlife Preservation Trust. *Dodo*, **20**, 48–52

Coimbra-Filho, A. F. & Mittermeier, R. A. (1977) Tree gouging, exudate eating and the 'short-tusked' condition in *Callithrix* and *Cebuella*. In *Biology and Conservation of the Callitrichidae*, ed. D. Kleiman, pp. 105–15. Washington, D.C.: Smithsonian Institution Press

Cross, J. F. & Martin, R. D. (1981) Calculation of gestation period and other reproductive parameters for primates. *Dodo*, **18**, 30–43

Dawson, G. A. & Dukelow, W. R. (1976) Reproductive characteristics of free living Panamanian tamarins, (*Saguinus oedipus geoffroyi*). *J. Med. Primatol.*, **5**, 266–75

Evans, S. (1983) Breeding of the cotton-top tamarin *Saguinus oedipus oedipus*: a comparison with the common marmoset. *Zoo Biol.*, **2**, 27–54

Frankel, O. H. & Soulé, M. E. (1981) *Conservation and Evolution*. Cambridge: Cambridge University Press

French, J. A. (1983) Lactation and fertility: an examination of nursing and interbirth intervals in cotton-top tamarins (*Saguinus o. oedipus*). *Folia Primatol.*, **40**, 276–82

Gengozian, N., Batson, J. S. & Smith, T. A. (1978) Breeding of marmosets in a colony environment. *Primate Med.*, **10**, 71–8

Hampton, S. H., Hampton, J. K. & Levy, B. M. (1972) Husbandry of rare marmoset species. In *Saving the Lion Marmoset*, ed. D. D. Bridgewater, pp. 70–85. West Virginia: Wild Animal Propagation Trust

Heltne, P. G., Wojcik, J. F. & Pook, A. G. (1981) Goeldi's monkey, genus *Callimico*. In *Ecology and Behavior of Neotropical Primates*, vol. 1, ed. R. A. Mittermeier & A. F. Coimbra-Filho, pp. 169–209. Rio de Janeiro: Academio Brasileira de Ciencias

Hershkovitz, P. (1977) *Living New World Monkeys (Platyrrhini)*, vol. 1. Chicago: University of Chicago Press

Kim, J. C. S. & Wolfe, R. J. (1980) Diseases of moustached marmosets. In *The Comparative Pathology of Zoo Animals*, ed. R. J. Montali & G. Migaki, pp. 431–5. Washington, D.C.: Smithsonian Institution Press

Kingston, W. R. & Muniz, J. A. P. C. (1983) Preliminary report on the establishment in Brazil of a breeding colony of marmosets (*Callithrix humeralifer humeralifer*). *Lab. Primate Newslett.*, **22**, 1–2

Kleiman, D. (1977) Characteristics of reproduction and sociosexual interactions in pairs of lion tamarins (*Leontopithecus rosalia*) during the reproductive cycle. In *The Biology and Conservation of the Callitrichidae*, ed. D. Kleiman, pp. 181–90. Washington, D.C.: Smithsonian Institution Press

Kleiman, D. & Evans, R. F. (1981) *International Studbook of the Golden Lion Tamarin* Leontopithecus rosalia rosalia. Washington, D.C.: National Zoological Park

Lunn, S. F. & McNeilly, A. S. (1982) Failure of lactation to have a consistent effect on interbirth interval in the common marmoset, *Callithrix jacchus jacchus*. *Folia Primatol.*, **37**, 99–105

Mallinson, J. J. C. (1984) Golden-headed lion tamarin contraband – a major conservation problem. *IUCN/SSC Primate Specialist Group Newsletter*, 4, 23–5

Omedes, A. (1981) A comparative study of social communication in two subspecies of the marmoset *Callithrix argentata*. Ph.D thesis, University College of Wales, Aberystwyth

Poole, T. B. & Evans, R. G. (1982) Reproduction, infant survival and productivity of a colony of common marmosets (*Callithrix jacchus jacchus*). *Lab. Anim.*, **16**, 88–97

Siegel, S. (1956) *Nonparametric Statistics for the Behavioral Sciences*. Tokyo: Kogakusha Co.

Stevenson, M. F. (1983) Effectiveness of primate captive breeding. In *Symposium on the Conservation of Primates and their Habitats, Vaughan Paper no. 31*; ed. D. Harper, pp. 201–32. Leicester: University of Leicester

Stevenson, M. F. (1984) The captive breeding of marmosets and tamarins. In *Management of Prosimians and New World Primates, Proceedings of Symposium 8, A.B.W.A.K.*, ed. J. Barzdo, pp. 49–67

Stevenson, M. F. & Rylands, A. B. (in press) The marmoset monkeys: genus *Callithrix*. In *Ecology and Behavior of Neotropical Primates*, vol. 2, ed. R. A. Mittermeier & A. F. Coimbra-Filho. Rio de Janeiro: Academia Brasileira de Ciencias

Thornback, J. & Jenkins, M. (1982) *The IUCN Mammal Red Data Book*, Part 1. Gland, Switzerland: IUCN

Vertovec, A. (1981) An observational study of socio-sexual behavior in a captive colony of pygmy marmosets (*Cebuella pygmaea*). Honors thesis, University of Wisconsin, Madison

Wolfe, L. G. & Deinhardt, F. (1978) Overview of viral oncology studies in *Saguinus* and *Callithrix* species. *Primate Med.*, **10**, 96–118

Wolfe, L. G., Deinhardt, F., Ogden, J. D., Adams, M. A. & Fisher, L. E. (1975) Reproduction of wild-caught and laboratory-born marmoset species used in biomedical research (*Saguinus* spp., *Callithrix jacchus*). *Lab. Anim. Sci.*, **25**, 802–13

VI.6

Management of reproduction in a breeding colony of bushbabies

M. K. IZARD AND E. L. SIMONS

Introduction

Most prosimian primates are either endangered, breed poorly in captivity, or are not readily available. Galagos, commonly known as bushbabies, are prosimian primates that are not endangered, have the potential for high reproductive output in captivity, and are easy to handle, feed and house. For these reasons, galagos are becoming increasingly important as models for the early stages of primate evolution. In addition they are being used in several diverse areas of biomedical research including studies of the visual system, Parkinson's disease, and contact lens research. In order to insure the future availability of bushbabies for biomedical research, while at the same time decreasing or hopefully eliminating dependence on wild stocks of *Galago*, domestic breeding colonies of *Galago* are a necessity.

With funding from the Division of Research Resources, National Institutes of Health, a breeding colony of three species of bushbabies was established at the Duke University Primate Center (DUPC). The objective of the project is to determine optimal husbandry techniques for the captive maintenance and propagation of these three *Galago* species.

Methods

Animals

The smallest of the three species of bushbaby in the DUPC colony is *G. senegalensis*. The subspecies *G. senegalensis moholi* was chosen for its high incidence of twin births. This subspecies weighs between 150 and 200 g. We have 90 individuals of which 45 are females.

G. garnettii, a medium-sized galago, was chosen for the breeding colony, because Duke already had a long-term captive colony, with some individuals in the eighth and ninth captive generations. They weigh between 900 and 1200 g. They usually produce only one infant at a time, with twins accounting for 11–15% of all births. We have 23 *G. garnettii*, of which 11 are females. In the literature, this species has been referred to as *G. crassicaudatus crassicaudatus*, but we adhere to the classification of Olson (1979), who put this type in a separate species.

The third species, the larger *G. crassicaudatus*, weighs between 1500 and 2000 g. We have three subspecies, *monteiri*, *argentatus*, and the South African subspecies, *umbrosus*. *G. crassicaudatus* has twin or triplet births about 60% of the time. We have 67 *G. crassicaudatus*, of which 30 are female.

Caging

G. senegalensis is housed in cages of vinyl-coated wire measuring 2 m × 1 m × 0.5 m, furnished with partitions, shelves, and nest boxes. One pair is housed on each side of the cage, which is partitioned down the middle. *G. garnettii* and *G. crassicaudatus* are housed in cages measuring 2 m × 1 m × 1 m, also furnished with partitions, shelves and nest boxes. Both types of cages have areas that can be closed off to separate cage mates if necessary.

Light cycle

G. senegalensis is kept on an LD 12:12 light cycle, under which conditions they are aseasonal breeders. This is true also for *G. crassicaudatus*. *G. garnettii* is housed under the fluctuating photoperiod for the latitude of Durham, North Carolina, which is about 36° North. They are also aseasonal breeders.

Diet

All three species are fed a diet consisting of chopped fruits and vegetables, Purina High Protein Monkey Chow and Purina Cat Chow, and crickets.

Management program

A major facet of the determination of optimal husbandry techniques has involved the development of a management program for reproduction. The management program for reproduction involves: (1) vaginal smears; (2) pregnancy palpations; (3) isolation of pregnant females 5 days prior to parturition; and (4) return of a male to the mother and infant when the infant is 30 days old.

Vaginal smears

To take a vaginal smear, a cotton swab, from which most of the cotton has been removed, is dipped in saline, inserted deep into the vagina, and gently twirled. Deep insertion is important because smears of the outer vagina are very difficult to decipher, and may be misleading. The cotton swab is rolled across a glass slide, then stained with a Diff–Quick stain set, which stains epithelial cells a deep blue.

For *G. crassicaudatus,* in which the vagina is always open, a diestrus smear is seen during pregnancy and anestrus, or between cycles. The criteria we use for diestrus is a smear containing predominantly leucocytes. Epithelial cells, if present at all, are small and rounded.

The vagina of a *G. senegalensis* or *G. garnettii* is sealed during pregnancy or between cycles. When open at a cycle, it normally shows a proestrus smear. During proestrus, leucocytes begin to decrease in number, and eventually disappear from smears. Epithelial cells increase in size, and are rounded.

Estrous smears consist predominantly of anucleate, cornified epithelial cells. Other cell types are absent. Often spermatozoa are present in large numbers.

During metestrus, the cornified epithelial cells begin to clump, and there is a gradual reinvasion of leucocytes until diestrus is reached, in which the leucocytes are the predominant cell type.

When we first established the management program, we did vaginal smears or inspections twice weekly, but as information accumulated on the length of the estrous cycle, we were able to decrease the number of smears, so that now we only spend about 45 min a day actually doing smears.

As a by-product of the management program, information has accumulated on the length of estrous cycles, the length of the stages of the estrous cycle, and the length of lactation.

The interval between successive estrous periods is 34.5, 38.6, and 51.3 days for *G. senegalensis*, *G. garnettii*, and *G. crassicaudatus,* respectively (Fig. 1). We have very small sample sizes for *G. garnettii* and *G. crassicaudatus* because in our colony, they rarely cycle twice in succession. *G. garnettii* usually conceives at the first cycle after parturition, and *G. crassicaudatus* cycles sporadically. Four of the six estrous cycle intervals depicted for *G. crassicaudatus* are from one female, who is reproductively abnormal, the other two intervals are about 35 days, comparable to the other two species.

The length of proestrus averages 2–3 days for *G. senegalensis moholi*, 3 days for *G. garnettii*, and 9 days for *G. crassicaudatus* (Fig. 1). Estrus lasts 1–2 days in *G. senegalensis*, about 5 days in *G. garnettii*, and 6 days in *G. crassicaudatus*. Metestrus, which for our purposes is defined as the length of time in which a vaginal smear can be distinguished from diestrus, is about 8 days for all three species.

Other information that has emerged as part of the vaginal smear program is information on the length of lactation. When a lactating female is smeared or inspected, she is also checked for lactation by expressing a drop of milk from one or more of her six nipples. *G. senegalensis* has a lactation length that averages 102 days (Fig. 2), which is significantly shorter than that of *G. garnettii* (146 days) or *G. crassicaudatus* (135 days). The length of lactation in *G. garnettii* and *G. crassicaudatus* does not differ significantly. Milk can be expressed for 8, 13, and 18 days after an infant dies, in *G. s. moholi*, *G. garnettii* and *G. crassicaudatus*, respectively, so the lactation lengths observed, in the absence of behavioral data, may overestimate the length of time the infant is actually nursing.

Fig. 1. Characteristics of the estrous cycle and cycle length in *Galago*. P = proestrus; E = estrus; M = metestrus; and D = diestrus.

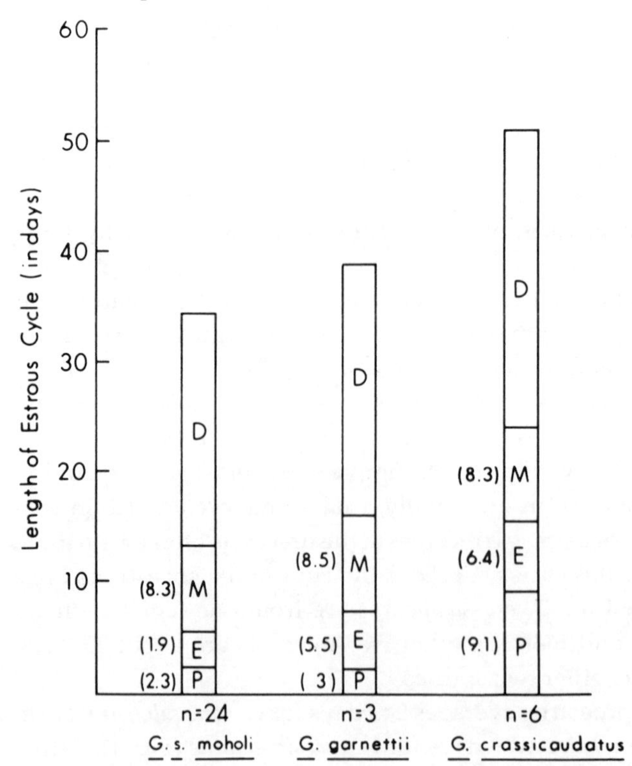

Pregnancy palpation

The second component of the management program is pregnancy palpation. A fetus can be detected by abdominal palpation at about 30–40 days of gestation, which is also about the time a second cycle can be expected. Palpations are confirmed with weights taken in the last third of gestation, in which weight gain is substantial.

Isolation of pregnant females

The third component of the management program is the isolation of pregnant females 5 days prior to parturition. The date of parturition is predicted based on gestation lengths calculated at DUPC. The gestation length is defined as the interval from the first day spermatozoa appear in a vaginal smear to the day of birth. Gestation lengths average 124 days for *G. s. moholi*, 129 days for *G. garnettii* and 134 days for *G. crassicaudatus* (Fig. 3). All three gestation lengths are significantly different from each other, an observation especially important in the comparison of gestation lengths between *G. garnettii* and *G. crassicaudatus*. As mentioned, *G. garnettii* has in the past been considered to be a subspecies of *G. crassicaudatus*, but here is one more piece of information to build a case for classifying these two forms as separate species.

Pregnant females are separated prior to parturition because of the following observation. Neonatal mortality rate declines significantly in all three species if females are isolated (Fig. 4). We went back through the records at the DUPC, covering almost 18 years in which

Fig. 2. The length of lactation in *Galago*, according to infant survival.

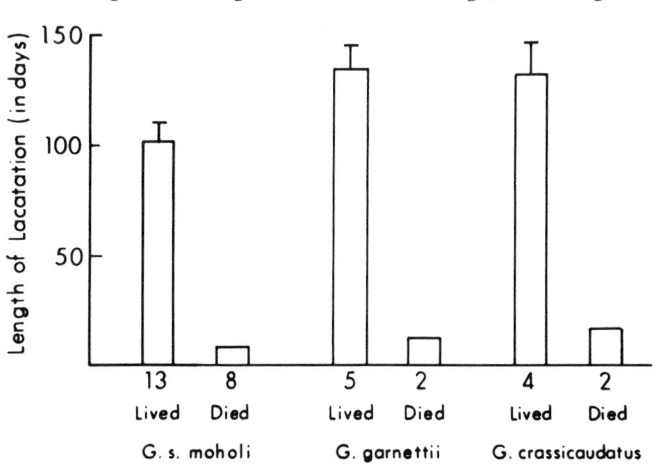

galagos have been kept there, and classified neonatal mortality rates according to whether or not females were isolated at parturition. In *G. senegalensis*, we observed a decline in neonatal mortality rate of about 45%. This decline is about 50% for *G. garnettii* and 32% for *G. crassicaudatus*. It is obvious what such large decreases in neonatal mortality rate can do for colony production. This is the major reason why we feel the time spent doing vaginal smears is worthwhile. If monitoring of reproductive cyclicity results in accurate prediction of parturition dates, so that females can be isolated prior to parturition, then the neonatal mortality rate can be kept low.

Reintroduction of the male

The fourth feature of the reproductive management progam is that males are put back in with females when infants are 1 month old. This assures that if a female cycles during lactation (which is not uncommon in our colony) a male with whom she is compatible is already with her. There are problems with the male and older infants getting along together if we wait much longer than 30 days for reintroduction of the male.

The influence of management factors on reproduction

There are three areas in which management factors can improve reproduction or have the potential to improve reproduction. The first, neonatal mortality rate, has already been mentioned. To test the hypothesis that management can improve neonatal mortality rate, we looked at neonatal mortality rates in the three species of *Galago* before and after the implementation of the management program.

Fig. 3. Gestation length in *Galago*.

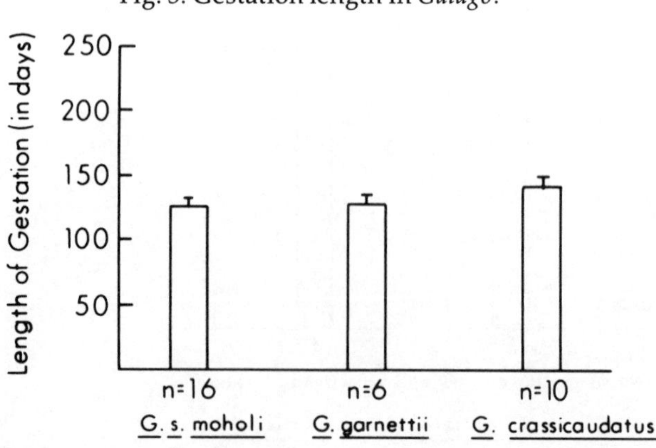

'Before' the implementation of the management program covers the period from August 1981 to October 1982. 'After' covers the period between October 1982 and March 1984. There was a significant decline in neonatal mortality rate in both *G. s. moholi* and *G. garnettii* (Fig. 4). Although neonatal mortality declined in *G. crassicaudatus*, the difference was not statistically significant. However, it is obvious that there is an advantage to be gained by isolating pregnant females.

Management has the potential to improve reproduction in a second area, that of interbirth intervals. Interbirth interval has declined with the implementation of the management program (Fig. 5). None of the differences depicted here is significant, largely because of the small sample sizes after the implementation of the management program, and the large standard deviations in the sample before the implementation of the management program. However, despite the lack of statistical significance, there is certainly a biological significance to these data. If interbirth interval decreases, production per female over time should increase. Interbirth interval, as might be expected, decreases if a female loses her infants (Fig. 5). A significant decline in interbirth interval is seen in *G. garnettii* and *G. crassicaudatus*. The decrease in *G. senegalensis* interbirth intervals approaches significance. This set of data suggests that if infants were removed for handraising, total colony interbirth interval would decrease. However, we are not prepared to invest the large amount of time necessary

Fig. 4. Neonatal mortality rate in *Galago*: the influence of management and isolation.

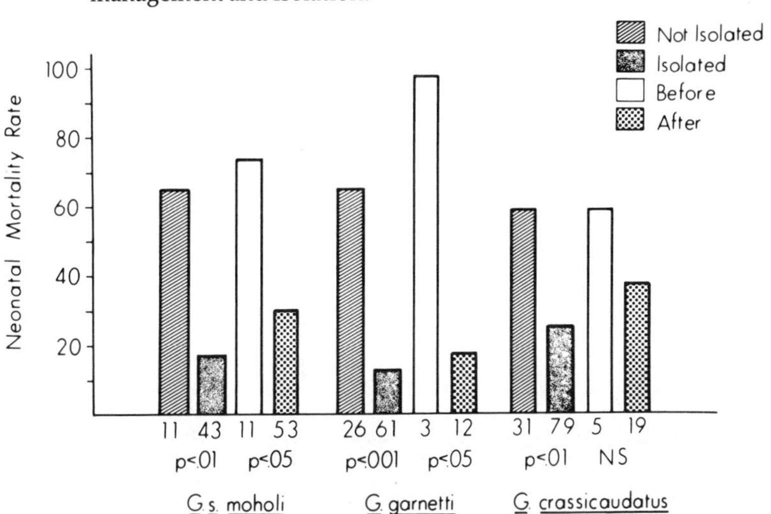

Fig. 5. Interbirth intervals in *Galago* under management and according to survival of the young.

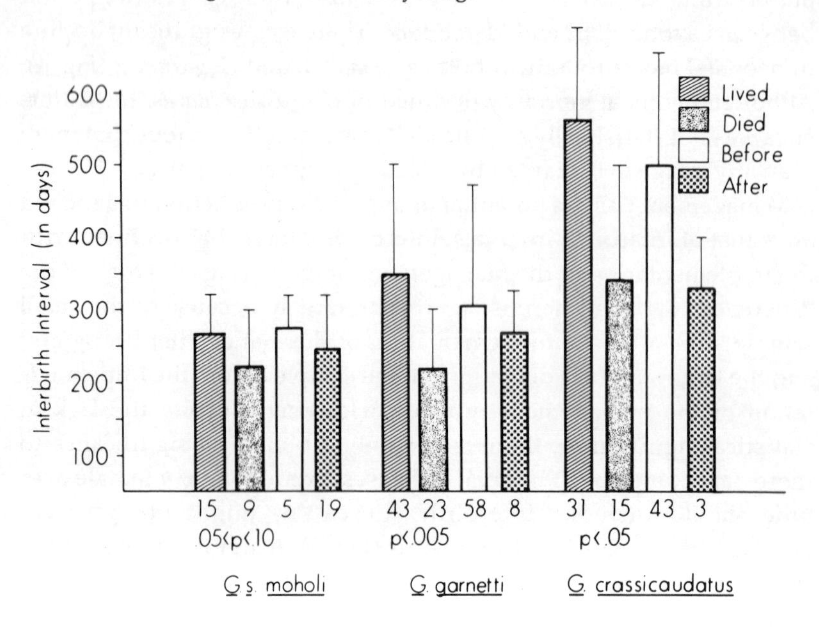

Fig. 6. Age at puberty in female *Galago*: influence of social housing with young or adult males.

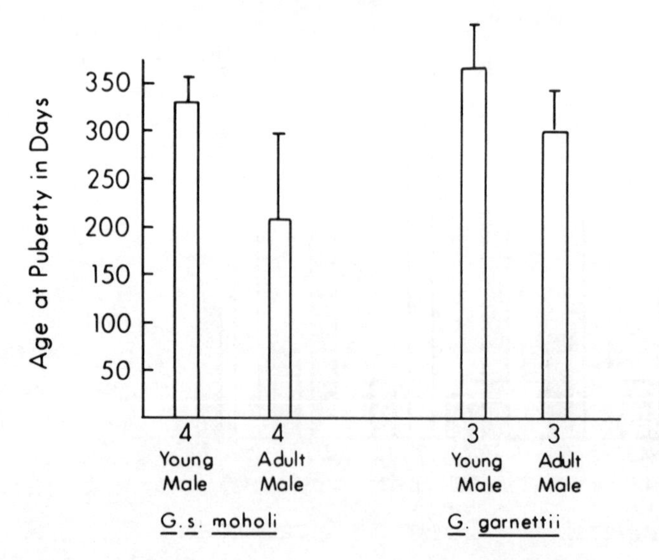

to handraise infants, because there may be anywhere from 15 to 20 infants in the colony at a time.

The third area in which management has the potential to improve reproduction is in pubertal age in females (Fig. 6). Our sample sizes are small at present, and the differences shown here are not statistically significant, but there is a clear trend for females housed with adult males to reach puberty at an earlier age than females housed with peer males.

Discussion and conclusions

The results presented above demonstrate that management factors can influence reproductive output in a breeding colony of bushbabies. By isolating pregnant females, neonatal mortality rate declines significantly, having a positive effect on colony production rates. Although our sample sizes are small at present, it appears that interbirth interval decreases as a result of the management program. This is probably due to the fact that close attention is paid to each individual female and that accurate records are kept. A decrease in interbirth interval increases a particular female's lifetime reproductive output. Finally, by housing young females with adult males rather than peer males, a decrease in the age at puberty results. This decrease in pubertal age will, over time, result in a decreased intergenerational interval, as well as an increase in the lifetime production of a particular female. Clearly, management factors can have a positive effect on the success of a breeding colony or any captive conservation effort.

Acknowledgements

This is Publication No. 278 of The Duke University Primate Center. This study was supported by a grant from the National Institutes of Health. We thank Nancy Wharton and Clyde Bittle for their expert technical assistance.

References

Olson, T. (1979) Studies on aspects of the morphology of the genus *Otolemur* Coquerel, 1859. Ph.D thesis, University of London

VI.7

A global overview of primate conservation

R. A. MITTERMEIER

Introduction

The past two decades have seen serious declines in the wild populations of most nonhuman primate species. Although often difficult to quantify, such declines pose a critical threat to the survival of a growing number of the world's 200 primate species.

The goals of this paper are to review briefly the factors contributing to the disappearance of primates and to present an overview of worldwide primate conservation. Major threats, priority regions, and the actions and efforts of those working to conserve primates are discussed.

Major threats to primate populations

There are three broad categories of major threats to primate populations in the wild: habitat destruction; hunting for food and other purposes; and live capture for export or local trade. The relative importance of each of these threats varies from species to species and in different regions, but one or more affects almost all existing primate populations.

Habitat destruction

Habitat destruction and modification is the single most important factor contributing to the decline of primate populations on a global basis. More than 90% of all primate species occur in the tropical forests of Asia, Africa, and South and Central America (Fig. 1). As these forests are exploited and disappear, so too do the species depending on them for survival. Tropical forests are destroyed for a variety of reasons, among the most frequent of which are conversion and use of land for agriculture and pasture, fuelwood gathering, poorly managed industrial logging, and massive hydroelectric projects.

Conversion of tropical forest lands to agriculture is the single most important cause of forest destruction (Fig. 2). Shifting agriculture, typically occurring in areas of low fertility and leading to cycles of degradation, abandonment, and the exploitation of new forest areas, is probably the greatest threat. Fuelwood gathering can account for as much as 80% of the wood removed from forests and can be a major cause of deforestation in some areas. An international trade in charcoal is now thought to be having a serious global effect on tropical forests (Office of Technology Assessment, 1984). Industrial logging, while in some cases selective, can cause secondary damage from heavy equipment traffic, road building, and poor management and replanting schemes. Flooding areas due to the construction of hydroelectric dams has caused massive losses in many forest reserve areas, which tend to be away from high concentrations of human populations and therefore suitable sites for dams.

At the root of these problems is the rapidly growing human population in tropical countries, adding millions of landless people to the already overtaxed traditional agricultural areas. Control of high growth rates for human populations and improved education for the better management of tropical forests are partial solutions. The needs

Fig. 1. The tropical rainforests of the world, showing the distribution of open and closed forests.

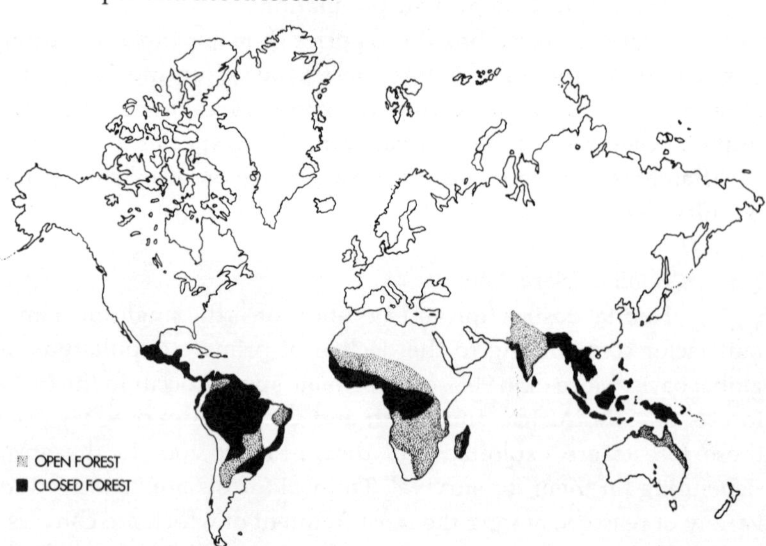

of the local people must be considered by conservationists, yet the needs of the human and the nonhuman primate populations are similar, since the survival of both ultimately depends on the survival of tropical forests.

Hunting for food and other purposes

The use of primates as a source of food probably has the greatest effect on those populations exploited by hunting, and again these effects vary between populations and regions. In at least three parts of the world, the hunting of primates for food is a major threat which may sometimes have an effect even greater than that of deforestation: the Amazonian region of South America (Fig. 3); and both West and Central Africa. In some places primates can account for as much as 25% of meat intake (Mittermeier, 1977). Large primates tend to suffer more than small species from hunting pressure and heavy hunting pressure can result in local extinctions even in areas of suitable habitat (Mittermeier & Coimbra-Filho, 1977; Soini, 1982).

Closely associated with meat hunting is the use of meat or body parts of primates for medicinal purposes. Although usually a minor factor in the decline of primate populations, this form of hunting can

Fig. 2. Destruction of tropical forests in southeastern Brazil. (Photo by R. A. Mittermeier.)

have serious effects of reducing the populations of already endangered species (e.g. the lion-tailed macaque). The use of primate body parts for ornamentation, for clothing or for trade (Fig. 4) generally has little effect in South America, but in other parts of the world (e.g. the case of the mountain gorilla in Rwanda and Zaïre) it can be a serious threat (Fossey, 1983).

Primates are occasionally killed to use as bait in traps for capturing other species. They may also be killed or maimed in traps set for other species, as is true for chimpanzees and gorillas (Ghiglieri, 1984). Sport and trophy hunting are relatively rare, but since they tend to concentrate on large, spectacular species (e.g. gorillas, Du Chaillu, 1930), they can have serious local consequences for species already reduced in numbers.

Fig. 3. Spider monkey (*Ateles belzebuth*) shot for food in Colombia. (Photo by F. Medem.)

Fig. 4. A large rug made of black and white colobus monkey (*Colobus querza*) skins on sale in tourist shops in East Africa, 1973. (Photo by R. A. Mittermeier.)

Primates can be considered a serious agricultural pest in some areas, and this conflict can lead to the destruction of local populations. It is difficult to assess the consequences of killing primates as agricultural pests on their populations, but it may be causing more losses than is generally recognised. Again as human populations increase and encroach on the marginal agricultural lands that are primate habitats, the level of the conflict will increase, posing a serious threat to crop-raiding species.

Live capture of primates

The third major threat to primate populations is live capture either for export or to serve a local pet or research market. Primates have been exported primarily for use in biomedical research, as pets and for zoos. Primate exports have been declining with the imposition of export bans by source countries, with import restrictions by user countries, and decreased demands for the use of primates in biomedical research (see Mack & Mittermeier, 1984, for a review). Trade in endangered or vulnerable species can have significant effects, although habitat destruction and hunting are probably more significant for most populations. The live capture of any species, and particularly that of the great apes for zoos and research, usually involves the death of several individuals for every one making it to the final destination. Trade figures do not reflect this wastage and are minimum numbers of primates removed from wild populations.

Regional review of conservation priorities

The Neotropical region

In this region, the highest-priority area is the Atlantic forest of eastern Brazil. Only about 1–5% of the original forest cover remains, and 75% of the 21 species and subspecies of primates (80% of which occur nowhere else) are endangered. Endangered species are listed by region in Table 1.

Amazonia as a whole must be considered as the next highest primate conservation priority as it is the world's largest tropical forest region and has the majority of Neotropical species. Since the area covers nine countries, primate and forest conservation programs must be organized on a country-by-country basis. Opportunities still exist in this area to conserve large forest tracts and their primates.

In the non-Amazonian forests, the status of primate populations and habitats varies greatly from highly disturbed to relatively untouched. Setting aside reserves and parks for highly endangered species found in these forests (e.g. *Lagothrix flavicauda* in the cloud forests of northern Peru; *Saguinus oedipus* in northern Colombia) is a top priority in these regions.

Africa

No mainland African primate genus can be considered endangered, but a number of species and subspecies definitely are. The mountain gorilla (*Gorilla gorilla beringei*) is probably the most endangered African primate (Harcourt, 1983). All the other great apes are of great conservation concern, and few facts are available on populations of these species. The African tropical forests are of critical importance in the conservation of the endangered species, having endemic and diverse primate communities. The islands of Zanzibar and Fernando Po and the East African ecological 'islands' surrounded by non-forest vegetation all contain distinctive communities.

Madagascar has a unique primate radiation (Petter, Albignac & Rumpler, 1977; Tattersall, 1982), and the island's primate fauna must be considered one of the world's highest conservation priorities. Hunting and habitat destruction resulting in primate extinctions have taken place since the arrival of humans on the island. Widespread habitat destruction is the major threat to primates on Madagascar. Within Madagascar, the eastern forest region probably requires the most attention.

Asia

In contrast to the Neotropics and mainland Africa, Asian primates are found in large numbers on both the sub-continent and the islands, and primates occupy several zoographical regions. About 30% of all Asian species are considered to be endangered. Species endemic to islands, the orang-utan and several gibbons are of particular concern. Conservation measures are being taken to ensure the survival of the lion-tailed macaque, while research and survey work on the status of the snub-nosed monkeys (genus *Rhinopithecus*) is currently the highest Asian primate conservation priority. Indonesia, with the highest primate diversity, is the most important country in the region.

Table 1. *The world's most endangered primates listed by species and region. Species are ranked first according to taxonomic uniqueness and then according to the degree of depletion of wild populations*

Genus/Species	Range	Comments
Endangered families		
Daubentoniidae		
Daubentonia madagascariensis	Formerly the eastern forests of Madagascar	A monotypic family; highly endangered by habitat destruction and killed on sight by villagers as evil omen; no population estimates available
Endangered monotypic genera		
Brachyteles		
Brachyteles arachnoides	Formerly the southeastern portion of Brazil's Atlantic forest region	A monotypic genus; highly endangered by habitat destruction and poaching; 200–300 remain
Allocebus		
Allocebus trichotis	Eastern forests of Madagascar	A monotypic genus; known only from four specimens, three of them collected in the last century; possibly extinct
Indri		
Indri indri	Part of eastern forests of Madagascar	A monotypic genus; endangered by habitat destruction within its small and patchy range; no population estimates available
Varecia		
Varecia variegata	Eastern forests of Madagascar	A monotypic genus; endangered by habitat destruction and hunting; no population estimates available
Simias		
Simias concolor	Mentawai Islands off Sumatra, Indonesia	A monotypic genus; endangered by habitat destruction and hunting and the most endangered of the four Mentawai primates; population estimated at 19000 (WWF, 1980)
Endangered polytypic genera		
Leontopithecus *Leontopithecus rosalia* *Leontopithecus chrysomelas*	Southeastern portion of Brazil's Atlantic forest region in states of Rio de	All three species highly endangered by habitat destruction and illegal live capture within their very small ranges; only a few hundred *L. rosalia*

Table 1. *Continued*

Genus/Species	Range	Comments
Leontopithecus chrysopygus	Janeiro, Bahia and São Paulo	and *L. chrysopygus* survive in the wild; *L. chrysomelas* also depleted, but no recent population estimates available
Rhinopithecus *Rhinopithecus roxellanae* (including *roxellanae*, *brelichi* and *bietii* as subspecies) *Rhinopithecus avunculus*	China and Vietnam	All four taxa endangered by habitat destruction and perhaps also hunting. *R. r. brelichi* down to 500 individuals; *R. r. bietii* to about 200; *R. roxellanae* to 3700–5700 (Poirier, 1983); *R. avunculus* known from only a handful of museum specimens, possibly extinct?
Endangered species and subspecies		
1. Species/subspecies numbering only in the hundreds		
Hapalemur simus	Now known only from the humid forest east of Fianarantsoa on the east coast of Madagascar	Although no population estimates are available, this species has been considered extremely rare and on the brink of extinction
Propithecus diadema (including *diadema*, *candidus*, *edwardsi*, *holomelas* and and *perrieri* as subspecies)	Eastern forests of Madagascar	Although no population estimates are available, the species as a whole is considered highly endangered because of habitat destruction, and all subspecies must exist at low population sizes; *P. d. perrieri* is considered the rarest (Tattersall, 1982)
Macaca silenus	Western Ghats of south India	Highly endangered by habitat destruction and occasional hunting; population estimated at 670–2000 (Ali, 1982)
Colobus badius gordonorum	Uzungwa Mts and Magombera Forest Reserve in Tanzania	Endangered by habitat destruction and heavy hunting pressure; Magombera Forest Reserve, thought to be its main stronghold, has a population of only 300 animals (Rodgers, 1981)

Continued overleaf

Table 1. *Continued*

Genus/Species	Range	Comments
Endangered species and subspecies (*cont.*)		
1. Species/subspecies numbering only in the hundreds (*cont.*)		
Gorilla gorilla berengei	Virunga Volcanoes of Zaïre, Rwanda and Uganda; Bwindi Forest in Uganda	Highly endangered by habitat encroachment and poaching; total population in the Virungas 255 animals, in Bwindi 95–130 (Harcourt, 1983)
2. Species/subspecies numbering in the low thousands		
Macaca pagensis	Mentawai Islands off Sumatra, Indonesia	Endangered by habitat destruction and hunting in very restricted range; population estimated at 39 000 (WWF, 1980)
Cercocebus galeritus galeritus	Tana River, Kenya	Endangered, very small population in fragmented habitat; total numbers estimated at 1200–1700 (Marsh, 1978)
Colobus badius rufomitratus	Tana River, Kenya	Endangered, very small population in fragmented habitat; total numbers estimated at 1400–2000 (Marsh, 1978)
Colobus badius kirkii	Zanzibar	Very small population in fragmented habitat; total numbers estimated at 1700 (Silkiluwasha, 1981)
Colobus badius preussi	Cameroon	Endangered by habitat destruction and hunting within very small range; total numbers estimated at less than 8000 (J. S. Gartlan, personal communication)
Presbytis potenziani	Mentawai Islands off Sumatra, Indonesia	Endangered by habitat destruction and hunting; population estimated at 46 000 (WWF, 1980)
Hylobates klossi	Mentawai Islands off Sumatra, Indonesia	Endangered by habitat destruction and hunting in very restricted range; population estimated at 36 000 (WWF, 1980)
Hylobates moloch	West Java	Endangered by habitat destruction in already fragmented habitat; total numbers estimated at 2400–7900 (Kappeler, 1981)

Table 1. *Continued*

Genus/Species	Range	Comments
Pan troglodytes verus	West Africa from southern Senegal to western Nigeria; now largely restricted to Guinea, Sierra Leone, Liberia and Ivory Coast	Endangered by habitat destruction, hunting and live capture for export; total numbers in known habitats estimated at 1500; in potential habitats 15 700 (Teleki & Baldwin, 1979)
Pan paniscus	Central Zaïre south of the Zaïre River	At risk from habitat destruction and hunting in some parts of its range; total numbers in known habitats about 2200; in potential habitats 13 000 (Teleki & Baldwin, 1979)
Pongo pygmaeus (including *pygmaeus* and *abelii* as subspecies)	Borneo and Sumatra	Threatened by habitat destruction and occasional live capture. 5000–15 000 still thought to exist in Sumatra, 3500 in Sabah and about 250 in Sarawak (Rijksen, 1978; Davies & Payne, 1982); no published estimate for Kalimantan; recent unpublished information indicates that the total population may be much higher than previously believed
Gorilla gorilla graueri	Eastern Zaïre	Endangered by habitat destruction and hunting; total numbers estimated at 4000
Gorilla gorilla gorilla	Western Africa (including Cameroon, Gabon, Central African Republic, Congo Republic, Equatorial Guinea and Angola)	Threatened by habitat destruction and hunting; total numbers unknown, but recent estimate for Gabon of 35 000 ± 7000 indicates that total population may be higher than previously believed (Tutin & Fernandez, 1982)
3. Species/subspecies for which population estimates not available, but known to be of conservation concern		
Mirza coquereli	Forests of western Madagascar	May be endangered
Lepilemur ruficaudatus	Forests of western Madagascar	May be endangered

Continued overleaf

Table 1. *Continued*

Genus/Species	Range	Comments
Endangered species and subspecies (*cont.*)		
3. Species/subspecies for which population estimates not available (*cont.*)		
Lepilemur leucopus	Southern Madagascar	Thought to be the most endangered member of the genus *Lepilemur*
Lemur macaco	Coastal areas of northwestern Madagascar and islands of Nosy Be and Nosy Komba	Probably endangered or vulnerable
Lemur mongoz	Northwestern Madagascar and Moili and Ndzouani in the Comores	Endangered both on the Madagascan mainland and the Comores (Tattersall, 1982)
Lemur rubriventer	Sparsely distributed in eastern forests of Madagascar	Poorly known, but probably endangered
Callithrix flaviceps	Atlantic forest region of eastern Brazil in southeastern Espirito Santo and adjacent parts of Minas Gerais	Endangered by widespread habitat destruction
Callithrix aurita	Atlantic forest region of eastern Brazil in São Paulo and adjacent parts of Minas Gerais and Rio de Janeiro	Endangered by widespread habitat destruction
Callithrix geoffroyi	Atlantic forest region of eastern Brazil in Espirito Santo and Minas Gerais	Endangered by widespread habitat destruction
Saguinus oedipus	Northwestern Colombia	Endangered by widespread habitat destruction and live capture for export
Saguinus bicolor bicolor	Vicinity of Manaus in Brazilian Amazonia	Endangered by habitat destruction in a very small range

Table 1. *Continued*

Genus/Species	Range	Comments
Callimico goeldi	Widely but very sparsely distributed in upper Amazonia	A rare species about which very litle is known
Callicebus personatus (including *personatus, melanochir* and *nigrifrons* as subspecies)	Atlantic forest region of eastern Brazil from southern Bahia to São Paulo	Endangered by widespread habitat destruction and hunting in some areas
Saimiri oerstedii	Western Panama and southern Costa Rica	Endangered by widespread habitat destruction
Chiropotes satanas satanas	Lower Brazilian Amazonia	Endangered by habitat destruction and heavy hunting pressure in its restricted range
Cacajao calvus calvus	Upper Brazilian Amazonia, between the Rio Japura, the Rio Solimoes and the Rio Auati-Parana	Possibly endangered
Cebus apella xanthosternos	Atlantic forest region of eastern Brazil, restricted to small area in southern Bahia	Endangered by widespread habitat destruction and heavy hunting pressure
Alouatta fusca (including *fusca* and *clamitans* as subspecies)	Atlantic forest region of eastern Brazil from southern Bahia to Rio Grande do Sul	Endangered by widespread habitat destruction and hunting; the norther subspecies *fusca* may be close to extinction
Lagothrix flavicauda	Cloud forest region of the northern Peruvian Andes	Endangered by habitat destruction and hunting
Ateles geoffroyi azuerensis	Azuero Peninsula of Panama	Endangered by habitat destruction and hunting; possibly on the verge of extinction?
Ateles fusciceps fusciceps	Pacific slope forests of northern Ecuador	Endangered by habitat destruction and hunting; possibly on the verge of extinction

Continued overleaf

Table 1. *Continued*

Genus/Species	Range	Comments
Endangered species and subspecies (*cont.*)		
3. Species/subspecies for which population estimates not available (*cont.*)		
Ateles belzebuth hybridus	Disjunct distribution in northern Colombia and northern Venezuela	Endangered by habitat destruction and hunting
Cercocebus galeritus sanjei	Uzungwa Mts, Tanzania	A recently discovered subspecies quite restricted in range (Homewood & Rodgers, 1981)
Cercopithecus erythrogaster	Southwestern Nigeria	Endangered by habitat destruction and hunting (Oates & Anadu, 1982)
Cercopithecus erythrotis erythrotis	Fernando Po in Equatorial Guinea	Thought to be endangered
Papio leucophaeus	Cameroon, eastern Nigeria and Fernando Po	Endangered by heavy hunting pressure
Colobus satanas	Cameroon, Equatorial Guinea (including Fernando Po), Gabon and Congo Republic	Endangered by habitat destruction
Colobus badius bouvieri	Congo Republic	Possibly endangered
Colobus badius pennanti	Fernando Po	Possibly endangered
Presbytis aygula	West Java	Endangered by habitat destruction
Pygathrix nemaeus	Laos, Cambodia, Vietnam, perhaps Hainan in China	A monotypic genus; poorly known, possibly endangered or vulnerable
Hylobates pileatus	Thailand, Laos and Cambodia	Endangered by habitat destruction and hunting in Thailand; status unknown in Laos and Cambodia

The future outlook

Although expanding human populations and increasing human needs for development make it inevitable that a proportion of the world's forests and the primates living in them will disappear, the role of primate conservationists is to minimise this loss by:

1. Protecting areas for particular endangered and vulnerable species
2. Creating large national parks or reserves in areas of high primate diversity or abundance
3. Maintaining parks and reserves already in existence and enforcing protective legislation
4. Creating public awareness of the need for primate conservation in source countries.

Special emphasis is being placed on conservation of habitat and the furtherance of conservation education. The long-term survival of primate populations will only be possible if they can co-exist with local people. Conservation-orientated breeding programs for endangered species offer the only hope for primates when all their remaining habitat has been destroyed. Such programs should not, however, replace efforts to save wild populations. Finally, eliminating illegal trade in primates and encouraging captive breeding for most, if not all, primates used in research will help ensure the survival of the great mammalian order to which we belong.

References

Ali, R. (1982) Update on the status of India's lion-tailed macaque. *IUCN/SSC Primate Specialist Group Newsletter*, 2, 21

Davies, G. & Payne, J. (1982) *A Faunal Survey of Sabah*. IUCN/WWF Project No. 1692. Kuala Lumpur: World Wildlife Fund – Malaysia

Du Chaillu, P. (1930) *Paul Du Chaillu: Gorilla Hunter*. New York: Harper Bros

Fossey, D. (1983) *Gorillas in the Mist*. Boston: Houghton Mifflin Co.

Ghiglieri, M. (1984) *The Chimpanzees of Kibale Forest: A Field Study of Ecology and Social Structure*. New York: Columbia University Press

Harcourt, A. H. (1983) Conservation and the Virunga gorilla population. *Afr. J. Ecol.*, **21**, 139–42

Homewood, K. M. & Rodgers, W. A. (1981) A previously undescribed mangabey from southern Tanzania. *Int. J. Primatol.*, **2**, 47–55

Kappeler, M. (1981) The Javan silvery gibbon (*Hylobates lar moloch*), habitat, distribution, numbers. Unpublished doctoral dissertation

Mack, D. & Mittermeier, R. A. (1984) *The Primate Trade*. Washington, D.C.: World Wildlife Fund – US

Marsh, C. (1978) Problems of primate conservation in a patchy environment along the lower Tana River, Kenya. In *Recent Advances in Primatology*, vol. 2, *Conservation*, ed. D. J. Chivers & W. Lane-Petter, pp. 86–7. London: Academic Press

Mittermeier, R. A. (1977) Distribution, synecology and conservation of Surinam primates. Unpublished doctoral dissertation, Harvard University

Mittermeier, R. A. & Coimbra-Filho, A. F. (1977) Primate conservation in Brazilian Amazonia. In *Primate Conservation*, ed. Prince Rainier & G. Bourne, pp. 117–66. New York: Academic Press

Oates, J. & Anadu, P. A. (1982) Report on a survey of rainforest primates in southwest Nigeria. *IUCN/SSC Primate Specialist Group Newsletter*, **2**, 17

Office of Technology Assessment (1984) *Technologies to Sustain Tropical Forest Resources*. Washington, D.C.: Congress of the United States

Petter, J. J., Albignac, R. & Rumpler, Y. (1977) *Faune de Madagascar 44. Mammifères Lemuriens (Primates Prosimiens)*. Paris: ORSTOM/CNRS

Poirier, F. E. (1983) The golden monkey in the People's Republic of China. *IUCN/SSC Primate Specialist Group Newsletter*, **3**, 31–2

Rijksen, H. D. (1978) *A Field Study on Sumatran Orang Utans* (Pongo pygmaeus abelii *Lesson 1827*). Wageningen, Netherlands: H. Veenam & Zonen, B. V.

Rodgers, W. A. (1981) The distribution and conservation status of colobus monkeys in Tanzania. *Primates*, **22**(1), 33–45

Silkiluwasha, F. (1981) The distribution and conservation status of the Zanzibar red colobus. *Afr. J. Ecol.*, **19**, 187–94

Soini, P. (1982) Primate conservation in Peruvian Amazonia. *Int. Zoo Ybk*, **22**, 37–47

Tattersall, I. (1982) *The Primates of Madagascar*. New York: Columbia University Press

Teleki, G. & Baldwin, L. (1979) *Known and Estimated Distributions of Extant Chimpanzee Populations* (Pan troglodytes *and* Pan paniscus) *in Equatorial Africa*. Special report to the IUCN/SSC Primate Specialist Group. Gland, Switzerland: IUCN

Tutin, C. & Fernandez, M. (1982) Preliminary results of a gorilla and chimpanzee survey in Gabon. *IUCN/SSC Primate Specialist Group Newsletter*, **2**, 19

World Wildlife Fund (1980) *Saving Siberut: A Conservation Master Plan*. Bogor, Indonesia: World Wildlife Fund – Indonesia

Part VII

Primate Conservation in the Broader Realm

VII.1

Introduction: primate conservation in the broader realm

D. WESTERN

Introduction

Why emphasize primate conservation in the broader realm? Partly because any taxon-specific approach, like any single-issue political advocacy, urges special concessions at the expense of other interests, such as those of other species or the biological integrity of ecosystems. Furthermore, primatology has historically been little integrated with other wildlife studies, conservation, or national policies. The traditional separation is breaking down fast, but the process is far from complete.

Species conservation is a necessary effort, but biologically and practically it is an insufficient conservation goal. We can now maintain zoo populations such as Père David's deer and Przewalski's horse long after a species has become extinct in nature. We can potentially resurrect a species by freezing sperm and ova, and by cross-species fostering after a species has become totally extinct. But saving such genomes is no guarantee of preserving their natural evolutionary continuity (Frankel, 1983). If the natural ecological realm of a species is eradicated in the meantime, its reintroduction to an alien environment constitutes ecological tinkering. In short, conserving species is no guarantee of maintaining natural ecological systems. Species conservation should aid, and result from, the preservation of natural ecosystems, and be managed when and where necessary for that purpose, not be divorced from it by becoming a stamp-collecting exercise.

If species diversity were the sole objective of conservation, our arguments for conserving large tracts of natural land could become self-defeating in a nano-second of evolutionary time once we could decode genetic programs and synthesize any base sequence. Come to that, if diversity were our only goal, genetic engineering could conceivably produce more species than we have space or patience for.

The ultimate goal of biological conservation is surely to maintain systems of sufficient ecological integrity that species can interact and evolve free of human intervention (Frankel, 1983). The distinction between the conservation of biological diversity and the maintenance of ecological integrity is not a trivial one. To the contrary, because little land is available for conservation, the two objectives can be diametrically opposed. Biological integrity demands large tracts of land undisturbed by humanity (Terborgh, 1975; Diamond & May, 1976) whereas we can feasibly include all species by designating numerous small, scattered parks (Western & Ssemakula, 1981). However, in the long term such small reserves would be biologically inviable (Wilcox, 1980).

My purpose is not to belittle taxonomic conservation but rather to urge a broader approach, one that includes an awareness of the biological costs. On a more practical level I want to stress, particularly in the Third World context, the difficulties of conserving all species within nature reserves, especially when policies and practices vary so much from country to country, and even from region to region.

So, what should be the broader conservation realm as far as the primatologist (or any other species preservationist) is concerned? It must at least consider primates in the context of their ecosystems, as stressed by Terborgh (Chapter VII.2). It must establish the minimum biological and material resources needed to maintain a species as an evolutionary entity independent of humanity. How those factors are affected by human goals and constraints and what can be done to justify conservation in specific Third World countries are addressed by Strum (Chapter VII.3) and Hearn (Chapter VII.4). Finally, the research we must undertake to understand the issues involved, and to formulate realistic conservation objectives, policies and practices, must ultimately be considered.

To introduce these specific topics, I will first consider the motivation for conservation, and the global coverage of protected areas, then the spectrum of issues in one primate region, East Africa. Here I look at its conservation history, priorities and accomplishments, listing the indigenous primate species and the extent of legal protection they receive. I will then consider how significant primates are ecologically relative to other larger mammals, and, finally, discuss what options exist for improving their conservation status.

Global conservation

Protected areas are set aside for many reasons, biological and otherwise. Until recently, most protected areas were established for reasons other than the biological (Nash, 1967; Runte, 1979); they seldom had the benefit of ecological design, and survive today for reasons of political, economic and recreational expediency, or because land threats are still insignificant. Among the many reasons for establishing natural parks and equivalent reserves are the esthetic, recreational, political (including international kudos and military defence), economic (including profitability and affordability) and cultural. Such ulterior motives are not inimical to biological conservation. Indeed they are the reasons why we have such an extensive network of protected areas (Miller, 1982). Only by understanding these complex motivations can we ally them to species and ecosystem conservation. Just as modern agriculture depends on resources and markets beyond the farm gate, so too must conservation take place beyond the protected areas, as I will later emphasize, for that is where the greatest potential exists (Western, 1982).

To make this point clear, an analysis of protected areas worldwide shows that only 3.7 million km^2, or 2.5% of the land surface, is set aside as national parks or equivalent reserves, and that both the number and average size of newly established areas is declining annually (IUCN, 1982).

The East African region (Kenya, Uganda and Tanzania), for which I elaborate issues of primate conservation in more detail, supports 10% of the world's primate species, and 47% of those found in Africa.

East Africa – its conservation and its primates

Wildlife preservation in East Africa, unlike most other early efforts (Nash, 1967), originated from the Victorian notion that species have rights independent of humanity (Western, 1983), and through colonial efforts to protect indigenous wildlife from 'the destruction which has overtaken wild animals in southern Africa and in other parts of the globe' (quoted in Simon, 1962). The first reserves, established in the late nineteenth century, were 'not for the gratification of the sportsmen, but for the preservation of interesting types of wildlife' (quoted in Simon, 1962). It was a concept new to the world (Thomas, 1983), alien to Africa and the precursor of modern biological rationale for national parks and reserves. Despite that auspicious beginning, it was another 40 years before East Africa's protected areas received much attention and were established on a more secure footing, though still for largely altruistic reasons. The colonials could well afford this since it was not their land which they so generously devoted to wildlife.

What has been the progress of East African conservation in recent decades, and what are the prospects for future expansion in natural reserves? The trends can be illustrated by plotting the conservation area allocated through time (Fig. 1). Two points are pertinent – fewer, smaller reserves have been established in the last 30 years, a pattern which amplifies the worldwide trend. What is not so obvious is that new reserves are legally, as well as biologically, less secure than national parks, for most are national reserves, which can permit alternative human uses, and often do.

Realistically, with population growing at 3.5–4% annually, with land at a premium for development (Myers, 1972), and with more than 7% of East Africa already under natural reserve status, we can anticipate little future expansion. Neither can we biologically redesign the system of reserves (Western & Ssemakula, 1981). Future conservation prospects will depend therefore, on five possibilities: (1) small additions to the network of natural reserves; (2) minor boundary modifications; (3) management intervention; (4) integrated management of reserves and adjoining lands; and (5) justifying conservation in non-protected rural areas. I will consider each case, but will first review

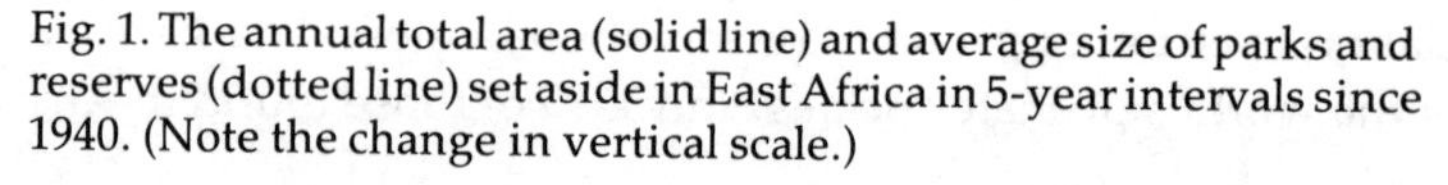

Fig. 1. The annual total area (solid line) and average size of parks and reserves (dotted line) set aside in East Africa in 5-year intervals since 1940. (Note the change in vertical scale.)

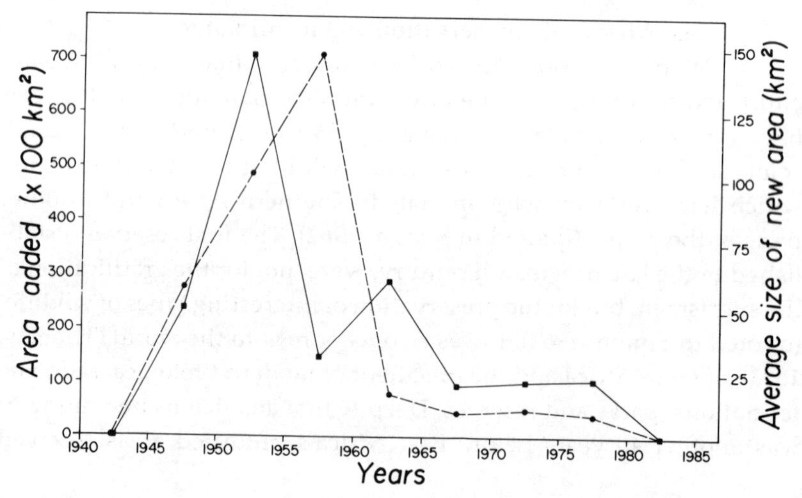

how well the existing protected areas cover the large-mammal fauna, including primates.

The implicit reason for protecting natural areas in East Africa has been to secure spectacular and diverse wildlife communities, such as Serengeti and Amboseli, representative habitats, such as montane (Mt Kilimanjaro), marine (Watamu), and savanna (Queen Elizabeth), or, exceptionally, to secure unique or endangered species (Kigezi Gorilla Sanctuary). Where land pressures denied complete ecosystem protection, emphasis was placed on dry season refuges (Western & Ssemakula, 1981). Even though not designed with the benefit of modern survey techniques and criteria for optimal reserve design (Terborgh, 1975; Diamond & May, 1976), how completely do the East African protected areas cover the large-mammal fauna?

All 55 East African ungulate species are included in one or more protected area, in all cases with populations exceeding 500 individuals (Western & Ssemakula, 1981), a relatively safe threshold for short-term survival in the face of inbreeding depression (Soulé, 1980). Two significant findings emerge from an analysis of East African reserves. First, there is no obvious species number–area relation (Miller & Harris, 1977; Western & Ssemakula, 1981), so area alone is not an adequate criterion for maximizing species diversity. Secondly, habitat diversity and specific resources, such as water, have a significant bearing on faunal diversity and migration, and underscore the need for site-specific analysis of optimal reserve boundaries. Considering land constraints and prevailing biological uncertainties when the reserves were established, they give remarkably good conservation coverage to the East African large ungulate fauna.

But what of the primates? It is worth first pointing out that primates, as measured by biomass production (Fig. 2), are usually less than 1% of the large-mammal biomass. In Amboseli National Park, for example, the total primate mass approximates one large elephant bull. Perhaps this emphasizes why primates have not been, and are unlikely to be, a priority conservation concern in East Africa, in contrast to the tropical forests especially in the New World (see Terborgh: Chapter VII.2). Despite their relatively tiny biomass, the same rough and ready conservation guidelines used in establishing the East African reserves have also covered the primate realm remarkably well.

According to Dorst & Dandelot (1970) and Kingdon (1971), East Africa has 21 primate species. Referring to the distribution maps and descriptions given in both sources, and Williams' (1967) list of primates found in parks and reserves, as well as information available from primatologists working in East Africa, it is evident that every species is included within at least one protected area. Census data are still too imprecise to tell whether populations large enough to avoid inbreeding depression are included in each case.

Primates and the need for inclusive conservation

Given the reality of human pressures in East Africa, and most nations, the waning prospects of new protected areas of substantial size, and the relative insignificance of primate biomass, what should be the future conservation priorities?

First, it is important to re-emphasize the possible conservation conflicts that arise between Third World efforts to maximize species diversity and those to secure ecological integrity, and to recognize that each additional conservation unit must exert maximum inclusive biological value. Both considerations point up the dangers to ecosystem conservation of single-species and single-issue advocacy. From the developing country's perspective, is it not better to lose some natural species and have the rest self-maintaining, than to have all dependent on extensive human intervention? Those are real choices which confront governments, if for economic and practical reasons alone, and primate conservation must therefore address the conflict of

Fig. 2. Annual production of all large mammals including livestock (open circles), wildlife (squares) and primates only (solid circles), plotted against annual rainfall. All data are compiled from East African ecosystems.

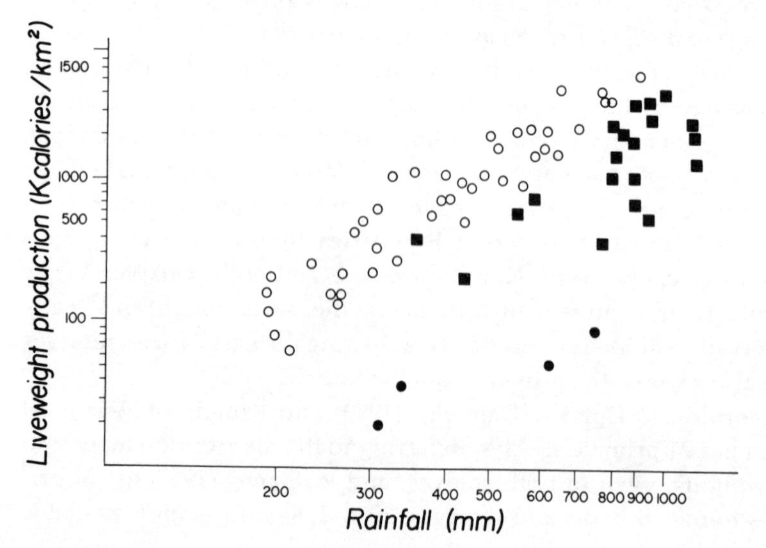

purpose. Each new small conservation area has an exorbitantly high cost–benefit ratio, and its start-up capital starves existing larger reserves of much-needed recurrent funds and staffing. Consequently, we should first question whether the existing reserves are adequate in scope, sufficient in design and security, and ultimately sustainable biologically and politically. If not, why should the addition of new areas do more than weaken the existing ones?

The five possibilities for improving conservation do not have equal potential. I have already stressed the limited role possible for new reserves and boundary realignments. Under Kenya's wildlife policy (Anon, 1975), these options are seen as filling the gaps in existing coverage, the gaps referring more to habitats than individual species. In either case, the logical approach is to review existing reserves, determine which species or habitats are safe, and which either omitted or vulnerable by virtue of inadequate area, numbers, or security. Reserve additions or boundary alterations can then be made according to the greatest inclusive value of all poorly protected species and habitats. By this criterion, species which are threatened at the edge of their range in one country but predominate and are securely protected in other countries, such as the De Brazza's monkey in Kenya, should not be considered a high priority. IUCN/SSC Elephant and Rhino Specialist Group has adopted criteria which rank the importance of all African populations of pachyderms according to biological characteristics, such as population size, genetic rarity and ecological significance of the region, as well as a series of political and socioeconomic criteria, including protectability, economic significance and opportunity cost (IUCN, 1984). Such criteria are not entirely satisfactory, but do help screen populations and direct international and national efforts to the ecologically and genetically significant and salvageable populations, rather than to saving every endangered population.

The three remaining categories have far greater conservation potential for both the maintenance of ecological integrity and species diversity, but they demand an imaginative approach to human and wildlife interactions, sophisticated technical and integrative management skills, and a broad and comprehensive approach to conservation.

Whereas the prospects of boundary readjustments are limited, a great deal more can be accomplished by integrated planning to buffer a reserve from the effects of insularization and ecological impoverishment. Although Kenya's parks and reserves occupy 7% of its land, the ecosystems on which they depend cover more than 25% (Western, 1983); it is biologically impractical to correct such ecological inadequacies. Kenya, recognizing that point, has an explicit policy of defining those 'dispersal areas' beyond the park boundaries vital for their maintenance and, through socioeconomic incentives, is encouraging landowners to accommodate the overspill. The policy of securing a functional ecosystem for existing parks takes precedence over the creation of new parks and reserves, based on the premise that effectively all permissible conservation land has been secured.

What I have stressed so far is the role reserves play in maintaining biological integrity and representative biogeographic realms, and their limited global potential. Such is not the case for the 95% of our land which is rural. On rural lands the conservation potential for species, as opposed to ecosystems, is enormous and largely ignored by the conservationist, even though most wildlife in developed nations, especially the United States, survives there and not in national parks. Need it be different in the developing world?

By far the majority of wildlife in the developing world also survives outside, rather than inside parks, and frequently in highly exploited regions, such as ranches and commercial forests. In short, human development need not be inimical to wildlife and many tropical forests may actually add species diversity or increase wildlife densities through shifting cultivation and selective timber exploitation (Levenson, 1981). Here species can survive, if the costs to humanity are sufficiently low or the benefits great enough. Hunting and commercial wildlife harvest can complement farm incomes (Mentis, 1978) and species such as giraffe and savanna primates, that compete minimally with livestock, survive well on ranches.

The new conservation challenge posed by diminishing natural lands is to evaluate land use patterns and trends, to identify compatible wildlife species, and those liable to disappear entirely beyond the national parks and reserves. Candidates for captive propagation or intensive *in situ* management should be those species incompatible with human activity beyond natural reserves and incompletely protected within them.

The final conservation option, applicable both to the maintenance of ecological balance and species, is management intervention. In theory there is no reason why small reserves should lose species as anticipated by island biogeographic theory. If inbreeding depression or disease threatens a species, we can exchange animals with other populations or restock from zoos. After all, the highest wildlife densities, and often survivorship rates, occur in zoos. But, although species can be saved by management intervention, the cost may be prohibitive; we can only afford to save so many species (Myers, 1979). New breeding techniques, such as artificial insemination, can greatly reduce the cost of gene transfer between populations, and no doubt will continue to expand the scope of management in natural reserves.

Most reserves are so small (75% of all parks and equivalent reserves are less than 1000 km^2) that they require management intervention to save some species. What we urgently require are genetic and biological guidelines for maintaining viable species populations (Soulé, 1980; Frankel & Soulé, 1981), and practical methods of monitoring key ecological parameters and applying corrective management.

Species conservation has advanced rapidly on both theoretical and practical planes to the point where many species have been rescued from extinction by measures taken in nature or in captivity, and to the stage where high-level governmental strategies (Anon., 1981), international trade conventions, such as the Convention for International Trade in Endangered Species and The IUCN World Conservation Strategy, focus on how to conserve our biological diversity. More international conferences (e.g. 'Primates: the road to self-sustaining populations', at San Diego Zoo in 1985) are planned in future to consider how to save primates using every available field, zoo and laboratory technique.

By contrast, awareness and concern for the maintenance of ecological integrity are rudimentary. Controversy still surrounds the issue of whether communities are assembled in some structured fashion or are random species aggregations (Diamond, 1975; Connor & Simberloff, 1979), and whether species in natural reserves can become locally overabundant and should be managed (Jewell & Holt, 1981). Even the largest reserves can be influenced by human activities, as in the case of concentrations of elephants in Murchison, Tsavo, Luangwa Valley, and other national parks, where their overabundance may destroy wooded habitats and threaten resident species, including primates (Laws, 1970).

Unless more attention is given to ecological integrity, we may inadvertently be burdened with more species rescue plans than we can handle or justify.

Acknowledgements

I am grateful to New York Zoological Society for sponsorship and to the participants in the conservation workshop of the IPS Congress.

References

Anon. (1975) *Statement on the Future of Wildlife Management in Kenya*. Sessional Paper No. 3. Nairobi: Government Printers

Anon. (1981) *Proceedings of the U.S. Strategy Conference on Biological Diversity*. Washington, D.C.: Department of State

Connor, E. F. & Simberloff, D. (1979) The assembly of species communities: chance or competition? *Ecology*, **60**(6), 1132–40

Diamond, J. M. (1975) Assembly of species communities. In *Ecology and Evolution of Communities*, ed. M. L. Cody & J. M. Diamond, pp. 342–444. Harvard: Belknap

Diamond, J. M. & May, R. M. (1976) Island biogeography and the design of natural reserves. In *Theoretical Ecology*, ed. R. M. May, pp. 163–86. Oxford: Blackwell

Dorst, J. & Dandelot, P. (1970) *A Field Guide to the Larger Mammals of Africa*. London: Collins

Frankel, O. H. (1983) The place of management in conservation. In *Genetics and Conservation*, ed. C. M. Schonewald-Cox, S. M. Chambers, B. MacBryde & L. Thomas. London: Benjamin/Cumming

Frankel, O. H. & Soulé, M. (1981) *Conservation and Evolution*. Cambridge: Cambridge University Press

IUCN (1982) *United Nations List of National Parks and Protected Areas*. Gland, Switzerland: IUCN

IUCN (1984) *The Status and Conservation of Africa's Elephants and Rhinos*. Gland, Switzerland: IUCN

Jewell, P. A. & Holt, S. (1981) (Eds) *Problems in Management of Locally Abundant Wild Mammals*. New York: Academic Press

Kingdon, J. (1971) *An Atlas of Evolution in Africa. Vol. 1*. London: Academic Press

Laws, R. M. (1970) Elephants as agents of habitat and landscape change. *Oikos*, **21**, 1–15

Levenson, J. B. (1981) Woodlots as biogeographic islands in southeastern Wisconsin. In *Forest Island Dynamics in Man-Dominated Landscapes*, ed. R. L. Burgess & D. M. Sharpe, pp. 14–40. Berlin: Springer-Verlag

Mentis, M. T. (1978) Economically optimal species – mixes and stocking rates for ungulates in South Africa. *Proc. First Int. Rangeland Congr.*, 146–9

Miller, K. R. (1982) Parks and protected areas: considerations for the future. *Ambio*, **11**(5), 315–17

Miller, R. I. & Harris, L. D. (1977) Isolation and expiration in wildlife reserves. *Biol. Conserv.*, **12**, 311–16

Myers, N. (1972) National parks in savannah Africa. *Science*, **178**, 1255–63

Myers, N. (1979) *The Sinking Ark*. Oxford: Pergamon Press

Nash, R. (1967) *Wilderness and the American Mind*. New Haven: Yale University Press

Runte, A. (1979) *National Parks: The American Experience*. Lincoln, Nebr.: University of Nebraska Press

Simon, N. M. (1962) *Between the Sunlight and the Thunder: The Wildlife of Kenya*. London: Collins

Soulé, M. E. (1980) Thresholds for survival: maintaining fitness and evolutionary potential. In *Conservation Biology*, ed. M. E. Soulé & B. Wilcox. Sunderland, Mass.: Sinauer

Terborgh, J. (1975) Faunal equilibria and the design of wildlife preserves. In *Tropical Ecological Systems*, ed. F. B. Golley & E. Medina, pp. 369–380. Berlin: Springer-Verlag

Thomas, K. (1983) *Man and the Natural World.* New York: Pantheon
Western, D. (1982) Amboseli National Park: enlisting landowners to conserve migratory wildlife. *Ambio,* **11**(5), 302–8
Western, D. (1983) *The Origins and Development of Conservation in East Africa.* Belkin Memorial Lecture. San Diego, Calif.: University of California
Western, D. & Ssemakula, J. (1981) The future for the savannah ecosystems: ecological islands or faunal enclaves? *Afr. J. Ecol.,* **19**, 7–19
Wilcox, B. A. (1980) Insular ecology and conservation. In *Conservation Biology,* ed. M. E. Soulé & B. Wilcox. Sunderland, Mass.: Sinauer
Williams, J. G. (1967) *A Field Guide to the National Parks of East Africa.* London: Collins

VII.2

Conserving New World primates: present problems and future solutions

J. TERBORGH

Introduction

In several parts of the Old World tropics we are down to a last-ditch effort to protect remaining primate populations from the combined ravages of overhunting and habitat destruction. Fortunately, over much of the New World tropics, the natural environment has not yet been so heavily exploited and there still remains time to protect adequate populations of most primate species, provided the governments of the countries involved take appropriate action. I do not imply that the Western Hemisphere is free of crisis situations; only that such situations are as yet the exception rather than the rule.

This state of affairs permits us the relative luxury of taking a broad view of the situation in the Western Hemisphere and asking what are the principal pressures on primate populations, how are these pressures likely to change in the forseeable future, and what measures might be taken to help assure the continued coexistence of primates and people in this part of the world? Particular trouble spots and problem species will be mentioned only briefly, as these have been covered by Mittermeier in Chapter VI.7.

Principal threats to New World primate populations

The principle threats to primate populations are the same the world over: (1) exploitation of forests for timber, pulpwood and construction materials; (2) transformation of wild lands to land that is intensively managed for crops, pasture, fuelwood, etc.; (3) direct slaughter of primates for food and pelts; and (4) removal of live animals for zoos, the pet trade and biomedical research. The severity of these threats varies by region according to the local state of development and the cultural practices and traditions of the people. It also varies between primate species according to their values as food or export commodities.

Of these four threats, habitat destruction has received the greatest attention, rightfully, because it leads to population losses that are essentially irreversible. Several New World areas are already severely affected by habitat loss, and a number of primate species and subspecies are consequently endangered. In decreasing order of urgency, the most affected regions are southeastern Brazil (*Brachyteles*, *Leontopithecus* and races of *Alouatta*, *Callicebus*, *Callithrix*, *Cebus*), northern Colombia and Pacific Colombia and Ecuador (forms of *Saguinus*, *Ateles*), and tropical Mexico (races of *Alouatta*, *Ateles* and *Cebus*). Elsewhere within the range of primates in the New World habitat loss has not yet reached the crisis stage, though numerous ongoing hydroelectric, roadbuilding and agricultural development projects presage problems in a number of additional areas in the near future.

Amazonia: a cause for alarm

The center of diversity of New World primates lies in central and western Amazonia (see Mittermeier: Chapter VI.7), a continent-sized region that has only marginally been invaded by modern technology. To this day one can fly for hours over unbroken primary forest without seeing a sign that man has modified the landscape. A seemingly limitless variegated green carpet stretches to the horizon, broken only by muddy, meandering streams. One could easily be lulled into complacency by viewing this awesome scene, unrivalled anywhere else on Earth. But what appears to be immense pristine wilderness is in reality an illusion. Virtually the entire Amazonian forest has been exploited for more than a century for a succession of commodities – first, medicinal plants, collectors' items, natural rubber, turtle eggs, then Brazil nuts, Cinchona bark, mahogany, and more recently, lesser timber species, cat and crocodilian hides, palm hearts and tropical fish. Through all these waves of exploitation the forest has survived virtually intact, because the generally poor soils do not support permanent agriculture. Nevertheless, men have penetrated almost everywhere under the verdant canopy, helping themselves to anything of value.

Among the valuable commodities of the Amazonian forest is wild game. In the near absence of agriculture, much of the population of the region depends on game and fish as the only available sources of protein. Relentless hunting pressure has brought about a collapse of the populations of many of the larger birds and mammals, including the principal primates. This disheartening development has almost entirely escaped the notice of the conservation community. To be sure, one sees an occasional lament in popular wildlife magazines over the slaughter of crocodilians, the pillage of turtle eggs and the plight of the Amazonian manatee, but these are only the most obvious symptoms of a wholly systemic condition. One seldom sees mention of the loss of forest-dwelling forms, including monkeys, peccaries, tapir, curassow and macaws. Is this lack of notice simply a reflection of indifference? Not entirely. It is more accurately a reflection of ignorance, because measuring the population densities of forest animals is a difficult proposition at best. The difficulty is exacerbated by a lack of adequately trained personnel to carry out censuses, and a startling deficiency of controls. Combine all these circumstances, and one can begin to appreciate the actual situation – an almost total lack of information on what the population densities of Amazonian forest mammals ought to be in the unperturbed state. Without the essential controls, it is impossible to demonstrate how bad the situation really is and to make a case for bringing it under control. Now, at last, there are some data, thanks to some very recently published censuses. What these censuses indicate is that the once abundant wildlife of the Amazon has vanished ahead of any visible development, just as the slaughter of the American Bison preceded the breaking of the sod in the Great Plains of North America.

Forest mammal communities in the Neotropics

Data on the natural abundance of forest mammals in the Neotropics are exceedingly scarce. Essentially, only two fully protected sites have been censused: Barro Colorado Island, Panama, and Cocha Cashu in the Manu National Park of Peru. Census results on the frugivorous mammal communities of the two localities are in close agreement as to overall biomass at 1300–1500 kg/km^2 (Table 1). In these forests, frugivores (broadly construed to include seed eaters) comprise the pre-eminent component of the mammalian community. (This is in contrast to many Old World forests where herbivores predominate.) In addition to the frugivores at these sites, but not included for lack of sufficiently reliable data, are carnivores, some specialized creatures such as armadillos and anteaters, and the herbivores, mainly sloths, deer and tapir. If one were to add in these latter groups, my guess is that the total mammalian biomass of Barro Colorado Island and Cocha Cashu would be around 2000 kg/km^2. From this estimate, crude as it is, we can appreciate that the forest environment does not sustain anything approaching the animal mass found on the African savannahs (commonly 10–20 tonnes/km^2 – Western: Chapter VII.1).

The roles that primates play in the forest and savannah ecosystems are strikingly different. Primates are an extremely minor element of the savannah community, contributing less than 1% of total biomass (Western: Chapter VII.1). But in the Amazonian forest at Cocha Cashu, primates contribute more to mammalian biomass than any other group (49% of frugivore biomass, and c. 33% of total mammal biomass, excluding bats). At Barro Colorado Island, primate diversity is lower than at Cocha Cashu (5 *vs* 11 species), and perhaps as a consequence, the prominence of primates in the frugivore community is rather less (21.9%). In any case, it is a telling commentary on our ignorance of tropical forest mammal communities that these data are all that exist to represent undisturbed sites in the Western Hemisphere.

This unfortunate situation has been somewhat relieved within the past year by the appearance of two new sets of censuses representing a broad selection of sites in Peru, Ecuador, Bolivia and Brazil. One set of censuses covers nearly all mammals other than bats, and the other covers primates only. Let us examine the more general data set first (Table 2).

The eight censused sites are all located in central and western Amazonia, but differ in at least three respects: rainfall, soil quality and

Table 1. *Biomass estimates for mammalian frugivores at Cocha Cashu, Peru, and Barro Colorado Island, Panama (kg/km^2)*

	Locality	
Mammal group	Cocha Cashu, Peru[a]	Barro Colorado Island, Panama[b]
Marsupials	60	110[c]
Primates	650	421
Rodents: large (>1 kg)	168	454
small (<1 kg)	114	186[c]
Procyonids	110	120[c]
Peccaries	230	153[c]
Total	1332	1444

[a] From census data of L. H. Emmons and others, as published in Terborgh (1983).
[b] Data from Eisenberg & Thorington (1973).
[c] Some species in group not included in census of Eisenberg & Thorington.

exposure to hunting. While it will not be possible to analyze the effects of all these variables here, inspection of the Table suggests that large primates (>1 kg adult weight), deer and tapir are badly depleted in localities exposed to hunting. Species that have little value as game, including small monkeys, small rodents and nocturnal forms such as marsupials and procyonids are affected much less or not at all. Here is our first indication that uncontrolled hunting has had a severe impact on game populations in Amazonia.

The second such indication comes from a series of censuses commissioned by the Pan American Health Organization to assess the status of wild primate populations in Peru and Bolivia (Table 3). Of 14 sites censused, 9 were judged to be exposed to heavy hunting pressure, 2 to moderate hunting and 3 to light hunting (the latter are all protected, one being Cocha Cashu). These data clearly show the devastating impact of hunting. The mean biomass of large primates (>1 kg) in the heavily hunted sites was only 3% of that in the protected sites. The species most affected are those in the genera *Ateles*, *Lagothrix*, *Alouatta*, *Cebus* and *Pithecia*. As in the previous set of censuses, the effect of hunting on small primates (<1 kg) is much less severe.

How representative are the data presented of the situation over the Amazonian forest as a whole? With no experience in Brazil, I cannot speak for that country, but for Ecuador, Peru and Bolivia I think they provide a reasonably accurate picture. None of the censused sites was near a major population center; some sites were near small villages while the rest were in fairly remote areas.

How can one explain the crash of primate populations in areas far removed from human settlement? The answer, I believe, lies in the widespread practice of market hunting, impelled by a demand for dried meat that can be kept without refrigeration. Teams of professional hunters work their way up tributary streams, systematically harvesting the game on both banks inland to a distance of 10 km or more. The flesh is then salted, dried in the sun and later sold in brittle sheets as *carne del monte*. Any large animal that is worth the price of a shotgun shell is included in the kill. Primates, being the most abundant and conspicuous larger animals of the forest, are harvested with the highest efficiency. In all the vastness of Amazonia, the animals have few strongholds. This is because streams navigable by small canoe are about 30 km apart on average, offering hunters access to all but the uppermost headwaters of nearly every river system. The loss of Amazonian game is thus, I fear, a general and not merely a local occurrence. The remedy, obviously, is to control hunting, but to do this in a huge, nearly empty frontier region is easier said than done.

Table 2. *Results of diurnal and nocturnal transect censuses: number of individuals (or troops) encountered per 10 km (from Emmons, 1984)*

	Cocha Cashu Madre de Dios, Peru	Barro Colorado Island,[a] Panama	Limoncocha Napo-Pastaza, Ecuador	Tambopata Madre de Dios, Peru	Yanomono Loreto, Peru	CMSE Amazonas, Brazil	Eastern Ecuador Napo-Pastaza, Ecuador	Mishana Loreto, Peru
Stream type[b]	White	White	White	White/black	White/black	Black	Black	Black
Hunting	None	None	Heavy	Light	Heavy	Light	Moderate	Heavy
Marsupials	4.8	5.8	8.0	2.6	2.7	0.6	1.8	2.2
Primates[c]								
large (>1 kg)	6.4	7.1	0.1	0.4	0.3	2.0	0.8	0.3
small (<1 kg)	3.3	0.3	3.8	2.8	3.1	1.0	2.4	3.1
Sloths	0.1	0.2	0.7	0.4				
Armadillos	0.3	3.5		0.8		0.6	0.6	

Rodents[d]								
large (>1 kg)	0.8	4.6	1.0	1.1		0.8	2.4	1.7
small (<1 kg)	30.4	30.5	18.4	13.1	7.3	2.1	12.8	10.0
Procyonids[e]	2.6		2.7	1.1	3.3	0.3	1.2	1.1
Felids	0.5			0.4		0.2		
Tapir		0.1		0.4				
Deer	0.2	1.8		0.4		0.5		

[a] Data for Barro Colorado Island from Glanz (1982)
[b] Stream color is a rough indication of soil quality, white water regions generally being characterized by more fertile soils than black water regions.
[c] Numbers for primates indicate troops, not individuals.
[d] A few species were not censused, including agoutis (*Dasyprocta* spp.) and peccaries (*Tayassu* spp.).
[e] Procyonids not censused at Barro Colorado Island.

Table 3. *Effect of hunting on the biomass of large (>1 kg) and small (<1 kg) primates at 14 sites in Peru and Bolivia (kg/km²)*[a]

Primate size class	Hunting pressure (no. sites)		
	Light (3)	Moderate (2)	Heavy (9)
Large	400	41	13
Small	48	1[b]	20
Total biomass	448	42	33

[a] Data from Freese *et al.* (1982).
[b] Both sites listed as being exposed to moderate hunting were in Bolivia beyond the geographic limit of *Saimiri sciureus*, the most abundant small primate.

Fig. 1. Seasonal variation in fruitfall at Cocha Cashu, Peru, 1976–1977 (lines represent two replicate sets of 50 fruit traps each). Note that variation as portrayed in the figure is damped by the logarithmic ordinate. From Terborgh (1983).

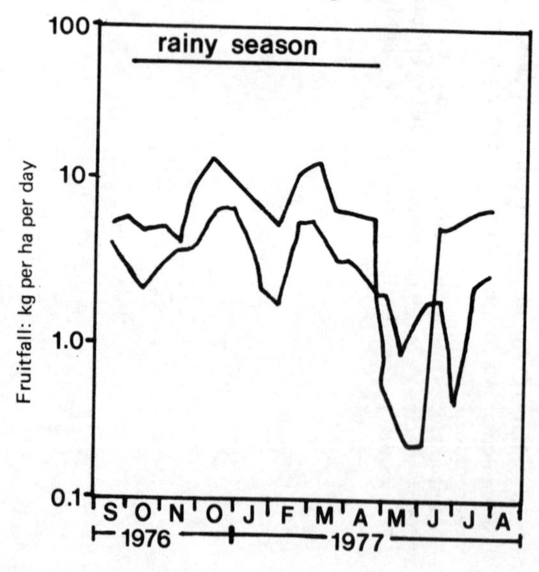

Resource fluctuations in a seasonal environment and their implications for forest management

I come now to the second main point of this chapter: how might tropical forests be managed in the future for commercial profit without eliminating all wildlife? For some preliminary clues, we must look at the resources that sustain mammal populations and the seasonal fluctuations in these resources.

Returning to Barro Colorado Island and Cocha Cashu as our two control sites, we note that they are situated, respectively, at 9°N and 12°S. Being located about halfway between the Equator and the latitudinal limit of the Tropics, the two sites can be considered representative of average conditions within the tropical belt of the Western Hemisphere. The climate in the two localities is similar. The mean temperature and rainfall on Barro Colorado Island are, respectively, 27°C and 2600 mm, while at Cocha Cashu the corresponding values are 24°C and 2050 mm (Dietrich, Windsor & Dunne, 1982; Terborgh, 1983). Seasonality is manifested primarily in the rainfall, both localities experiencing a dry season of 4–5 months' duration in which the precipitation is less than 10 cm per month. The alternation between wet and dry periods drives the annual flowering and fruiting cycles of the forest with the consequence that fruiting is concentrated in the rainy period (Fig. 1; Terborgh, 1983). Although the data shown are for Cocha Cashu, the seasonal pattern at Barro Colorado Island is quite similar, as is total annual fruitfall: 2180 kg/ha (BCI) and 1990 kg/ha (Cocha Cashu). Using our measurements of fruit falling from the canopy, and rough estimates of the daily food requirement of approximately 1400 kg/km^2 of frugivores, it is easily calculated that fruit is available in superabundance for about 8 months of the year (Terborgh, 1983). During the remaining 4 months fruit is in short supply, and for 2 or 3 of these months the amount available is actually less than needed to meet the metabolic requirements of the frugivores (including primates). During this annual period of scarcity, mammalian frugivores must get by on other resources such as leaves, insects, nectar, hard nuts, tree exudates, pith and meristem (Terborgh, 1983). A number of the avian frugivores simply migrate out of the area. In any case, every primate at Cocha Cashu modifies its diet in the late wet/early dry season to cope with the scarcity of fruit.

From an ecological standpoint, the annual food crisis is the most interesting time of year, for it is then when interspecific competition is most intense and ecological roles become most sharply defined. I have shown elsewhere that among Cocha Cashu primates interspecific dietary overlap is less in the dry season than during the period when fruit is abundant (Terborgh, 1983). Nevertheless, a surprisingly small number of plant species supplies the bulk of the non-leaf vegetable matter consumed during the dry season (Table 4). The 10 species listed accounted for more than 90% of feeding time (on plant materials) of five primate species (two *Cebus* spp., *Saimiri sciureus* and two *Saguinus* spp.), and we know from other observations that the same plants constitute critical resources for other mammals and birds as well (Janson, Terborgh & Emmons, 1981). The 10 species include three palms (hard nuts), three figs (soft fruit), a vine bearing 1-cm drupes,

Table 4. *Critical dry season plant resources in the Cocha Cashu flora (from Terborgh, 1983)*

Plant species	Resource	Period available	Eaten by
Palms			
Astrocaryum sp.	Seed	May, June	*Cebus* spp., peccaries, rodents, macaws
Scheelea sp.	Mesocarp, seed	Throughout	*Cebus* spp., Saimiri (mesocarp); squirrels (seed)
Iriartea ventricosa	Exocarp, seed	May–July	*Callicebus, Cebus, Ateles* (exocarp); peccaries (seed)
Vines			
Celtis iguanea	Fruit (a drupe)	March–Aug.	*Saguinus* spp., *Callicebus*
Combretum assimile	Nectar	July	*Cebus* spp., *Saimiri*, parakeets, 7 spp. primates, marsupials, procyonids, many birds
Trees			
Ficus perforata	Fruit (a fig)	Irregularly throughout season	Most primates, procyonids, marsupials, many birds
Ficus erythrosticta	Fruit (a fig)		
Ficus killipii	Fruit (a fig)		
Quararibea cordata	Nectar	Aug.–Sept.	*Saguinus* spp., *Cebus* spp., *Saimiri*, procyonids, marsupials, many birds
Erythrina ulei	Nectar, flowers	July–Aug.	*Cebus* spp., *Ateles*, parrots and other birds

and three nectar sources (two trees and a liana). Some of these plants occur only in early successional forest (in areas disturbed by a meandering river) and others in mature forest. What is remarkable is that the list of species is so short, encompassing less than 1% of the known plant diversity of the area.

This fact is of special significance, for it contains a key to the multiple-use management of tropical forests in the future. Management practices that were to eliminate crucial dry season resources would greatly diminish the carrying capacity of the forest for animals, especially primates. On the other hand, the plant species in question occupy at most only a few per cent of the space in the forest, so could be left standing at very little cost. But for animals to persist at anything approaching normal abundance in managed forests, it is essential that they be accorded a significant place in management planning. Otherwise we face a future in which tropical wildlife will hardly exist outside of scattered parks and reserves.

I should like to close by remarking on the fact that man's use of the tropical forest has changed over time. Aboriginal man obtained all his needs from the forest: food, building materials, tools, cooking utensils, weapons, adornments, condiments, medicines, even pets. Lately, the indigenous aborigines have been supplanted by peasant colonists who live partly on the renewable resources of the forest, and partly on an exploitative cash economy. They obtain fewer of their necessities from the forest, and, perhaps consequently, use it more harshly. Finally, we are now entering a period in which tropical forest lands are being subjected to intensive management for single products – cattle, palm oil, pulpwood, and others. The first two phases of exploitation (aboriginal and peasant) are (or can be) compatible with the continued existence of diversity and wildlife, but they are viewed as being obsolete. Nevertheless, these two types of exploitation are demonstrably sustainable, but the third, which is regarded by planners as the wave of the future, has not yet proven to be sustainable. In most places where large-scale intensive management has been attempted in the humid tropical lowlands, the end result has been seriously degraded soils and drastic loss of productivity. Might we thus, in our contemporary passion for intensive management, be going a step too far? Perhaps by backing up a little and returning to the notion of extensive management of the forest for all its manifold resources, we would reap a greater benefit over the long run. At the very least we should give it a try, for not to try is to admit defeat.

Acknowledgements

I am most grateful to Louise H. Emmons for providing me with a prepublication copy of her *Biotropica* article. I thank the Peruvian Ministry of Agriculture for authorization to conduct research in the Manu National Park. Supported by National Science Foundation grant DEB 76-09831.

References

Dietrich, W. E., Windsor, D. M. & Dunne, T. (1982) Geology, climate, and hydrology of Barro Colorado Island. In *The Ecology of a Tropical Forest: Seasonal Rhythms and Long-term Changes*, ed. E. C. Leigh, Jr, A. S. Rand & D. M. Windsor, pp. 21–46. Washington, D.C.: Smithsonian Institution Press

Eisenberg, J. F. & Thorington, R. W. (1973) A preliminary analysis of a Neotropical mammal fauna. *Biotropica*, **5**, 150–61

Emmons, L. H. (1984) Geographic variation in densities and diversities of non-flying mammals in Amazonia. *Biotropica*, **16**(3), 210–22

Foster, R. B. (1982) The seasonal rhythm of fruitfall on Barro Colorado Island. In *The Ecology of a Tropical Forest: Seasonal Rhythms and Long-term Changes*, ed. E. C. Leigh, Jr, A. S. Rand & D. M. Windsor, pp. 151–72. Washington, D.C.: Smithsonian Institution Press

Freese, C. H., Heltne, P. G., Castro, R. N. & Whitesides, G. (1982) Patterns and determinants of monkey densities in Peru and Bolivia. *Int. J. Primatol.*, **3**, 53–90

Glanz, W. E. (1982) The terrestrial mammal fauna of Barro Colorado Island: censuses and long-term changes. In *The Ecology of a Tropical Forest: Seasonal Rhythms and Long-term Changes*, ed. E. G. Leigh, Jr, A. S. Rand & D. M. Windsor, pp. 455–68. Washington, D.C.: Smithsonian Institution Press

Janson, C. H., Terborgh, J. & Emmons, L. H. (1981) Non-flying mammals as pollinating agents in the Amazonian forest. *Biotropica*, Suppl., 1–6

Smythe, N. (1970) Relationships between fruiting seasons and seed dispersal methods in a Neotropical forest. *Am. Naturalist*, **104**, 25–35

Terborgh, J. (1983) *Five New World Primates: A Study in Comparative Ecology*. Princeton, N.J.: Princeton University Press

VII.3

Activist conservation: the human factor in primate conservation in source countries

S. C. STRUM

Introduction

Primate conservation takes place in Third World countries whose cultural, political and economic realities are different from the background of the foreign conservationists interested in developing and implementing conservation and management policies. The inappropriateness of the perspective of 'outsiders' and subsequent policies is greatest in its lack of consideration and concern for the 'human factor' in conservation and management. To complicate matters, often source country attitudes are not explicitly defined, vary greatly and are evolving, although implicit goals and attitudes still guide government decisions powerfully.

Amidst all the variation, there is a major shift in the orientation of source countries towards sustainable development of biotic resources with an emphasis on the 'role of protected areas in sustaining society' (McNeely & Miller, 1984). Despite this groundswell, the only country which has explicitly formulated these policies is Kenya. A cogent wildlife management policy exists, stating goals, priorities and future directions. This policy candidly considers the 'human factor', taking as its rationale the goal of 'maximum returns from the land; optimum returns from wildlife'. Although the inescapable conclusion is that 'wildlife must pay its way', this principle is flexible. In practice, solutions vary with circumstances and policy differs for the National Parks and Reserves and for private land.

The Kenyan policy is instructive. It provides a framework for considering primate conservation in the 'broader realm' of wildlife conservation in source countries (see Western: Chapter VII.1) and highlights several critical issues. This paper will address these issues in two ways, first, by summarizing the position paper (Sessional paper

No. 3, *Statement on Future Wildlife Management Policy in Kenya*, Republic of Kenya, 1975) which crystalizes the Kenyan view, and secondly, by briefly reviewing specific primate cases in Kenya and elsewhere using the framework of evaluation provided by Kenyan policy. The latter incorporates my personal experience with the management and conservation of one population of savannah baboons. Taken together, Kenya's policy and personal experience provide suggestions for future directions in primate conservation in source countries.

Sessional paper No. 3

Sessional paper No. 3 is a thoughtful and sophisticated management document. Since few people have access to the actual publication and to avoid misinterpretation, I will quote directly from it rather than paraphrase its meaning. The paper is divided into sections, the most relevant for the current question of primate conservation are: Goals and Objectives; Policy on Parks and Reserves; Benefits to People; Control of Animals and People; and The Future.

Goals and Objectives

1. The Government's fundamental goal with respect to Wildlife is to optimize the returns from this resource, taking account of returns from other forms of land use.
2. Returns include not only the economic gains from tourism and from consumptive uses of wildlife, but also intangibles such as the aesthetic, cultural and scientific gains from conservation of habitats and the fauna within them.
3. Wildlife is only one resource on the land, and proper wildlife management is, therefore, but one aspect of land use planning and management, designed to maximize the returns from land. Land use conflicts involving wildlife are frequently debated in terms of "people versus animals". This statement of the problem is illegitimate. Rather, where conflicts exist, they are between different groups of people specifically, between those who believe that wildlife should be conserved in specific areas, versus those who wish to follow land uses which they believe are inconsistent with the survival of wildlife.
4. The Government's basic policy problems with respect to wildlife are, therefore, to –
   a. identify the best land uses (or combination of uses) for specific areas of land, in terms of their long and short run benefits to people;

b. to ensure implementation of those uses; and
c. ensure a fair distribution of benefits of those uses.

5. On 5 per cent of Kenya's area, exclusive wildlife use promises the highest returns, mainly through supporting the tourist industry . . . but also by maintaining examples of the main types of habitats found in this country . . . [National Parks]
6. Maximum returns from another 80 per cent of the land can only be secured through proper utilization of wildlife in combination with other forms of land use, mainly ranching, but also forestry [3 per cent of Kenya is made up of gazetted forest].
7. Finally, wildlife management has an important role to play, even in the approximately 15 per cent of Kenya currently suitable for high density arable agriculture, where the aim of policy is efficiently to minimize damage to crops by wild animals. This area may of course expand with the introduction of new crops or extension of irrigation schemes in drier areas.
8. The main future emphasis of wildlife policy will be upon finding means to secure optimum returns from the wildlife resource, and upon implementation of those means for the benefit of landowners and the nation generally. Policing and control activities, directed to preservation of wildlife wherever possible, will cease to be the main content of policy although policing and control activities will of course continue to be important components of policy.
9. The main objective of wildlife policy in the past was to preserve as many animals as possible . . . This policy was correct for its time. There were enormous uncertainties about the techniques which might be adopted for securing high returns from wildlife . . . Under traditional pastoralism, the carrying capacity freed by higher wildlife off-takes would have been occupied by expansion in domestic stock numbers, resulting in the elimination of wildlife as a usable resource. In these circumstances, "preservation wherever possible" was the only rational policy, and has had the great benefit of ensuring that Kenya now possesses a wildlife resource worth managing.
10. Recent and continuing changes in land ownership and the scale and nature of range development, both require and permit a more activist approach to wildlife management. If wildlife is to continue (in pastoral areas), it must yield returns to the ranchers, which are at least equal to the returns from the livestock, which could replace it.

12. Since the rate of progress on implementation of improved livestock management differs in different parts of the country, the specific wildlife policies followed in different parts of Kenya must differ . . . . to secure the fundamental goal – maximum returns from land, optimum returns from wildlife – in the presence of differing institutional, management, and social situations.

Thus Kenya's policy recognizes that wildlife will play a varying role in different areas of the country depending on the existence of other resources and their value. In some areas, wildlife promises to produce the highest returns from the land, through tourism, while in other areas wildlife utilization or elimination will be a necessity. More important is the projected shift in management policy from policing and control to the implementation of schemes to secure optimum returns from wildlife.

Policy on Parks and Reserves

Kenya has set aside 6% of its land for National Parks (Western, 1984*a*). These National Parks and Reserves represent *the* major investment in conservation and preservation (compare the 6–10% of land area set aside as National Parks by East and Central African countries to the 1–2% of land devoted to National Parks in the United States and Europe (IUCN, 1982)). Protected Areas have specific goals and rationales for wildlife including the primates within them.

46. National Parks are State Lands which are managed exclusively for the following objectives –
    a. to preserve, in a reasonably "natural" state, examples of the main types of habitats which are found within Kenya for aesthetic, scientific and cultural purposes;
    b. to provide educational and recreational opportunities for Kenyans;
    c. to provide an attraction for tourists, and so serve as a major basis for Kenya's economically profitable tourist industry; and
    d. to sustain any other activities which are not in conflict with the above (e.g. water catchment, commercial photography, etc.).
47. National Reserves may be on any type of land . . . Their main objectives are similar to those for National Parks, with the exception that other land uses may be specifically allowed . . .
61. . . . Protection Areas are the most important innovation in the new legislation . . . are absolutely critical for the continued

survival of the Wildlife herds which attract visitors to Nairobi National Park . . . etc. . . . There must therefore be provision for payment of some returns from tourism within the Park/ Reserve to these landowners.

It is clear from these provisions that the legitimacy of wildlife within Parks and Reserves does not depend on specifically financial considerations. However, human factors do play a role. 'Protection Area' is a new concept which recognizes that many Park populations, even in the best Parks, will not be able to survive if isolated from their traditional dispersal range which necessarily exists outside of the Parks. Although primates are not migratory, Park boundaries do not necessarily coincide with either the distribution of primates in the area or the home ranges or territories of any one group. Therefore the relationship of Parks to surrounding areas is of relevance to primates as well as migratory large mammals.

Benefits to People and Control of Animals and People

What are the benefits that wildlife is to provide to people and how are they to be dispensed? Furthermore, how are the costs of wildlife to be assessed? And how are human rights and wildlife rights to be integrated?

56. . . . Local Authorities and local inhabitants stand to benefit from the economic activity generated by Parks . . . (employment, new markets for agricultural produce, hotels and lodges located outside Parks, and licenses and cesses on these activities).
60. The landowner used by wildlife herds from specific Parks and Reserves should, at a minimum, not suffer any net damage from supporting these . . . on their land. To this end, Government shall pay to landowners . . . grazing fees scaled to the costs . . . imposed on them, after taking account of any direct benefits received . . . in the form of revenues from hunting, cropping, and tourist facilities on their land . . . Such agreements shall be gazetted, and the areas . . . be called Protection Areas.
79. [Control of Animals] The range of available techniques includes deterrence (through use of thunderflashes, night fires, dogs, shooting of one or two members of a herd), erection of game proof barriers (which is very expensive and can only be countenanced where game damage is likely to be large and where it is in the interest of sound land use management that wildlife be excluded from the area), translocation

(which will only be applied when the value of the animal at its destination, whether in Kenya or abroad, is judged to be sufficiently large to cover the costs involved), and extermination (via poisoning, shooting, or destruction of habitat. This method will be carried out where the animals must be excluded from an area).

86. [Control of people] . . . [the government] will take active steps to eliminate illegal killing or sale of wild animals and products made from them, both through activities in the field to apprehend poachers, and in the towns to intercept trade in resulting products . . .

82. Members of the public shall have the right to kill wild animals in cases of immediate danger to life . . . in cases of immediate threats of substantial damage to property . . . The onus will be on the person concerned to prove that he killed the animal out of necessity.

Thus, Kenya compensates certain landowners for sustaining wildlife on their land, and protects animals from illegal or improper exploitation, while acknowledging the right of the public, in other areas, to protect themselves and their property from wildlife damage. Yet, wild animals are 'innocent until proven guilty' and human detractors must make a legal case for wildlife destruction.

The Future

Future directions in Kenya's policy include both additional Parks and Reserves and also the preservation of wildlife on private land by making wildlife of direct benefit to the landowner. Land for the expansion of Parks is limited and additional Parks will be oriented towards specific goals outlined in existing policy. Utilization clearly plays a role in arguments for wildlife on private lands.

51. The Government shall seek the creation of new National Parks and Reserves . . . in accordance with its over-riding objective to optimize returns from wildlife, and in accordance with specific objectives of Parks and Reserves as stated . . .

52. To this end . . . will carry out careful surveys of potential Parks and Reserves within Kenya, identify "gaps" in the system of Parks and Reserves, and seek the establishment of Parks and Reserves which are likely to be beneficial . . .

65. . . . shall explore the possibilities for, induce and establishment of, and regulate consumptive forms of wildlife utilization, wherever these uses will make a net contribution to Kenya's economic and social development.

36. . . . to inform landowners about the possibilities for securing higher long run returns from wildlife on their land, advising them upon particular courses of action to be followed to that end, and assisting them to carry out those actions.
39. . . . It is important that many of these issues be handled by way of regulations rather than legislative provisions, in order that regulations may be flexible in light of changing circumstances.

The issues

Given the rationale of Kenya's policy, particularly the emphasis on benefits from wildlife and the rights of people, certain primate conservation issues emerge. Specifically:

1. How do primates fit into the wildlife goals as outlined and are there special benefits derived from primates?
2. What are the problems in the management of primate populations and are they compatible with general wildlife management?
3. What is the future of primates within wildlife policies as currently formulated?

For primates, as for other wildlife, there is no one answer to these questions. Because of changing circumstances and responding to ecological, economic, social and political conditions in different regions of the country, there are minimally three sets of answers. These are appropriate to the three different types of land use: Parks and Reserves, where wildlife, including primates, need not pay its way, private land used for domestic stock, and arable agricultural land, both situations where maximum returns from the land incorporating wildlife conservation could require direct economic benefit from wild animals.

Are there special benefits from primates?

Primate biomass represents a small fraction of wildlife biomass in most habitats in Kenya, although its place is increasingly important in some forested areas. Generally, then, primate conservation has been part of a broader concern with habitat and faunal preservation and not specifically focused on the primates themselves. Given this situation, are there special benefits to be derived from nonhuman primates which might enhance their position in conservation policies?

In accordance with Kenya's policy, the benefits of primates in the Parks and Reserves include their role as part of the ecosystem, their place in the preservation of natural fauna, and their aesthetic value.

The latter has more importance in modern-day Kenya because urbanization, a growing African middle class and the emergence of institutions like the Wildlife Clubs have created both a need and an appreciation for wildlife (see Western, 1983). In addition, although foreign scientists may treat primate studies as distinct, primate research is valued under existing policy as an aspect of wildlife research. All of these nonmonetary benefits accrue to primates *as wildlife* and are not in any way unique to them.

But a special value is found in at least two areas. Culturally, primates have a singular place because of interest in human origins, particularly in countries where the fossil evidence for primate and human evolution is found. Medically, because of this biological affinity between human and nonhuman primates, primate research has special value to the extent that the focus is on issues relevant to source country needs.

Economics

Economic benefits are often the most persuasive arguments for the conservation of wildlife, nationally and locally. In a country like Kenya where tourism plays a central role in the economy, it is unfortunate that primates occupy a shadowy position as a tourist attraction. Attitudes toward the more common primate species are, at best, ambivalent, both among indigenous people and among visitors. Yet, the attractiveness of primates and their role in tourism could be enhanced through education and proper presentation. In certain areas, like the Tana River Reserve, limited tourism could develop around the resident primate community.

There are also potential economic advantages to private landowners who are less impressed by aesthetic, cultural and scientific merits of wildlife. In pastoral lands where primate numbers are at a minimum, economic gains would reflect the extent to which both tourism and utilization could focus on primates and be cost effective. Where arable agriculture now occupies previously choice primate habitats or is still surrounded by these, any direct benefits from primates would likely come from utilization or very specialized tourism.

In summary, the list of benefits derived from nonhuman primates greatly resembles that for wildlife generally. Economic considerations often predominate, particularly outside of the Parks, but primates can and do have aesthetic, cultural and scientific value. Any 'special' status for nonhuman primates depends, to a great extent, on the attitudes of their human relatives, both indigenous and foreign. Fortunately, attitudes have the potential to change. Both better education and specialized tourism would enhance the position of nonhuman primates *vis à vis* other wildlife in the assessments of tourists and policy makers.

Are there problems in the management of primate populations?

Niche overlap between human and nonhuman primates and shrinking natural habitats bring human and nonhuman primates into direct contact with increasing frequency. Because nonhuman primates are intelligent, have great manual dexterity, and are often opportunistic, maintaining them within an ecosystem may have special costs.

When nonhuman primates find themselves near human activity they are quick to exploit new opportunities. Primate pests are more difficult to control, more difficult to contain and more difficult to circumvent than pests of other taxa. Techniques that routinely work with nonprimate species have limited effectiveness with primates. This creates serious management problems for both rare and populous primate species.

We can conclude that while the benefits of primates generally resemble those of other wildlife, the costs of maintaining nonhuman primates within an ecosystem may be greater than for other taxa; so much so that these costs actually tip the balance against primate species both within protected areas and on private land.

A few concrete examples will help to illustrate the situation.

Examples

Nairobi National Park, despite being at the edge of the capital city, has a variety of wildlife seasonally present in high densities. The protection extended to those primates within Nairobi Park is as members of a valued ecosystem. But two species, baboons and vervet monkeys, create problems within the Park. Or, rather, they are part of a serious problem. Tourists find the monkeys appealing and, although directly and legally admonished from doing so, feed them. In the process groups of dangerous monkeys are created. Other troops which live on the edge of the Park scale fences that deter other large mammals, and invade gardens, houses, and lodges looking for food. As pests, these primates pose a serious threat to both the wildlife attitudes of local people and to tourist enjoyment. As a result, in order to safeguard general wildlife interests, groups of primates often have to be destroyed.

A related issue is illustrated by Laws' (1970) discussion of elephant activity in Murchison National Park, Uganda. The pressure of surrounding human populations compressed the indigenous elephant population leading to major changes in the landscape, including total deforestation. Laws argued for the containment of elephants in order to protect woodlands and maintain animal diversity within the Park. Although primates are present in significant numbers, and Laws' recommendations would protect them, it is the welfare of other species, not primates, which is seen as important to the source country and hence to policy. The Murchison situation is not unique. While the conservation focus is usually on the large mammal community which includes primates, primate species never figure in either rationales or policy decisions.

Thus the fate of primates even within Protected Areas is delicately balanced against general wildlife interests. Despite the obvious value of primates and their legal and ecological 'right to life' in Kenya's policy, (1) primate interests often have lower priorities than other taxa, or (2) the costs of maintaining them outweigh the benefits. Conserving primates for their own sake or as a focus of habitat conservation is limited in Kenyan Parks because the majority are savannah/woodland areas. There is greater potential for primates in forest Parks but these have their own problems such as low tourist visitation and therefore low economic potential to policy makers.

Primates living on private land fare worse under Kenyan policy since they do not have the special status of those in protected areas. Generally there are few nonhuman primate species in pastoral regions and direct interaction between human and nonhuman primates is minimal, as is conflict between nonhuman primates and domestic stock or rangeland management. The fate of primate species in these pastoral lands depends, to a great extent, on pre-existing attitudes which have historical or cultural origin. Economic benefits from wildlife, through tourism that includes primates and through other means such as utilization, must be tailor-made to each set of circumstances if they are to be used with any force as rationales for primate conservation in these areas.

My own familiarity is with recently settled marginal areas. A brief review of the events of the last 8 years on Kekopey Ranch, near Gilgil, Kenya, where I have studied baboons since 1972 (Strum, 1981, 1984; Strum & Western, 1982), may be instructive.

This high-altitude savannah with an annual average rainfall of 600 mm is ideal cattle country. As an extremely productive cattle and sheep concern, Kekopey Ranch had a large complement of wildlife, including baboons. The land was divided for smallhold agriculture in early 1977 and settlement began in 1978. August, 1979, witnessed the first conflict between baboons and humans over crops, which assumed serious proportions within a few months. Major crop loss and major baboon mortality occurred during the next 9 months. Beginning in late 1980, the Baboon Project initiated a program aimed at understanding the development of crop-raiding among the baboons and investigating methods of containment and control. Concurrently we hoped to explore and institute a rural development scheme that was both realistic and sound for Kekopey.

Research and education were important components of our efforts. It quickly became apparent that we could not expect local farmers, who were suffering from wildlife depredations, to have altruistic attitudes towards wild animals. So, the Baboon Project sought ways to change farmers' opinions by optimizing wildlife benefits and maximizing returns from the land in this low rainfall area that was clearly of limited value agriculturally. The benefits included assistance with social services, schools, various educational activities, and aid with appropriate agricultural, energy and water technology, that we hoped would be seen as associated with wildlife. Plans also included future direct benefits from wildlife through some form of tourism.

Taking our cues from existing Kenyan policy, we sought alternative or supplemental economic activities independent of agriculture. A successful wool project was started in 1982. Members of many of the local farming families are currently involved in woolcraft, preparing, spinning and dying wool with local natural dyes and then weaving the wool into rugs, mats and tapestries. The wool project is an economic boon to the farmers far in excess of the maximum agricultural profits they could hope to realize on Kekopey.

All of these activities, taken together, reduced the conflict of interests between wildlife and agriculturalists and made wildlife (including primate) conservation potentially desirable.

The interests of primates are more difficult to champion in prime agricultural areas. Here multiple land use schemes in which wildlife is an important contributor are less feasible. Kenyan policy recognizes that the major activity in these regions (roughly 18% of Kenya in 1975) is control management, although preservation of forests for water catchment (and with the forests, primates) takes high priority. Thus the fate of primates in these areas rests with forest preservation policies, but the underlying principles of ecosystem conservation parallel those already outlined for wildlife conservation in general.

The exercise of putting primates into perspective in different land-use situations highlights an important question. *How much profit* and *what kind of benefit* is needed to make primates an attractive and important part in conservation and management plans, particularly on private land? There will be different answers for different land-use regimes but all have in common the necessity of carefully considering the economy and politics of the area, that is, the 'people' factor, and devising a realistic scheme that is appropriate to the area (Western, 1976, 1982, 1984*a*,*b*; Western & Henry, 1979). In some places, for example, social services might be more appealing than money, while obtaining government assistance with animal husbandry techniques might be favored by others. In some cases only time and influence, not money, are needed. Yet financial considerations can be very modest. I have been impressed in my work in a marginal agricultural area and in a pastoral area with how little profit from wildlife, including baboons/primates, is required to make an appreciable impact on landowners' attitudes. Reasonable research fees in one area added 25% to the net income for those landowners, while the addition of limited tourism focused on primates will double or triple the income (at no cost to the landowner) that is made from domestic stock in another area. In marginal and pastoral areas, the costs and problems created by primates are small, compared to areas of higher rainfall, and it seems that the benefits can more easily be made significant.

Conclusions

Third World countries are source countries for nonhuman primates. The current shift towards sustainable development of biotic resources and its consequent emphasis on the role of protected areas in sustaining society makes the 'human factor' of critical importance in every stage of primate conservation in Kenya and elsewhere. As Kenyan policy points out, although conflicts involving wildlife are frequently couched in terms of 'people *versus* animals', they are really differences of opinion between people who want to conserve wildlife in certain areas and those who want to use that land in a different manner. Therefore, any effective conservation policy must address a basic question: how can people be convinced to act in the interests of wild animals?

The Kenya situation is atypical in several ways that are relevant to generalizations about primate conservation. Wildlife tourism is a major industry and a top earner of hard currency for the country, so attitudes towards wildlife conservation are generally good. But most National Parks and Reserves in Kenya are wildlife 'spectacles' of savannah ecosystems where the primate contribution is minimal. Thus, primates as a taxa have relatively low priority in management decisions. The situation in many source countries where forests predominate is different. As primate biomass increases, relative to overall animal biomass, and primates provide some of the major spectacles in an ecosystem, as happens in forested regions, the rationale for primate conservation fares better.

Still, even with its unique characteristics, in many ways Kenya offers conservation's 'best of all possible worlds'. The political stability of the country, the well-articulated wildlife policies and the crucial role that wildlife plays in tourism, enhance the potential for conservation. (Forest tourism is more difficult to develop and currently less appealing than visitation to nonforested areas.) Yet the lessons learned from the Kenyan examples, should, in principle, apply in other countries where the primary primate habitat, forests, predominates. Perhaps the most important lesson is that there is no one solution or principle to be used in primate conservation. Different circumstances call for different combinations. Since it is humans who must do the conserving, policies must be pragmatic: maximizing returns from the land, optimizing returns from wildlife, providing protected areas where wild animals have priority, providing direct benefits to private landowners who tolerate wildlife on their property, having a flexible approach to the integration of human and wildlife interests, and above all, addressing human concerns in conservation and management decisions and policy. The value and valuing of wildlife can be improved through education and in some circumstances primate conservation problems are really only problems of education.

How can people be convinced to act in the interest of wild animals? Currently there are two general responses to this question: *active* conservation and *activist* conservation. The most outspoken of recent proponents of 'active conservation' is Dian Fossey in her book *Gorillas in the Mist* (Fossey, 1983).

Active conservation, as Fossey defines and illustrates it, has quite a narrow and immediate focus. It ignores human issues except to devise ways to motivate individuals to be more efficient policers or to control forcibly the interaction between wildlife and people. This position is very critical of what Fossey calls 'theoretical conservation' (essentially the *activist* approach outlined in Kenya's Sessional Paper and reviewed below) which is characterized by her as an abstract approach emphasizing the wrong things: wildlife education and direct economic benefits to local people from wildlife.

In contrast, *activist conservation* (what Fossey calls theoretical conservation) takes a cold hard look at the realities of source countries: economic, political and social. This approach strives to find a flexible way of integrating wildlife and human perspectives, making it possible for people to conserve wild animals out of enlightened self-interest, a powerful motivator.

The future of primates seems problematic under a regime of the Fossey type 'active' conservation, whether it is baboons at Gilgil or gorillas in the Virungas. There is simply not enough money or manpower available to entrust the fate of primate populations to physical (armed) protection. More important, such protection is merely a cosmetic treatment of symptoms and does not address the basic underlying problems such as land pressure of local populations.

The future under 'activist' conservation holds more promise, it seems to me. Education, the economics of tourism and other wildlife benefits given to people change attitudes and, more importantly, change people's lives. Even for Fossey's gorillas in the Virunga, education, increased tourism, and getting economic benefits to the local people have helped to stem the tide of poaching far more effectively than the armed patrols that were instituted earlier (Harcourt, 1984; von der Becke, 1984).

What should be apparent by now to those interested in primate conservation in source countries is that the 'people factor' cannot be ignored or dismissed. The benefits and costs of nonhuman primates to source countries, as just discussed within the framework of Kenyan policy, suggests that the future of primates depends on innovative plans which maximize potential benefits derived from primates and minimize the costs of primates within the larger wildlife conservation and management picture.

Problems remain; some are highlighted by the Kenyan policy. For example, future Parks and Reserves in Kenya will be instituted because of 'gaps' in the existing system. These gaps, in most cases, refer to habitat and ecosystem considerations; primates will certainly not be a main focus of future conservation planning in Kenya. Even in countries with extensive forests, the rationale for future conservation areas will more likely revolve around forests themselves than their primate residents. Therefore arguments for primate conservation will not carry much force.

Primate habitats are those most affected by the expansion of agriculture, an ever-present phenomenon in source countries. Even if habitats are not completely destroyed, nearby human settlement will offer tempting agricultural resources for opportunistic primates to exploit. Management of primate populations may overshadow primate conservation interests.

Given such realities, the future of primates will depend on better applied research, better techniques of nondestructive control of pests, and better understanding of how isolated populations can be managed or consolidated. But even this will not ensure a future for primates in source countries. Activist conservation with pragmatic solutions to human issues offers the greatest potential for primate conservation.

The Kenyan philosophy, a flexible policy 'maximizing returns from land and optimizing returns from wildlife', should be of great help to those concerned with primate conservation in other source countries where the shift towards sustainable development of biotic resources is equally important although less cogently formulated.

Acknowledgements

The insights for this paper come from fieldwork in Kenya from 1972 to the present. This research has been sponsored by the Office of the President, the Institute of Primate Research, the National Museums of Kenya and the University of Nairobi. Funding has been provided by the New York Zoological Society, L. S. B. Leakey Foundation, National Geographic Society, World Wildlife Fund, U.S., H. F. Guggenheim Foundation and University of California, San Diego. This paper draws heavily from the work of David Western and I am grateful to him for many helpful discussions.

References

Fossey, D. (1983) *Gorillas in the Mist*. New York: Houghton Mifflin Co.

Becke, J.-P. von der (1984) Mountain Gorilla Project. *Wildlife News*, **19**, 2

Harcourt, A. (1984) Mountain gorilla population report. *Wildlife News*, **19**, 2

IUCN Commission on National Parks and Protected Areas (1982) *United Nations List of National Parks and Protected Areas*. Gland, Switzerland: IUCN

Laws, R. M. (1970) Elephants as agents of habitat and landscape changes in East Africa. *Oikos*, **21**, 1–15

McNeely, J. A. & Miller, K. (1984) *National Parks, Conservation and Development: The Role of Protected Areas in Sustaining Society*. Washington, D.C.: Smithsonian Institution Press

Republic of Kenya (1975) *Statement on Future Wildlife Management Policy in Kenya*. Sessional Paper No. 3. Nairobi: Government Printing Office

Strum, S. C. (1981) Processes and products of change: baboon predation at Gilgil, Kenya. In *Omnivorous Primates*, ed. G. Teleki & R. Harding. New York: Columbia University Press

Strum, S. C. (1984) The Pumphouse Gang and the great crop raids. *Anim. Kingdom*, **87**, 36–43

Strum, S. C. & Western, J. D. (1982) Variations in fecundity with age and environment in olive baboons. *Am. J. Primatol.*, **3**, 61–76

Western, D. (1976) A new approach to Amboseli. *Parks*, **1**, 1–4

Western, D. (1982) Amboseli National Park: enlisting landowners to conserve migratory wildlife. *Ambio*, **11**, 302–8

Western, D. (1983) *The Origins and Development of Conservation in East Africa.* The Fifth David Marc Belkin Memorial Lecture. San Diego, Calif.: University of California

Western, D. (1984*a*) Conservation based rural development. In *Sustaining Tomorrow*, ed. F. Thibodean & H. Field. Hanover: University Press of New England

Western, D. (1984*b*) Struggling to save East Africa's wildlife. *Anim. Kingdom*, **87**, 32–5

Western, D. & Henry, W. (1979) Economics and conservation in Third World parks. *Bioscience*, **29**, 414–18

Western, D. & Van Praet, C. (1973) Cyclical changes in the habitat and climate of an East African ecosystem. *Nature*, **204**, 104–6

VII.4

Physiological aids in the conservation of primates

J. P. HEARN

Introduction

Research in physiology and biochemistry has naturally tended to develop separately from studies in ecology and behaviour. Normally, it is only in the laboratory, under closely controlled conditions, that specific questions related to animal behaviour can be asked using physiological techniques. However, in the past few years, there have been a number of developments that provide hope for extending the use of physiological technology to studies in the wild. This is not before time, as there are a number of ways in which physiological methods may be useful in future for promoting conservation of animals in the wild.

According to the moderate predictions of the World Bank (McNamara, 1984), the human population will rise from its present 4.5 billion to just under 10 billion by the year 2050. Much of this increase will be in developing countries which currently afford some space for exotic species. It follows that the wild environments currently available will diminish, resulting in smaller populations of wild animals, often isolated and sometimes of insufficient genetic diversity to survive. It will be necessary to manage wild populations to a greater degree than is now possible. In addition, exotic species in captivity and in the wild will need to be seen as one stock, the knowledge gained from captive animals being applied to improve the management of wild populations.

In this brief communication, a few areas are noted where recent development in technology may make available new methods that may help in future to conserve wild populations. Some of these methods may assist in studies of species in the wild, helping to monitor reproductive status or levels of stress. Others may provide practical ways of ensuring gene flow between restricted populations.

Non-invasive methods

A key that is required to open many possibilities is the further development of non-invasive methods. It is now possible to measure sex steroids and some gonadotrophins in urine and to a lesser extent in faeces. Recent developments in enzyme-linked immunosorbent assays should result in these quick, simple, non-isotopic assays being made available as field kits. Consequently, it should be possible, by collecting a few drops of urine, to measure pregnanediol or oestrogen metabolites that will give some idea of the reproductive status of an individual. A study of this nature in wild vervet monkeys has recently been reported (Andelman *et al.*, 1985). The availability of such technology, while easier to apply in some species than others, should give a physiological back-up to the behavioural scientist who is attempting to unscramble complex interactions using a pair of binoculars.

Another related area that is developing rapidly is that of telemetry, whereby small indwelling monitors can signal information on heart rate, respiration, body temperature, etc., to the observer. With a little ingenuity, weight recorders can be placed at nest sites to gather data on body weights of adults or young. Radiotracking has been a feature of ecological studies for many years and advances in electronics and communications technology now make possible far smaller and longer lasting transmitters that increase the numbers of species that can be investigated. All of these methods increase the ways in which more precise, quantifiable data can be collected. Furthermore, portable electronic keyboards, on which behavioural data can be recorded for subsequent transfer and analysis on a base computer, greatly reduce the time and the risk of transfer errors inherent in studies depending upon handwritten record sheets.

In addition to non-invasive methods, in recent years there have been considerable improvements in the application of rapid sedating and anaesthetic drugs. When applied properly, through use of darting, this enables the capture, sampling, and possible translocation of an animal without greatly disturbing the social group. Further developments of this nature are essential if animal populations are to be consolidated in smaller areas, or population structures manipulated to improve the chances of future survival.

Physiological monitoring

Using a combination of non-invasive and invasive methods, it should be possible in future to sample either urine or blood, and to monitor either individual or population status from the points of view of genetics, nutrition and disease. Genetic analysis will allow the planning of introduction of new stock. Nutritional analysis should give data that are of value in monitoring the availability of adequate diets in the wild and also provide, through comparisons of blood samples of animals in the wild and in captivity, an indication of the adequacy of nutrition of captive animals. Studies of disease, although perhaps less amenable to intervention in the wild, should help identify problems before they become a threat to the population. There is only limited need for such technology at present, but if in future more translocation and intensive management of species in smaller areas is to be seen, the monitoring of disease may become of far higher priority, particularly where a disease can affect more than one species.

It should be noted that before many of these procedures can become applied in any routine fashion, a lot of basic research is necessary to establish species differences and to validate the techniques with new species. Application before basic studies are completed will result in a confusing and meaningless mass of data with little practical relevance. The initial stages of development of new technology and its application and validation to novel species can best be worked out on captive populations. It is here that studies on animals in zoos can be of importance when seen in the wider perspective.

Gene transfer

As a consequence of managing captive and wild populations as a whole, it may be advisable in future to consider artificial methods of reproduction, insofar as they lend themselves to gene transfer between populations. Advances in this field over the past few years, both in domesticated stock such as cattle and sheep as well as in the human, have been remarkable (for a review, see Hearn & Hodges, 1985). The result has been a revolution in the management of domesticated stock, with extensive commercial and financial development, and a great improvement in the successful treatment of human infertility. As yet, little application of this technology has taken place with exotic species, partly because the basic studies required have yet to be done, and partly because the problem has not yet arisen. However, the problem is arising and will be seen within the next 20 to 30 years, so that much of the basic research necessary needs to be done now, with some urgency.

The technology concerned, which will not be dealt with in detail in this short communication, includes the artifical maturation of gametes, the freezing, storage and transfer of eggs, sperm and embryos, and the suppression by reversible contraception of individuals whose productivity in the gene pool is becoming overbalanced. At present, it is still possibly only to freeze the sperm of a few domesticated species, and extension to endangered species requires considerable further work, as does the freezing and transfer of embryos (Hearn, in press; Holt, in press).

Many of these techniques are still complicated and laboratory based. However, in most cases there is a trend to simplify methods that should before long be available in the field.

Neonatal loss

An area that has received very little attention to date is the possible prevention of the very high levels of neonatal loss, seen both in captive and wild populations. Where this is a consequence of lactation failure or of pathological behaviour in the wild, little can be done. In captivity, procedures for hand-rearing and the development of synthetic diets are making slow progress. Further development of synthetic diets and a greater knowledge of normal milk compositions are necessary before further progress can be made.

Conclusions

The brief discussion above notes some areas in which a physiological dimension might assist in the conservation of primate and other exotic species. No one would suggest the introduction of technology for its own sake, and it should be emphasised that the application of these techniques should only be attempted when there are clear reasons for doing so, either in the investigation of ecological questions or in the practical advancement of conservation. A greater interaction between physiological and ecological studies should lead to the advancement of both, creating a novel field of physiological ecology that allows a more quantitative approach to animals in the field. It is probably fair to say that in the past, schools employing a physiological approach, such as endocrinology, gamete biology or developmental biology, or indeed those studying exotic species in captivity, have not had a great interaction with field biologists and ecologists. In addition, the difficult tasks faced by field biologists in quantification of data and the long-term nature of their studies may not always be appreciated by the laboratory worker. In the interests of more efficient conservation, and the time constraints facing us due to the threats of expanding human populations, a far greater multidisciplinary collaboration is essential in future.

References

Andelman, S. J., Else, J. G., Hearn, J. P. & Hodges, J. K. (1985) The non-invasive monitoring of reproductive events in wild vervet monkeys (*Cercopithecus aethiops*) using urinary pregnanediol-3α-glucuronide and its correlation with behavioural observations. *J. Zool., London (A)*, **205**, 467–77

Hearn, J. P. (in press) Artificial acceleration of reproduction. In *Primate Conservation*, ed. K. Benirschke. Berlin: Springer Verlag

Hearn, J. P. & Hodges, J. K. (1985) (Eds) *Advances in Animal Conservation*. (*Symp. Zool. Soc. Lond.*, **54**). Oxford: Oxford University Press

Holt, W. V. (in press) Assessment, storage and transfer of sperm. In *Primate Conservation*, ed. K. Benirschke. Berlin: Springer Verlag

McNamara, R. S. (1984) *The Population Problem: Time Bomb or Myth*. Washington, D.C.: World Bank

INDEX